REGENERATIVE AGROURBANISM

REGENERATIVE AGROURBANISM

Experiencing edible placemaking transforming neglected or damaged landscapes, lives, and livelihoods

Robert L. France

– A web-enhanced book –

Wageningen Academic
Publishers

EAN: 9789086863860
e-EAN: 9789086869381
ISBN: 978-90-8686-386-0
eISBN: 978-90-8686-938-1
DOI: 10.3920/978-90-8686-938-1

First published, 2022

'The redemption business: healing the land, healing the food, healing the economy, and healing the culture.'

Reclaiming our food: how the grassroots food movement is changing the way we eat,
Tanya Denckla Cobb (2012)

'In virtually all the academic literature it is quite striking how disembodied written landscapes become. This is because virtually everything written about landscape is not only written on paper; it is principally derived from paper. Landscape is not bodily experienced ... Bodies remain at the desk rather than in the field ... What we are left with is paper landscapes, paper perspectives.'

The materiality of stone: explorations in landscape phenomenology,
Christopher Tilley (2004)

For my ancestral instructors and anecdotal cicerones

Peter Mazur (re: Chapter 3's 'garden stand entrepreneur'), Marilyn France, née Mazur (re: Chapter 3's 'gastronomic tourism'), Josephine Mazur (re: Chapter 4's 'food confraternity'), Barry France (re: Chapter 5's 'edible cottage gardens'), Ann France (re: Chapter 6's 'farming forbearers'), and William France (re: Chapter 7's 'POW gardeners versus Nazis').

Table of contents

Preface

Situating urban agriculture tours in time and 'splace'

The literature on urban agriculture has been expanding voluminously. A graph in France and Mougeot (2016) shows the dramatic climb in the cumulative total of English-language books available from online shops, and a table therein highlights, by title, the growing literature up to 2010, when the conference was held that gave rise to the subsequent publication. An infographic in Bohn and Viljoen (2016) illustrates how the source repository of reports about urban agriculture has likewise flourished over the same period. In the last two decades, and coincident to these trends continuing, there has also been a gradual shift in focus in the type of books being published. Ten years ago, as publications on the subject were in their 'salad days,' at the stage of transitioning from a century of scholarship about the establishment and operation of allotments and community gardens for sustenance, toward those recognising the arrival of a discipline of professional practice and academic study in contemporary urban planning and design (France 2016a), many of the books took on the form of first-person accounts of visiting sites with the intent to introduce and showcase the rich diversity of urban agriculture projects around the world (select examples of such titles are listed in France and Mougeot (2016)). Today, scholarship on urban agriculture has moved on from simply describing, to critically examining, the nuanced, nuts-and-bolts operations of case studies, as well as their underlying assumptions and much-touted claims. Such research has appeared in hundreds of journal articles and dozens of recent books.

The present book, with its deliberate focus on the experience of visiting projects, as described through the lens of narrative scholarship and rooted in phenomenological landscape research, is almost unique in the field of urban agriculture. Indeed, this book can be said to have been designed intentionally as a counterpoint to the existing literature on urban agriculture. What was remarkable to me in reviewing the rapidly expanding literature is the emphasis that continues to be given to urban gardens, food markets, educational farms, commercial enterprises, heritage landscapes, and the like, as conceptual spaces to be studied from a distance rather than as experiential *places* to be engaged in situ. The lack of any embodied perspective in much of the urban agriculture literature means that, as Tilley (2004: 28) found with respect to archaeology publications, 'they can only provide us with abstract models for thinking landscapes rather than models *of* landscapes that are sensuously lived.' The result is – echoing Tilley's criticism expressed in the epigraph for the present book – an academic literature in which urban agriculture projects are presented as little more than sensorially impoverished 'paper landscapes.'

The purpose of this book is to emphasise that 'agrourbanism' (see Chapter 1) is about placemaking in the urban fabric. 'Space' and 'place,' as geographers contend, are *not* the same thing (Creswell 2004; Tuan 2001). Abstract space, when humanised, becomes value-laden, and is transformed

DOI 10.3920/978-90-8686-938-1_0, © Wageningen Academic Publishers 2022

in the process to being place (Tilley 1994). Places are therefore centres of felt value and meaning that began as undifferentiated space (Tuan 2001). This is why when one refers to the idiosyncratic ambiance of a particular location, it is in regard to the 'spirit of a place' rather than that of a space.

The present book, with its emphasis on describing the agrourbanism experience, may at first blush seem somewhat anachronistic. This is reinforced by the fact that many of the tours took place a decade ago following on from the international conference I organised on the subject, 'Planning Urban Agricultural Systems for the 21st Century,' and preceding its consequent publication, *Integrated urban agriculture: precedents, practices, prospects* (France 2016b). It had been my intention at the time to describe these tours in a chapter in that volume, but as the manuscript grew in size due the scholarship of others, it became necessary to pare back my own contributions to the whole. And so, publication of the research endeavour remained stillborn, while periodically, over the interceding years, further site visits were undertaken in order to round out the specific types of projects in which I was interested. It is, therefore, this material, collected over the decade between 2012 and 2022, which forms the content of the present book.

Another way in which the present book is different is in the type of projects that were visited and described herein. Indeed, of the 74 projects from a dozen countries covered in this regard, fully 78% have not, to my knowledge, been discussed in any detail in the titled literature before. The reason for this is a consequence of the very specific nature of the projects that comprise the present book. Whereas some are community gardens and commercial operations – in other words, the sort of projects that customarily subsume the bulk of urban agriculture books – most of the projects are markedly different. This is because the overall focus of the present book is on the more holistic field of *agrourbanism* rather than the more narrowly focused field of urban agriculture per se. And so, in addition to a handful of 'the usual suspects,' other sites which are described include those receiving less coverage in the traditional urban agriculture literature: food markets, educational farms, aesthetic foodscapes, heritage farms, and regenerative landscapes.

Whereas academics revel in comprehensive and extended descriptions of case studies, practitioners often champion those that tend toward succinctness over long-windedness. This book presents its case studies modelled on the 'short-'n-sweet' format used in other landscape architecture texts, wherein the context of each case study is considered through no more than a dozen paragraphs spread over a handful of pages and supported by a limited number of cited references (De Meulder and Shannon 2008; Dreiseitl *et al.* 2001; Earth Pledge 2005; McGuire and Hanson 2019; Pinkham 2000; Rottle and Yocom 2009). Case studies are arranged into ten typologies within five chapters, each of which concerns transforming or 'regenerating' (see Chapter 1) an aspect of agrourbanism. These are expressed as Chapter 3's 'healing economies' (business and tourism), Chapter 4's 'healing connections' (hope and education), Chapter 5's 'healing art' (beautifully grown food), Chapter 6's 'healing heritage' (restoring agri-culture), and Chapter 7's 'healing sites and societies' (regenerative landscape design).

Each chapter section is composed of four elements. A brief, personal anecdote sets the stage. This is followed by an overview of a book that addresses the theme, sometimes in an unexpected or tangential fashion, of the particular section. For all but one of the books so considered, this is the first time that they have been discussed in detail in the urban agriculture literature. The core of each chapter section is composed of first-person accounts of the perceptual and sometimes other sensorial experiences of visiting a host of case studies. The approach used is that of 'layered' narratives, augmented by photo-essays, and undertaken through phenomenological 'thick descriptions,' as described in Chapter 2. The emphasis throughout, and unlike the general urban agriculture literature, is on regarding these locations as inhabited places rather than as inert spaces. It is this focus more than anything else that contributes to the present book being remarkably different in both tone and substance from others in the oeuvre of urban agriculture. Moreover, the present book accomplishes this by spotlighting the ecopsychological perception, or 'sight and sensibility' (Sewall 2002), of the visited sites, doing so through an accompanying bespoke website that contains hundreds of additional photographs (https://doi.org/10.3920/978-90-8686-938-1). What I have found to be surprising is that although many dozens of urban agriculture books have been published, most include hardly any – and some cases, absolutely no – photographs of the gardens, farms, and other sites that are presented. The result of this visual impoverishment of the agrourbansim literature is that we have an incomplete comprehension of the multifaceted nature and true value of the projects being described. This is because sight and thought are closely related, as in the expression 'I see,' signifying 'I understand' (Tuan 2001). Again, this derives from the primary focus of the literature being on urban gardens and farms as biophysical spaces in which to grow food, rather than on them as humanised places in which to grow experience. The present book, in contrast, is in the same vein of that fraction of the overall corpus that *does* acknowledge and sometimes even celebrates the visual attributes of edible placemaking (CITIES 2016; Cobb 2012; Gorgolewski *et al.* 2011; Hanson and Marty 2012; Hou *et al.* 2009; Philips 2013; Rich 2012). As such, the professional genre into which this book can be situated is that of landscape architecture or cultural geography rather than empirical science or numerical urban planning. Finally, each chapter section concludes with a point-form listing of learned lessons compiled from the case studies considered.

Two other points concern the placement of the case studies within the organisational structure of the book. Firstly, and what will be immediately apparent, is that some of the case studies could be comfortably situated within several different chapters due to the complex nature of their displayed attributes. And so, a judgement call was made as to which particular category was the best fit based on the most prominently exhibited features. Secondly, the sequence of particular case studies covered within any chapter section progresses roughly in concordance with a personal ranking of the demonstrative fulfilment of their claimed ends as well as, especially, a 'take-home' assessment of the overall experiential quality of what the specific site visits offered.

Finally, from a personal perspective, the writing of this book has been a conscious effort to exercise my right-side brain of creative holism rather than my left-side brain of analytical reductionism; in other words, drawing upon my experience as a social scientist with an interest in writing thick descriptions about ecocultural landscapes (France 2014, 2020) rather than that of my experience as a physical scientist scripting technical papers focused on quantification (France and Peters 1992; France *et al.* 1992, 1994).

References

Bohn, K. and A. Viljoen. 2016. CPUL repository. In: Viljoen, A. and K. Bohn (eds.) Second nature urban agriculture: Designing productive cities. Routledge, London, UK.

CITIES (ed.) 2016. Farming the city: food as a tool for today's urbanisation. Trancity, Haarlem, the Netherlands.

Cobb, T.D. 2012. Reclaiming our food: how the grassroots food movement is changing the way we eat. Storey Publishing, North Adams, MA, USA.

De Meulder, B. and K. Shannon. 2008. Water urbanisms. Idea Books, Amsterdam, the Netherlands.

Dreiseitl, H., D. Grau and K.H.C. Ludwig. 2001. Waterscapes: planning, building and designing with water. Birkhauser, Basel, Switzerland.

Earth Pledge. 2005. Green roofs: Ecological design and construction. Schiffer Books, Atglen, PA, USA.

France, R.L. 2014. Along the way: pilgrimage scenes from the Camino Frances to Santiago de Compostela. Libri, Farringdon, UK.

France, R.L. 2016a. The once and future city farm. In: France, R.L. (ed.) Integrated urban agriculture: precedents, practices, prospects. Libri, Farringdon, UK.

France, R.L. (ed.) 2016b. Integrated urban agriculture: precedents, practices, prospects. Libri, Farringdon, UK.

France, R.L. 2020. Waymarking Italy's influence on American environmental imagination while on pilgrimage to Assisi. Cambridge Scholars Publishing, Newcastle upon Tyne, UK.

France, R.L. and L.J.A. Mougeot. 2016. Reading urban gardens and farms: literature, layering and largesse. In: France, R.L. (ed.) Integrated urban agriculture: precedents, practices, prospects. Libri, Farringdon, UK.

France, R. and R. Peters. 1992. Temporal variance function for total phosphorus concentration. Canadian Journal of Fisheries and Aquatic Science 49: 975-977.

France, R., D. McQueen, A. Lynch and M. Dennison. 1992. Statistical comparison of seasonal trends for autocorrelated data: a test of consumer and resource mediated trophic interactions. Oikos 65:45-51.

France. R., R. Peters and Y. Prairie. 1994. Adjusting chlorophyll estimates through temporal weighting based on the seasonal development of phytobiomass. Aquatic Sciences 56/2: 106-114.

Gorgolewski, M., J. Komisar and J. Nasr. 2011. Carrot city: creating places for urban agriculture. Monacelli Press, New York, NY, USA.

Hanson, D. and E. Marty. 2012. Breaking through concrete: building an urban farm revival. University of California Press, Berkeley, CA, USA.

Hou, J., J.M. Johnson and L.J. Lawson. 2009. Greening cities, growing communities: learning from Seattle's urban community gardens. University of Washington Press, Seattle, WA, USA.

McGuire, M.P. and J.M. Henson. 2019. Fresh water: design research for inland water territories. Applied Research + Design, Novato, CA, USA.

Philips, A. 2013. Designing urban agriculture: A complete guide to the design, construction, maintenance and management of edible landscapes. Wiley, Hoboken, NY, USA.

Pinkham, R. 2000. Daylighting: new life for buried streams. Rocky Mountain Institute, USA.

Rich, S.C. 2012. Urban farms. Abrams Books, New York, NY, USA.

Rottle, N. and K. Yocom. 2010. Basics landscape architecture 02: Ecological design. AVA Publishing, Worthing, UK.

Sewall, L. 2002. Sight and sensibility: the ecopsychology of perception. Tarcher, New York, NY, USA.

Tilley, C. 1994. A phenomenology of landscape: places, paths and movements. Berg, Oxford, UK.

Tilley, C. 2004. The materiality of stone: explorations in landscape phenomenology. Berg, Oxford, UK.

Tuan, Y-F. 2001. Space and place: the perspective of experience. University of Minnesota Press, Minneapolis, MN, USA.

Acknowledgements

Financial support for site visits was provided by a travel grant from the Social Sciences and Humanities Research Council of Canada and a legacy fund from the estate of Barry France. The Dalhousie library system is thanked for acquisition of published research materials. To be especially thanked are those individuals who took time off from their day-to-day operations to help guide me through their projects and to answer my questions arising therefrom. If there are any misinterpretations or errors included in the case study descriptions, they remain mine own. The purposeful dedication, imaginary vision, and overt enthusiasm displayed by those participating in the regenerative agrourbanism projects visited were truly inspirational to behold. At a time when more than half of the world's population resides within cities, it is these individuals who are and will be the stewards for setting a course forward based on ensuring the continued wellbeing of not just 'place – plants – people' as explored herein, but also that of the fourth alliterative 'p': the planet itself. Throughout my career I have been fortunate to be able to meet, interact with, and learn from leaders whose well-known scholarship on the subject matter of the present book needs no further elaboration beside the following: Jane Jacobs (cities are places of people), Ian McHarg (landscapes are places of layered use), Michael Hough (cities are places of ecosystemic function), Carl Steinitz (landscapes are places to be experienced visually), and Wes Jackson (farms are places to be managed sustainably).

Part 1
Background

Chapter 1
Place – Plants – People: the paradigm of regenerative agrourbanism as waymarked through ten formative books

1.1 Agrourbanism: *'talkin' 'bout ... [re]generation'*

Food has been grown in and around cities since the dawn of civilisation on the floodplains of ancient Mesopotamia (France 2011). For, as I wrote in the Introduction to *Integrated urban agriculture: precedents, practices, prospects* (France 2016a:8):

> Scratch the surface of any 'new' green technology championed as part of sustainable design ... and one will often find a long-established and richly-diverse history of the effective use of the same technology at some time in the past. Much of sustainable living and design therefore amounts to reinventing the tried and proven effectiveness of that old wheel through the application of modern, efficient methods (spokes?). The same is true for growing and distributing food in cities.

Nowhere has the 'everything old is new again' perspective of sustainable design in general, and its constituent urban agriculture in particular, been expressed with more historical resonance than with the rediscovery of the concept of 'regeneration.' The Neolithic Revolution, where agriculture began, was characterised by widespread belief in regeneration (Garrow and Wilkin 2022; Lewis-Williams and Pearce 2005), a mindset that extends back even further to at least the Mesolithic and the myth of 'the eternal return' (Elidae 1991). Recently, and mimicking the cyclical processes of natural systems, regenerative design has been embraced by green building practice (Cole 2012), from where its ideals and ideas have expanded spatially to include urban design (Wansborough and Mageean 2010) and new ways in which to think about and undertake the planning of entire cities (Battisti 2021; Schurig and Turan 2022; Zari 2018). Given agriculture's pivotal role in Neolithic notions of rebirth and renewal, it is no surprise that linkages are being made between regeneration and urban agriculture, as for example, in relation to ecological mimicry (Nawfal and Mudares 2017; Rashed 2018), sustainable food systems (Parker 2010; Yamin 2019), circular economy (Mohan *et al.* 2020), heritage preservation (Levadna 2018), green infrastructure (Russo and Cirella 2020; Shidqina 2021), and social wellbeing (Swensen *et al.* 2012). It is within this multidisciplinary zeitgeist that the present work is situated.

This chapter is based on the idea that the description and development of a scholarly thesis can be aided through a process of 'literary waymarking' (France 2020). In the present case, salient elements from ten texts are absorbed into forming the paradigm of regenerative agrourbanism. This is illustrated through two infographics, wherein the nucleus of core concepts is represented

DOI 10.3920/978-90-8686-938-1_1, © Wageningen Academic Publishers 2022

by a couplet of books about holistic urban agriculture and a second couplet of books about holistic environmental restoration (Figure 1.1). Added to this and contributing to the whole are the radial constituents represented by six other books, each containing the word 'regenerative' in its title (Figure 1.2).

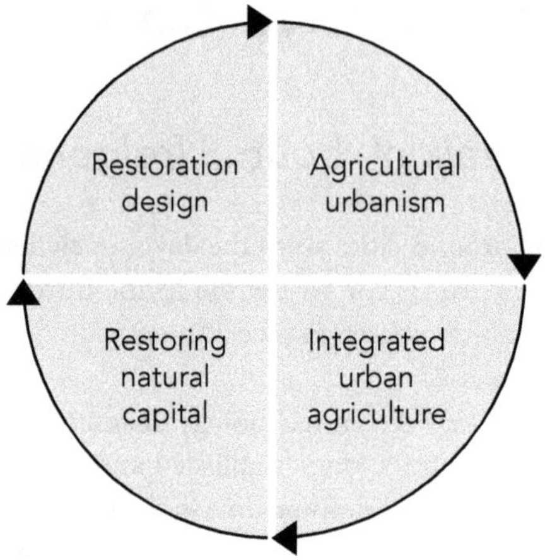

Figure 1.1. The core concepts that underlie regenerative agrourbanism.

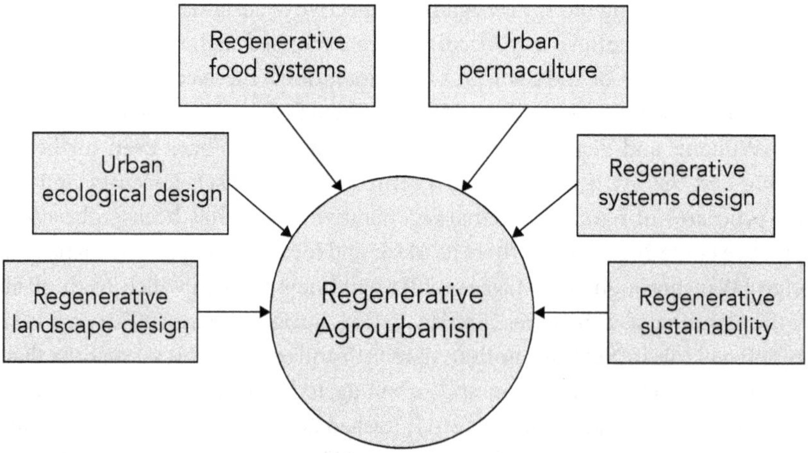

Figure 1.2. Regenerative agrourbanism and its constituent disciplines.

1.2 Agrorestorative nucleus

1.2.1 Agriculture

Regenerative agrourbanism is based on the contention that urban agriculture provides a diverse set of direct and indirect socio-ecological services to city dwellers and their associated physical environments and civic institutions. Therefore, more than merely the nuts-and-bolts operations involved with producing food, urban (and, henceforth, also peri-urban) agriculture should be conceived of as a central core of activities – production, processing, marketing, distribution, and consumption – along with a wide array of services and outcomes that support the total population. Urban agriculture, then, in addition to functioning as provenance for material goods, is a citizen-centric endeavour in the spirit described in Thomas Lyson's seminal *Civic agriculture: reconnecting farm, food, and community* (1994).

Book 1: *Integrated urban agriculture: precedents, practices, prospects*, edited by myself (2016b), is part of a trend that has recognised (Lovell 2010; Zasada 2011) and continues to champion (Orsini *et al.* 2020; Specht *et al.* 2021) the multi-functionality of urban agriculture. As itemised in *Urban and agricultural communities: opportunities for common ground* (CAST 2002), a book whose message was ahead of its time, these affiliated services and outcomes include: business entrepreneurship and economic vitality, individual and community health and wellbeing, landscape beautification, and environmental restoration and remediation; to which can be added education to improve literacy about food and farming.

The point is made in the book's Introduction that urban agriculture is a specialised form of environmental management, and as such, can learn much from related disciplines. For example, the shift from the engineering-only, techno-fix mindset of watershed management to the more holistic, socially-inclusive working tenets of *integrated* watershed management offers a precedent for the transition of growing food via urban agriculture to growing communities via *integrated* urban agriculture.

Integrated urban agriculture, by its very existence, challenges prevailing myths about human-nature relationships with respect to the false separation of nature and culture, particularly with respect to the capability of cities to harbour productive landscapes. Chapter contributions address the historical development of urban agriculture and its role in supporting sustainability. Public access to foodstuffs is emphasised as essential to ensure dietary health and security, just as is the growing and marketing of produce to ensure wealth and financial security. These are issues whose importance soars during times of societal austerity and crisis in the Western developed world, while being omnipresent in the Global South.

The landscape architecture of high-visibility, edible gardens of beautiful form in urban cores, in contrast to functional yet unattractive suburban farms, is noted to parallel the evolution of treatment wetlands into treatment wetland parks in innovative stormwater management. In addition to addressing concerns related to greenspace planning and commercial enterprises, integrated urban agriculture educates about agrarian heritage and food literacy. Finally, transforming derelict land into productive spaces is an essential step toward establishing resiliency in cities, just as is creating 'food hubs' for enabling sustainable food systems through what can be called 'agricultural urbanism.'

~

Regenerative agrourbanism is based on the contention that integrated urban agriculture is part of comprehensive community planning and development. Recognising this, some have sought to find a new nomenclature in which to better describe the holistic enterprise of landscapes as foodscapes. The year 2005 brought two contenders in this regard – 'agropolis' and 'CPULs' – with the publication of *Agropolis: the social, political, and environmental dimensions of urban agriculture*, by Luc Mougeot, and of *Continuous productive urban landscapes: designing urban agriculture for sustainable cities*, edited by André Viljoen (with Katrin Bohn and Joe Howe). In the concluding chapter of the latter, Viljoen and Bohn liken CPULs – the combination of people and place with cultivation and circulation – with 'tracks of an agrarian landscape' through a city setting (p. 268). Three years later, Carolyn Steel, in *Hungry city: how food shapes our lives*, coined the term 'sitopia,' linking the ancient Greek words for food (*sitos*) and place (*topos*), with, however, no specific urban focus, although the term would reappear in the subsequent foreword of a book about urban agriculture (Steel 2016).

Next, as Ladner (2011) recounts in his chapter 'Agriculture as the new golf: farming as a development amenity,' comes the introduction of several terms used in reference to the design of agricultural subdivisions in conservation communities that integrate residential housing with farming – what the Urban Land Institute would later refer to as 'agrihoods' (*Agrihoods: cultivating best practices* – Matthew Norris and Daron Joffe 2018). These two closely related monikers are 'agricultural urbanism,' from Janine De la Salle and Mark Holland's *Agricultural urbanism: handbook for building sustainable food & agriculture systems in 21st century cities* (see below) – whose publication I facilitated in 2010 – and 'agrarian urbanism,' from Andres Duany's *Garden cities: theory & practice of agrarian urbanism*, published the following year. Agrarian urbanism is described in the latter book as 'a concept that involves food not as a means of making a living, but as a basis for making a life' (p. 2). The distinction is made that whereas the word 'agricultural' pertains to 'the technical aspects of growing food, the term agrarian emphasises the *society* involved with all aspects of food' (p. 2). Although this may be true, Duany goes on to make great hay about how his concept is more holistic than that of De la Salle and Holland's, which he incorrectly (see below) marginalises as being only about 'settlements equipped with a working farm' (p. 8), an inaccuracy that is repeated by Ladner (2011).

The most recent suggestion of a candidate moniker for the emerging discipline of regarding urban agriculture through a wider-angled lens, is 'agrourbanism,' from *Agrourbanism: tools for governance and planning of agrarian landscape*, edited by Enrico Gottero (2021). Essentially a convenient, shorthand expression fusing 'agrarian' and 'urbanism,' Gottero, in contrast to Duany and Ladner, makes no distinction between these terms and De la Salle and Holland's and Duany's earlier phraseology, using all three versions interchangeably. He elaborates that agrourbanism is that which considers 'all activities related to agricultural production and urban uses, especially in order to underscore mutual benefits in terms of ecosystem and socio-economic services ... [through offering] the opportunity to rethink unsustainable models and to improve environmental conditions, as well as a chance to improve human health, quality of life and wellbeing both in cities and in the countryside' (p. 3).

Book 2: *Agricultural urbanism: handbook for building sustainable food & agriculture systems in 21st century cities*, by Janine De la Salle and Mark Holland (2010), along with *Urban agriculture: growing healthy, sustainable places*, by Kimberley Hodgson, Maria Campbell and Martin Bailkey, which came out the following year, represent a significant milestone in terms of urban planning finally awakening to consider food systems, a shortcoming that had been noted (Pothukuchi and Kaufman 2000), yet largely, but not completely, ignored for much of the first decade of the millennium (France 2016b). Notable exceptions to the profession's ambivalence to urban agriculture during this time included *Continuous productive urban landscapes: designing urban agriculture for sustainable cities* (Andre Viljoen, with Katrin Bohn and Joe Howe 2005) and *Agriculture in urban planning: generating livelihoods and food security* (Mark Redwood 2008).

For De la Salle and Holland and their contributors, agricultural urbanism is about much more than producing greens in backyard and community gardens or buying foodstuffs at farmers' markets; it represents a new way in which to build sustainable food systems through a strategic framework of not only the processing, marketing, and selling local food, but also in developing education and training programmes, as well as in celebrating the culture around food. All of this is approached through the disciplines of urban planning and design as a 'way of building a place around food' (p. 9).

The book is intended as a manifesto and manual, both rooted in achievable practicalities. The principles that are used to inform and guide decisions and actions include: taking an integrated, food-and-agriculture perspective; building a food and agriculture economy as part of a sustainable food system; increasing access to, and education about, food; providing sustenance and habitat for other species; constructing sustainable infrastructure for food and agriculture; and forging partnerships with organisations to maintain the food system, particularly in the face of climatic and social instability. Significantly, underlying all, is the emphasis on creating 'a rich experience of food and agriculture [through] use of placemaking design strategies to make food visible and to enhance the experiences that involve growing, harvesting, tasting, selling, buying, learning about, cooking, and sharing food' (p. 31).

1.2.2 Restoration

Regenerative agrourbanism is based on the contention that integrated urban agriculture represents a specialised form of 'restoration design' (see below). This makes perfect sense given that the process of environmental restoration has itself often been conceptually likened to the physical act of gardening nature. Dozens of compelling case studies have demonstrated the role played by urban agriculture in the remediation or restoration of damaged sites. However, far fewer examples exist wherein the processes of agriculture are specifically linked to various restoration goals. When agriculture is referred to as being restorative, it often means in the sense of undertaking farming in nature's image, as for example, *Restoration agriculture: real-world permaculture for farmers,* by Mark Shepard (2013), or of adopting site-specific remedial technologies, as in *Restoration agriculture*, by Saket Kushwaha (2021). Urban agriculture, when it has been specifically coupled with restoration, has been concerned with technical issues such as re-establishing gardens lost over time to encroaching development (Dobson *et al.* 2020), or in expanding horticultural production for purposes of food security and livelihood self-sufficiency (Gwetsay *et al.* 2016). On a more theoretical perspective, Handel (2016) has called for a partnering in the work of agriculturalists and restoration ecologists toward a goal of reimagining city greenspaces.

Book 3: *Healing natures, repairing relationships: new perspectives on restoring ecological spaces and consciousness*, edited by myself (2007), is part of a movement emphasising that restoration needs to expand beyond its reductionist focus in tinkering with the broken bits and pieces of ecosystems by including human perspectives. Important precedents in this philosophical development of a more holistic discipline of restoration include: *Environmental restoration: ethics, theory, and practice* (Throop 2000), *Restoring nature: perspectives from the social sciences and humanities* (Gobster and Hull 2000), *Nature by design: people, natural process, and ecological restoration* (Higgs 2003), and *The sunflower forest: ecological restoration and the new communion with nature* (Jordan 2003).

Humans exist in a mindset of disseverance from nature and desperately need to find positive ways in which to reconnect with their environment. This can be accomplished by physically engaging in the actual process of both ecological restoration and urban agriculture. The act of restoring, remediating, in other words, *healing,* degraded landscapes is an act of reciprocity, important not only for improving the quality of the outside environment of nature, but also that of the internal environment of the psyche, or human nature. The paradigm of 'restoration design' described in the book is the process by which participants creatively develop physical and conceptual relationships to engage repaired nature through the architectural transformation of their inhabited ecological space as well as their internal consciousness. Furthermore, restoration design shares much in common with integrated urban agriculture in that both firmly believe that the *process* of restoration and cultivation may be just as or even more important than the products that ensue from those efforts. In particular, restoration design is based on three major subject categories that encompass and structure the discipline: engagement, relationships, and design.

Chapters explore restoration design as a means to engage nature, either physically as an act of ceremony and ritual or as pageantry and performance, or conceptually as a positive act of mutualism and reciprocity with nature or sometimes potentially as a negative act reinforcing our dominance over the non-human world. For other contributors, restoration design is a way of establishing relationships, such as that of a human healer to the healed landscape, that of a resident's life to the inhabited re-greened home, and/or that of individuals to a lineage of people and place. Finally, for some, restoration design is a design process in which humans can express their artistic creativity in the environment as a vehicle for cultural awareness and education, or in exercising their impulse to reconfigure the landscape as part of defining their relationship with it.

~

Regenerative agrourbanism is based on the contention that urban agriculture, through being part of integrated food systems, purveys capital. Although generally considered from the limited perspective of this capital being solely financial in scope, with numerous publications attesting to this, holistic restoration of agrarian landscapes recognises that farming also provides a diversity of socio-ecological services which can be thought of as 'natural capital' (e.g. Crookes 2012; France *et al.* 2019).

Book 4: *Restoring natural capital: science, business, and practice,* edited by James Aronson, Suzanne J. Milton, and James N. Blignaut (2007), is based on the tenet that ecology must be practiced as if people matter, and that economics must be practiced as if nature matters. The book is important for being part of the shift in thinking wherein restoration is expanded beyond a focus on ecological form to one on social-ecological function. Other books in the same spirit include *Reclaiming nature: environmental justice and ecological restoration* (Boyce *et al.* 2007) and *Social innovation in Latin America: maintaining and restoring social and natural capital* (Calvo and Morales 2021).

The restoration of natural capital includes ecological restoration but goes further to include the socioeconomic interface between humans and the natural environment. Chapters demonstrate that in managed ecosystems, such as those involved with the production of food, the broader term of 'restoring natural capital' is more applicable than 'ecological restoration' per se. The book is rooted in the idea of restoration design (France 2007), with a focus on fulfilling relationships to nature rather than merely fixing pieces of nature; in other words, the interlinking of natural and social capital. The point is made for the need to move beyond sustainability toward resiliency as an end-goal. And the way to bring this about is through a multidisciplinary approach that involves the community in collaborative learning and management. A problem is that in practice, restoration decision-making processes can become complicated due to vagaries of the political, informative, and social landscapes that continually influence the procedure, which can lead to a reiterative, and at times even cyclical, suite of actions being taken.

Aronson *et al.*'s book has four major themes: the importance of valuing both the loss of natural capital and the restoration activities needed for it to be regained; the challenge of establishing realistic targets toward which progress can be measured; the need to prioritise and plan restorative activities at various scales; and the importance of collaboration and community support in order for projects to be judged as being successful. The latter is important since the editors and contributors are of the mind that restoring natural capital has a unique appeal to a wide public audience through articulating a new cultural relationship of people with nature. In this light, an emphasis on restoring natural capital can be a key to 'mainstreaming' restoration, doing so by engaging community participation; again, a theme raised in other, contemporaneous books (France 2007, 2008).

1.3 Regenerative constituents

1.3.1 Place

Book 5: *Handbook of regenerative landscape design*, edited by myself (2008), belongs to the oeuvre derived from a series of paradigm-defining conferences held at the Harvard Design School in the first decade of the twenty-first century that were significant for their focus on the reparation of abused and neglected urban landscapes for use by people. Other publications included: *Manufactured sites: rethinking the post-industrial landscape* (Kirkwood 2001), *Waterfronts in post-industrial cities* (Marshall 2001), *Healing natures, repairing relationships: new perspectives on restoring ecological spaces and consciousness* (France 2007), and *Designing to reclaim the landscape* (Berger 2008).

The handbook contains a series of case studies from around the world which offer a message of hope in terms of enabling better decisions to be made today to reduce the burdens for generations to come. The case studies demonstrate that in order to create meaningful and supporting urban places, it is necessary to view economic, social, and environmental challenges as interdependent parts of the same system, and that solutions must be grounded in people, not technology, and in practice, not abstract theory. Unlike ecological restoration, then, regenerative landscape design is just as much about people as it is about the nonhuman environment. '*Regenerative* [my emphasising here and henceforth in other quotations in this chapter] landscape design is a multifaceted strategy based on recognition that the wish of humans to reside in liveable, environmentally healthy communities can only come about through direct public participation in many phases of the overall reparation process' (p. 429). In particular, '*regenerative* landscape design, while incorporating elements from a family of related professional disciplines such as environmental engineering, ecological restoration, low impact development, and sustainable development, is unique in terms of the relative emphasis it places upon these different attributes. Also, *regenerative* landscape design, in building upon these previous reparative strategies, is often more holistic in its overall outlook, balancing the needs of both nature and people' (p. 427).

One of the most important elements of regenerative landscape design is its explicit endorsement of, and frequent reliance on, a 'bottom-up' process of civic engagement. This is demonstrated in three case studies from the book which focus on agriculture. One deals with renovating an ancient water supply system to olive orchards in a Syrian community, and demonstrates how site history and an appreciation of the idiosyncrasies of local politics are essential to informing decisions about reparations. A second case study focuses on the contributing role of edible landscapes in the overall revitalisation of downtown Detroit (this example is discussed in Chapter 7 in the present book). The third includes urban agriculture as part of a comprehensive project of regenerating and repurposing a degraded landscape in a northern California city to create a health services centre for Indigenous people. The overall mandate of holistic healing for both land and people is achieved through highlighting Native American permaculture practices, building a community garden and orchard, developing programs for nutrition education, hands-on training of greenhouse propagation techniques, and the harvesting, construction, and marketing of heritage crafts products from wetland plants. These, and other, non-agricultural, case studies demonstrate that regenerative landscape design projects are often: (a) bottom-up, grassroots initiatives; (b) link environmental with individual and cultural healing in the rebuilding of sustainable, socially conscious communities; and (c) blend engineering principles with design aesthetics to create multipurpose landscapes of high civic quality in terms of providing space for recreation, relaxation, and reflection.

~

Book 6: *Urban ecological design: a process for regenerative places*, by Danilo Palazzo and Frederick Steiner (2011), presents a holistic, interdisciplinary, and systematic process of infusing ecological thinking into urban design. Singular precedents with respect to the concept of cities as ecological systems include: *The granite garden: urban nature and human design* (Spirn 1984) and *Cities and natural process: a basis for sustainability* (Hough 2004).

The goal of Palazzo and Steiner is explained at the start: 'increasingly, we seek to create urban places that are at a minimum sustainable and we hope even *regenerative*' (p. 6). Their focus is on the maturation of urban design as a bone fide discipline based on making and managing the built forms of buildings and open- and green-spaces in a way that is rooted in objective science, responds to the ecology of the region and the urban location, and is flexible to future changes. It is in this manner through which knowledge can be connected to action. Critical to the process is the importance of undertaking site visits to gather knowledge about the critical experiences of residents and visitors to understand the sense of place or genius loci; in other words, how the sounds, colours, smells, shapes, and combinations influence receptivity to places. In particular, use is made of walking tours and user interviews to illustrate how people experience sites in a social context.

Urban ecological design, according to Palazzo and Steiner (2011:123), is based on recognising the importance of the environment in framing answers to Francis' (2001) case study methodology for evaluating landscape architecture vis-à-vis addressing questions such as: What does the project look like? How does it work? How does it feel? How is the place perceived and valued? What problem(s) is the project trying to solve? How is the place used? Who uses it? What are the key design concepts? How is the community served by the project? What is its social impact? How is the environment served by the project? What is its contribution to sustainability? And what are the site-specific lessons learned in comparison to the more general lessons? Tangential mention is made in the book of the role of urban agriculture in one case study example.

1.3.2 Plants

Book 7: *The permaculture city: regenerative design for urban, suburban, and town resilience*, by Toby Hemenway (2015), is based on regarding cities through the lens of permaculture, the paradigm in which edible landscapes are considered to be functional ecosystems. The book is important for its urban adoption of the concept of permaculture that was originally developed for rural settings, including *Permaculture: principles and pathways beyond sustainability* (Holmgren 2002) and *Restoration agriculture: real-world permaculture for farmers* (Shepard 2013).

The point is made right from the start that urban permaculture is only slightly about gardening and really much more about people, i.e. it is based on the idea of taking lessons gleaned from the growing of food and adapting them to the human experience. In this regard, reference is made to 12,000-year-old Göbekli Tepe, an archaeological site that has transformed understanding of the onset of the Neolithic Revolution. Hemenway emphasises that rather than being a discipline per se, permaculture is a universal design tool: 'what permaculturalists have learned by applying whole-systems thinking to food production and habitat restoration has profound and immediate application to the human ecology of the urban environment' (p. xi). In other words, permaculture creates holistic designs modelled on nature's wisdom, a strategy that is in sharp contrast to the piecemeal approach that so often characterises traditional urban design.

A permaculture flower diagram, modified from Holmgren's original, is shown that contains ten petals, each symbolic of a basic human need: food, energy, water, shelter, waste, health, spirit, community, justice, and livelihood; and all integrated in an organisational framework from personal, through local, to regional perspectives. The foundations of the permaculture design process are the core principles for ecological design: observe, connect, catch, and store energy and materials; make each element perform multiple functions, each being supported by multiple methods; make the least change for the greatest effect by starting with small-scale/ intensive systems; optimise edges, collaborate with succession, and use biological resources before technological ones; turn problems into solutions to obtain a yield; and recognise that lack of creativity is the greatest limit to learning from mistakes. It is these principles that are part of 'the *regenerative* toolbox' (p. 231) which serves as 'the touchstones for *regenerative* design' (p. 25).

The book begins by focusing on designing gardens for urban homes and communities before moving on to addressing physical issues related to water and energy and then social issues concerning livelihood and wealth. Recognising that at its heart sustainability is a social, psychological, as well as political, problem, permaculture design is motivated by a desire to empower communities through placemaking. Healthy communities need to be more than simply well-designed physical spaces, however; Hemenway suggests that equal attention also needs to be directed toward creating well-designed relationships and social systems. Placemaking in this light is about 'the ways that the physical infrastructure of the city affects what people do and feel' (p. 263). Because growing food ecologically builds diverse links to people, policies, and nature, permaculture design expands the work of sustainability in terms of 'moving human beings onto the path of a *regenerative* culture' (p. 240).

~

Book 8: *Routledge handbook of sustainable and regenerative food systems,* edited by Jessica Duncan, Michael Carolan, and Johannes S.C. Wiskerke (2021), is part of flurry of recent publications that reiterate the need to reconsider agriculture from a wider perspective of social-ecologicals system theory: *Regenerative agriculture* (Wafullah and Munene 2021), *Regenerative agriculture: farming with benefits profitable farms, healthy food, greener planet* (Landzettel 2021), *Regenerative agriculture: What's missing? What do we still need to know?* (Dent and Boincean 2021), and *Agroecology and regenerative agriculture: sustainable solutions for hunger, poverty, and climate change* (Shiva 2022). Where Duncan *et al.*'s approach is singular is in its comprehensive consideration of the concepts of regeneration as applied to *all* facets of the field-to-fork, production-to-consumption food system, rather than just the growing of food.

The book begins by making the claim that there is a need to move beyond sustainability and its preoccupation with efficiency, towards regeneration: 'it is clear that we need better concepts and stories that position us as part of nature; not as sustainers of nature, but as active participants in an integrated cycle of *regeneration*' (p. 21). Regeneration in this light is a holistic practice that moves beyond environmental impacts towards the creation of mutually supportive symbioses across the entire food system in both rural and urban settings. With respect to food production per se, '*regenerative* agriculture is a system of farming principles and practices that increases biodiversity, enriches soils, improves watersheds, and enhances ecosystem services' (p. 329).

Case studies highlight the important benefits of multifunctional agriculture. Chapters introduce various aspects such as the concepts of 'care-farming' and 'social agriculture' and how these contribute to community resilience; the need to abandon the false duality of nature and society as being separate; and the importance of circular food economies for regenerative design. Other chapters address the urban context of regenerative food systems, including food sharing in co-op distribution centres; linkages between peri-urban farms and both farmers' markets and food banks; racial equality and the new 'repeasantization' of households in Detroit; issues of food miles with respect to concerns about social justice and zero carbon footprints; and the utility of intensive urban farming practices such as hydroponics.

One important message is that 'planning *regenerative* working landscapes' (p. 376), as one chapter title phrases it, necessitates regarding agricultural practices and food systems from a landscape perspective based on human engagement and the spatial dimensions of placemaking. Furthermore, and with respect to sustainable city planning, it is essential to recognise that 'food is a fundamental city infrastructure which pervades, and shapes, all four key urban functions – dwelling, work, recreation, and transportation – codified by modern city planners' (p. 410). Only in this way will urban planners perform their jobs effectively and without failure.

1.3.3 People

Book 9: *Regenerative design for sustainable development*, by John Tillman Lyle (1994), is held in as high esteem today among progressive designers as when it was first published. It is regarded as a seminal work in the recognition and development of holistic or systems thinking, in the spirit of such classic, earlier books as *General systems theory* (Von Bertalanffy 1969) and *Operating manual for spaceship Earth* (Fuller 1969).

Based on recognition that sustainability requires that supply systems for energy and materials must be continually self-renewing or regenerative, '*regenerative* design means replacing the present linear system of throughput flows with cyclical flows at sources, consumption centres, and sinks' (p. 10). Regenerative design accomplishes this by adopting and making operational a set of demonstrable natural processes for human purposes; in other words, through using natural means of replenishment to maintain a functional wholeness. Landscapes are therefore more than physical contexts of ecosystems; they are dynamic processes. Regenerative systems can encompass every level of technology from the very simple to the highly complex, with the commonality being that they continue to evolve *after* their initial design and implementation, such that management also becomes a creative activity.

Strategies for regenerative design are enmeshed in complex natural and social processes, including letting nature do the work, considering nature as both model and context, aggregating rather than isolating elements, seeking multi-functionality and common solutions, and prioritising for sustainability. As Lyle correctly states (p. 11), '*regeneration* has to do with rebirth of life itself, thus with hope for the future.' As such, regenerative systems are not new, as for example, the concept of permaculture wherein crop production areas are regarded as functional ecosystems rather than outdoor factories. Nevertheless, a cautionary approach is needed given that 'of all *regenerative* practices and technologies, those related to agriculture are most enmeshed in uncertainty' (p. 213).

~

Book 10: *Designing for hope: pathways to regenerative sustainability*, by Dominque Hes and Chrisna du Plessis (2015), with its emphasis on participatory engagement and hope as guiding principles for regenerative design and restorative healing, follows from ideas developed earlier in *Healing natures, repairing relationships: new perspectives on restoring ecological spaces and consciousness* (France 2007) and *Handbook of regenerative landscape design* (2008).

A key point in this new retelling is of the need to explore mindsets and practices that go beyond notions of sustainability, so as to 'contribute to the well-being, nourishment and *regeneration* of the world and all its communities' (p. 11). This, the authors correctly note, is a much larger undertaking then Lyle's (1994) earlier focus on metabolic flows in biophysical systems. Echoing Palazzo and Seiner (2011), the authors state that '*regenerative* design emphasises the need to be open to all the ways of knowing a site, so that solutions can be developed that will enhance the potential and *regenerative* capacity of the site (p. 41). To this end, individual chapters reiterate concepts made earlier by others, including reconnecting to nature (i.e. 'nature deficit disorder'), patterning design on nature (i.e. 'biophilia'), the permaculture flower motif, and the analogy of the designer as a master gardener (holistic restoration). Rather than originality, the strength of the book is in its integration of these and other contributing components. As well, the important point is made that regenerative sustainability means more than merely reinvigorating or reviving a system, or even restoring it to a better state or condition; it means changing the system to something that is different and better, doing so through a co-evolutionary process of reconnecting human aspirations and activities with natural systems.

Agriculture figures prominently in the authors' discussion of regenerative sustainability. Besides giving credence to the over-arching importance of permaculture, mention is made of a handful of interesting urban agriculture projects, including an art-filled community garden, a waterfront redevelopment scheme, edible plantings at a sustainable building centre, a green grocery store co-op, and a u-pick garden along a sidewalk.

Hes and du Plessis' work is based on stewardship and the need to acknowledge the importance of relationship to place; 'place' being, from a regenerative point of view, 'far more than topography, climate, light and tectonic form ... [but rather] a construct which emerges from the entire network of ecological and cultural systems and their interactions within a geographic area' (p. 117). Quoting from a design organisation called Regenesis, the authors aptly state that '*regeneration* is first and foremost 'a process of engagement with the purpose of healing living systems and birthing a new spirit to consciously participate in expanding the healing process' (p. 120).

References

Aronson, J., S.J. Milton and J.N. Blignaut. 2007. Restoring natural capital: science, business, and practice. Island Press, Washington, DC, USA.

Battisti, A. 2021. Green dreams: regenerating cities through nature. In: Sayigh, A. and A. Trombadore (eds.) The importance of greenery in sustainable buildings. Springer, Berlin, Germany.

Berger, A. 2008. Designing to reclaim the landscape. Routledge, London, UK.

Boyce, J.K., S. Narain and E.A. Stanton. (eds.) 2007. Reclaiming nature: environmental justice and ecological restoration. Anthem Press, London, UK.

Calvo, S. and A. Morales. 2021. Social innovation in Latin America: maintaining and restoring social and natural capital. Routledge, London, UK.

Cole, R.J. 2012. Regenerative design and development: current theory and practice. Building Research Information 40: 1-6.

Council for Agricultural Science and Technology (CAST). 2002. Urban and agricultural communities: opportunities for common ground. Butler, A. and B. Maronek (eds.) Task Force Report 138, CAST.

Crookes, D.J. 2012. Modelling the ecological-economic impacts of restoring natural capital, with a special focus on water and agriculture, at eight sites in South Africa. Ph.D. Thesis. Stellenbosch University, South Africa.

De la Salle, J. and M. Holland. 2010. Agricultural urbanism: handbook for building sustainable food & agriculture systems in 21st century cities. Green Frigate Books/Libri, Farringdon, UK.

Dent, D. and B. Boincean 2021. Regenerative agriculture: what's missing? What do we still need to know? Springer, Berlin, Germany.

Dobson, M.C., J.L. Edmondson and P.H. Warren. 2020. Urban food cultivation in the United Kingdom: quantifying loss of allotment land and identifying potential for restoration. Landscape Urban Planning 199: 34-39.

Duany, A. 2011. Garden cities: theory & practice of agrarian urbanism. Duany Plater-Zyberk & Co., Miami, FL, USA.

Duncan, J., M. Carolan, and J.S.C. Wiskerke. (eds.) 2021. Routledge, London, UK handbook of sustainable and regenerative food systems. Routledge, London, UK.

Elidae, M. 1991. The myth of the eternal return, or, cosmos and history. Princeton University Press, Princeton, NJ, USA.

France, R.L. (ed.) 2007. Healing natures, repairing relationships: new perspectives on restoring ecological spaces and consciousness. Green Frigate Books/Libri, Farringdon, UK.

France, R.L. (ed.) 2008. Handbook of regenerative landscape design. CRC Press, Boca Raton, FL, USA.

France, R.L. 2011. Restorative redevelopment of devastated ecocultural landscapes. CRC Press, Boca Raton, FL, USA.

France, R.L. 2016a. The once and future city farm. In: France, R.L. (ed.) Integrated urban agriculture: precedents, practices, prospects. Libri, Farringdon, UK.

France, R.L. (ed.) 2016b. Integrated urban agriculture: precedents, practices, prospects. Libri, Farringdon, UK.

France, R.L. 2020. Waymarking Italy's influence on American environmental imagination while on pilgrimage to Assisi. Cambridge Scholars Publishing, Newcastle upon Tyne, UK.

France, R., J. Campbell and K. Sherren. 2019. Enhancing the social and natural capital of Canadian agro-ecosystems through incentive-based 'Alternative Land Use Services' (ALUS) programs: recurring themes and emerging lessons. Journal of Geoscience and Environmental Protection 7: 139-162.

Francis, M. 2001. A case study method for landscape architecture. Landscape Journal 20: 15-29.

Fuller, R.B. 1969. Operating manual for spaceship earth. Lars Muller Publishers, Baden, Switzerland.

Garrow, D. and N. Wilkin. 2022. The world of Stonehenge. The British Museum, London, UK.

Gobster, P.H. and R.B. Hull. 2000. Restoring nature: perspectives from the social sciences and humanities. Island Press, Washington, DC, USA.

Gottero, E. 2021. Agrourbanism: tools for governance and planning of agrarian landscape. Springer, Berlin, Germany.

Gwetsayi, R.T., L. Dube and C. Mashapa. 2016. Urban horticulture for food security and livelihood restoration in Mutare City, Eastern Zimbabwe. Greener Journal of Social Science 6: 56-64.

Handel, S.N. 2016. Greens and greening: Agriculture and restoration ecology in the city. Ecological Restoration 34: 1-2.

Hemenway, T. 2015. The permaculture city: regenerative design for urban, suburban, and town resilience. Chelsea Green Publishing, White River Junction, VT, USA.

Hes, D. and C. du Plessis. 2015. Designing for hope: pathways to regenerative sustainability. Earthscan, London, UK.

Higgs, E. 2003. Nature by design: people, natural process, and ecological restoration. MIT Press, Cambridge, MA, USA.

Hodgson, K., M.C. Campbell and M. Bailkey. 2011. Urban agriculture: growing healthy, sustainable places. American Planning Association, Chicago, IL, USA.

Holmgren, D. 2002. Permaculture: principles and pathways beyond sustainability. Holmgren Design Serv, Hepburn Springs, Australia.

Hough, M. 2004. Cities and natural process: a basis for sustainability. Routledge, London, UK.

Jordan III, W.R. 2003. The Sunflower forest: ecological restoration and the new communion with nature. University of California Press, Berkeley, CA, USA.

Kirkwood, N. 2001. Manufactured sites: rethinking the post-industrial landscape. Spon Press, Abington, UK.

Kushwaha, S. 2021. Restoration agriculture. Delve Publishing, Burlington, Canada.

Ladner, P. 2011. The urban food revolution: changing the way we feed cities. New Society Publishers, Gabriola Island, Canada.

Landzettel, M. 2021. Regenerative agriculture: farming with benefits. Profitable farms. Healthy food. Greener planet. World University Service, Wiesbaden, Germany.

Levadna, A. 2018. Regeneration of Pozniaky village: giving new life to the hidden heritage through urban agriculture. M.A. Thesis, Scuola di Architetture Urbanistica Ingegneria delle Costruzioni.

Lewis-Williams, D. and D. Pearce. 2005. Inside the Neolithic mind. Thames & Hudson, London, UK.

Lovell, S.T. 2010. Multifunctional urban agriculture for sustainable land use planning in the United States. Sustainability 2: 2499-2522.

Lyle, J.T. 1994. Regenerative design for sustainable development. John Wiley& Sons, Hoboken, NY, USA.

Lyson, T.A. 2004. Civic agriculture: reconnecting farm, food, and community. Tufts University Press, MA, USA.

Marshall, R. 2001. Waterfronts in post-industrial cities. Spon Press, Abington, UK.

Mohan, S.V., M. Hemalatha, K. Amulya, G. Velvizhi, P. Chiranjeevi, O. Sarkar, A.N. Kumar, K.V. Krishna, J.A. Modestra, S. Dahiya, D.K. Yeruva, S.K. Butti, J.S. Sravan, S. Chatterjee, and R. Kona. 2020. Decentralized urban farming through keyhole garden: a case study with circular economy and regenerative perspective. Material Circular Economy 2: 1-24.

Mougeot, L.J.A. 2005. Agropolis: the social, political, and environmental dimensions of urban agriculture. Earthscan, London, UK.

Nawfal, M. and A.L. Mudares. 2017. Urban agriculture as one of the ecological applications of the regenerative city. Journal of Engineering 9: 64-81.

Norris, M. and D. Joffe. 2018. Agrihoods: cultivating best practices. Urban Land Institute, Washington, DC, USA.

Orsini, F., G. Pennisi, N. Michelon, A. Minelli, G. Bazzocchi, E. Sanye-Mengual and G. Gianquinto. 2020. Features and functions of multifunctional urban agriculture in the Global North: a review. Frontiers of Sustainable Food Systems 4: 562513.

Palazzo, D. and F. Steiner. 2011. Urban ecological design: a process for regenerative places. Island Press, Washington, DC, USA.

Parker, T.L. 2010. The case for urban agriculture: regenerative, human-scale food production systems in urban landscapes. M.A. Thesis, Washington State University.

Pothukuchi, K. and J.L. Kaufman. 2000. The food system: a stranger to the planning field. Journal of the American Planning Association 66: 113-1124.

Rashed, R. 2018. Urban agriculture: a regenerative urban development practice to decrease the ecological footprints of cities. International Journal of Environmental Science and Sustainable Development 8: 85-98.

Redwood, M. (ed.) 2008. Agriculture in urban planning: generating livelihoods and food security. Earthscan, London, UK.

Russo, A. and G.T. Cirella. 2020. Edible green infrastructure for urban regeneration and food security: case studies from the Campania region. Agriculture 10: 358-373.

Schurig, S. and K. Turan. 2022. The concept of a 'regenerative city': how to turn cities into regenerative systems. Journal of Urban Regeneration and Renewal 15: 161-175.

Shepard, M. 2013. Restoration agriculture: real-world permaculture for farmers. Acres USA, Greeley, CO, USA.

Shidqina, L. 2021. Farm the gaps! Regenerative residual spaces in between infrastructure for agro-civism to support urban food security in Kuala Lumpur, Malaysia. M.A. Thesis, Scuola di Architetture Urbanistica Ingegneria delle Costruzioni.

Shiva, V. 2022. Regenerative agriculture: sustainable solutions for hunger, poverty, and climate change. Synergetic Press, Santa Fe, NM, USA.

Specht, K., J. Schimichowski and R. Fox-Kamper. 2021. Multifunctional urban landscapes: the potential role of urban agriculture as an element of sustainable land management. In: Weith, T., T. Barkmann, N. Gaasch, S. Rogga, C. Straub and J. Zscheischler (eds.) Sustainable land management in a European context: a co-design approach. Springer, Berlin, Germany.

Spirn, A. W. 1984. The granite garden: urban nature and human design. Basic Books, New York, NY, USA.

Steel, C. 2008. Hungry city: how food shapes our lives. Vintage Books, New York, NY, USA.

Steel, C. 2016. Sitopia and the urban paradox. In: Farming the city: food as a tool for today's urbanisation. CITIES.

Swensen, G., V.E. Stafseng and V.K. Simon Nielsen. 2012. Visionscapes: combining heritage and urban gardening to enhance areas requiring regeneration. International Journal of Heritage Studies 28: 511-537.

Throop, W. 2000. Environmental restoration: ethics, theory, and practice. Humanity Books, Amherst, NY, USA.

Viljoen, A. (ed.) 2008. Continuous productive urban landscapes: designing urban agriculture for sustainable cities. Elsevier, Amsterdam, the Netherlands.

Von Bertalanffy, L. 1969. General system theory: foundations, development, applications. George Braziller, New York, NY, USA.

Wafullah, N.T. and K.C. Munene. 2021. Regenerative agriculture. Delve Publishing, Burlington, Canada.

Wansborough, M. and A. Mageean. 2000. The role of urban design in cultural regeneration. Journal of Urban Design 5: 181-197.

Yaman, S.D. 2019. Urban agriculture for sustainable regeneration: Nine Elms, London. M.A. Thesis, Scuola di Architetture Urbanistica Ingegneria delle Costruzioni.

Zari, M.P. 2018. Regenerative urban design and ecosystem biomimicry. Routledge, London, UK.

Zasada, I. 2011. Multifunctional peri-urban agriculture – A review of societal demands and the provision of goods and services by farming. Land Use Policy 28: 639-648.

Chapter 2
Narrative scholarship and peripatetic phenomenology as landscape research

2.1 Sensorial embodied experience

The experience of nature: a psychological perspective (Kaplan and Kaplan 1989) is regarded as a seminal publication in the field of behavioural geography, ecopsychology, and landscape studies. Much of its content concerns the visualisation of mundane urban sites in concordance with Gibson's (1979) idea of 'affordance,' or that which the environment offers to a perceiver. The satisfaction or pleasure obtained from 'deep immersion' (France 2003a) in a particular landscape is part of a recreational opportunity spectrum, as for example, the recognition that gardeners 'derive much more than vegetables and flowers from their efforts' (Kaplan and Kaplan 1989: 119). The book refers to both restorative environments and restorative experiences, and considers the importance of nature and wellbeing that can result from experiencing an environment through the simple act of walking and allowing one's senses to get the 'feel' of a place. None of this is new. Back in the nineteenth century, Thoreau would famously go out of his way to use not only his eyes (France 2001) but his entire body as an instrument in which to perceive and experience, and from that, to comment upon, nature (France 2003b). Today such approaches are referred to as 'phenomenology,' for which many definitions have been suggested.

Brown and Toadvine (2003) equate phenomenology with a return to the elements or 'things' of the environment themselves rather than theoretical abstractions of such. Rooted in the philosophy of, among others, Merleau-Ponty and Heidegger, phenomenology regards the world as it is directly experienced. As Tilley (2004: 1) elaborates, 'the aim is not to explain the world ... but to describe that world as preciously as possible in the manner in which human beings experience it;' in short, it is the interdisciplinary examination of human relationship with nature. Tilley's (2004: 16) summary is apt: 'the real world is the perceived world is the phenomenological world.'

For Wylie (2012), phenomenology concerns the lived and immediate, not pre-meditated, experience of utilising the senses to acquire knowledge about the world. Encounters are 'based on the sensory perception of the human body' (p. 10) and it is important to move beyond ocular-centric dominance to consider the whole-body experience. In support of this contention, he cites Tilley (1994), an archaeologist notable for interpreting the past through qualitative analyses of the present, vis-à-vis interacting with the landscape in a way that uses the body rather than just the eyes, as for example, the sight, touch, sound, and smell of the constituent elements of rock menhirs in northern France (Tilley 2004). Both scholars share a criticism that most academics have tended to reduce landscapes to visual scenes and not bodily experience, and that although photos alone can offer passive appreciation of a site, they may be but a poor representation of everything that was experienced.

DOI 10.3920/978-90-8686-938-1_2, © Wageningen Academic Publishers 2022

Phenomenology, for Bannon (2016), is a relational philosophical method based on examining experience and engagement with the world as we actually encounter it through perception. It is about being at home in the world through seeing with deliberate intention more than through visual impressions, and is based on the meaning and value inherent in the body's relation – perceptual, emotional – to its surroundings. For example, beauty, to a phenomenologist, is not a trait possessed by the environment but is rather a quality of the *relationship* between the perceiver and that which is perceived.

Others refer to phenomenology as being the geography of the 'lifeworld' (Seamon 1980) or 'life space' (Wilson 1980), with the notion of body as subject but also as instrument to map out body-and-place choreographies or 'place-ballets.' The idea of using the experience of embodied engagement with a landscape (Unwin 2008) to create sensuous maps of socially embedded aesthetics or urban 'sensecapes' has been explored by Degen (2008). 'I would now like to invite the reader on a sensuous journey ...' (p. 41), encapsulates her phenomenological approach. Agreeing with Thibaud (2015), Degen and Lewis (2020) consider atmosphere to be a useful concept for describing the relational dynamics of people and place. In particular, sensory dimensions make up what we experience as the atmosphere of place, as for example, its vibrancy or its eclectic character. Phenomenology is therefore that which resonates with us when we get 'a feel' or 'a sense' of a place. But here too, as has been previously noted, there is often a visual bias that takes precedence over the other senses. To counter this, Degen (2008: 46) specifically mentions that 'tastescapes play an increasingly major role in the sensuous landscape of regenerated neighbourhoods' because, increasingly, an identity of place is often portrayed through its gastronomic reputation. Moreover, despite sight being the most immediate sense, it is often the least memorised, especially in comparison to smell (Tuan 1990).

In the end, phenomenology uses the crafts of wordsmithing and photography to craft sensuous stories that describe the perceptions and emotions of bodily engaging with nature. As Tilley (2004: 26) describes, 'from a phenomenological perspective, language flows from the body rather than the mind, or rather, from a mind that is embodied, bound up with the sensorial world.' And the way to capture this flow of bodies, words, and images is through use of 'thick descriptions.'

2.2 Thick descriptions

Throughout his distinguished career, the eminent landscape scholar, Denis Cosgrove, called for the need to discover and exploit alternative forms of writing and representing the intertwined – inner and outer – geographies of people and place. For many geographers, landscapes have become disembodied, inert entities that are studied from afar rather than through direct immersion (Wylie 2012). In contrast to richly textured, carnal descriptions of interactions with an environment, the standard fare in virtually all academic journals are sensorially impoverished, analytical accounts (Tilley 2004). A counter to this approach is 'thick description,' whereby clear pictures of the environment or situation are conveyed through self-reflective essays whose goal are to make one's experiences concrete for readers through use of evocative language that shows rather than merely tells (Anonymous 2012; Geertz 1973).

Thick description is a qualitative research approach in social science that is based on providing detailed narratives or vignettes and interpretations of situations observed and experienced by a researcher, as well, where applicable, supplemented with background information (Ponterotto (2006). The approach moves beyond presenting facts and overt appearances, and instead provides details, context, emotions, and underlying meanings and intentions. Whereas thin descriptions are limited to what is observed, thick descriptions may comment on cultural significance, emotional interpretations, and finer details to such a degree as to provide the reader of the account with impressions of verisimilitude as if they were corporeally, not only, conceptually, present. Therefore, rather than avoiding subjectivity as in empirical research, thick description research embraces personal thoughts and feelings and observations in its attempt to describe the evocative interactions of a person and a place, making the experiences of the former visible to a reader so that s/he can better appreciate the latter. Thick descriptions have become a standard tool in the discipline of 'auto-ethnography' (Adams *et al.* 2015; Humphries 2005), in which the researcher uses self-reflection and writing to explore anecdotal and personal experience and to connect this personal narrative to wider cultural and social meanings and scholarly understanding.

At its core, thick description is 'sensemaking,' the process of giving meaning to experiences (Ponterotto 2006). It can take a variety of forms, as, for example, switching between first-person and second-person narration. This is what is referred to as 'layered accounts,' qualitative research in which the researcher writes from the perspective of more than a single voice (Hermann 2012). Interpretive archaeologist Christopher Tilley pioneered such a phenomenological approach to landscape studies through developing an alternative mode of scholarship which incorporates the importance of thick description and sensuous engagement emphasising participant observation, with referencing the observations of preceding scholars (Tilley 1994, 2004). This expanded form of phenomenology (Wylie 2012) therefore combines immediate embodied experience with other forms of analysis, such as archival research and critical evaluation of the pertinent literature. Although not mentioned by these previous scholars, the critical evaluation aspect is similar to that used in case study analyses by landscape architects (Francis 2001) and agricultural land-use planners (France and Campbell 2015) through the 'W5' (i.e. who, what, when, where, why) protocol (France 2013).

Because thick descriptions endeavour to paint a clear picture of the environment or situation (Anonymous 2012), the research method of intertwining of photo images and words is this regard is part of the discipline of 'visual ethnography.' Photos are open documents that provide viewers layered meanings (Kharel 2015), and as such, contribute to layered accounts. The subjective understanding provided by participant observation research of visually augmented thick descriptions conveys not only facts but also commentary and interpretation. However, unlike documentary films, photos need voice, the challenge being to organise words and images in such a way to describe the *story* of the experience (Geertz 1973).

The use of photography in social science and anthropology has a long-established tradition (Harper 1987). In particular, ethnographers have often created visual diaries to capture elements deemed interesting or significant (Johnsen *et al.* 2008), just as phenomenologists have employed cameras to 'reveal the dynamic relationship between a perceiver and their environment' (Unwin 2008:8). The back cover of Tilley's (1994) seminal book describes it as 'an extended photographic essay about topographic features of the landscape.' These approaches have newfound purpose in visual narratives designed to record the experience of urban walks (Degen and Rose 2012; Mason and Davies 2009; Middleton 2010). As Coles (2014) describes, 'the sequenced images provoke phenomenological and narrative modes that go beyond individual images to frame a collection of narratives ... [that encapsulate] the inherent complexity of place.' This is part of the critical spatial practice referred to as topographic 'site-writing,' wherein an assemblage of photos and words 'utilise different types of writings, and more broadly, 'texts', to engage with place and place-making and to (re)present the interrelations between its 'surfaces' and 'structures.'' And for many, the implicit reflexivity of visual/textual narratives for examining the feelings, reactions, and motives that influence one's impression of a place, is best undertaken through the simple act of pedestrianism.

2.3 Walking methodology

In the eighteenth and nineteenth centuries, walking came to be regarded as a way in which to experience rather than simply to traverse through a landscape (Ingold 2004; Wunderlich 2008), something that was celebrated in both poetry and prose (Coverly 2012; Jarvis 1997; Solnit 2002; Wallace 1993). Today, walking is regarded as the key means through which to feel, and from that, to understand, a landscape (Howard *et al.* 2013; Ingold 2004, 2011).

As a modern social science methodology (Bates and Rhys-Raylor 2017), walking makes urban places both commendable as well as comprehendible (Aoki and Yashimizu 2015; Kusenbach 2003; Middleton 2010, 2011; Rendell 2006; Vergunst 2010). Urban walking as aesthetic practice (Careri 2017) has its roots in a series of mid-twentieth century Parisian ideas and practices of avant-garde artists and political activists who built upon the concept of the *flâneur* (Jenks and Neves 2000), and which has more recently been reborn in the experiential field of 'psychogeography' (Basset 2004; Richardson 2015). The 'walk-' or 'go-along' method is a qualitative ethnographic research tool based in phenomenology that uses walks buttressed by opinions of others garnered through interviews or through readings to explore everyday experience related to environmental perception, spatial practices, and social expressions of landscape form and function (Doring and Ratter 2021; Kusenbach 2003). Degen (2008), for example, begins her book about 'participant pedestrianism,' in which observations and interpretations are described through researchers' discourses, with the words: 'Walk down any street in any city ...' (p. 5).

By engaging our senses, walking is integral to our perception of an environment (O'Neill and Roberts 2020). It enables researchers to tune into their sensory experiences through relational descriptions, and in so doing, enables place-learning (Springgay and Truman 2018). The field of scholarship in this regard is expanding rapidly. For example, a conference held in the United Kingdom on 'walking stories' and 'walking ethnography,' was based on idea that the land evokes conversations, reflections, and narrations that take place only through movement. As well, the 'The Nature of Cities' conference, held during the pandemic in 2021, had a series of online field trips in which international attendees could watch live-narrated video recordings of actual walks along urban riverscapes in London, Beirut, and Melbourne, city forests in India, British Columbia, and Oregon, and neighbourhoods, parks, playgrounds, canals, and landfills in England, Scotland, New York, California, Hawaii, and Argentina. Walking offers possibilities for more evocative and creative forms of academic writing while also based on the idea that the relation of an individual's experiences to a landscape provides an opportunity to comment upon wider questions (Wylie 2012). Place-learning through walking involves sensory inquiry and embodiment of experience that connects mind, body, and the environment (Springgay and Truman 2018), and is becoming increasingly utilised in the professional literature (Clarke and Jones 2001; Edenson 2000; Sidaway 2009; Wylie 2002, 2005).

More than providing transportation, walking is therefore an elemental way of perceiving place (Wunderlich 2008), and is part of the relational study of landscapes (Doring and Ratter 2021). As Tilley (2004: 26) states, 'places and landscapes are created and experienced through mobility.' In this regard, landscape is not a physical constant but is something that is created through relationships to its inhabitants through their perceptions and embodied experiences (Ingold 2011). In short, through walking, Tilley (1994) explains, landscapes are woven into life while in turn lives are woven into the landscape. Walking, in this light, is the process of appropriating topography, with the act of moving through a landscape being just as important as that of arriving to it. The sensate and kinaesthetic attributes of walking allow it to a placemaking practice that showcases how we interpret our surroundings and our position within it (O'Neill and Roberts 2020). In this light, walking, as a method of exploring a city, is an aesthetic process that creates place from space (Careri 2017). The ensuing 'walkscapes' can be likened to being architectural creations based on perceiving and conceiving part of the world. In a sense, walkscapes are art projects traced upon the topography of a landscape through use of the body in much the same way as a painter would use a brush upon a canvas.

'Mind-walking' (Ingold 2011) creates 'paths of observation' wherein perception is a function of movement (Wunderlich 2008). As Gibson (1979: 46) states, 'cognition should *not* be *set off* from locomotion along the lines of a division between head and heels, since walking is itself a form of circumambulatory knowing.' Once this is recognised, he continues, 'a whole new field of inquiry is opened up in which our knowledge of the environment is altered by techniques of footwork.'

For Gibson (1979), perception derives, *ipso facto*, from movement. Ingold (2002, 2004) argues that, more than cognition, it is locomotion that underscores and facilitates perception and is a 'form of circumambulatory knowing.' Although walking developed into the visual activity of scanning, Ingold (2004) argues that it is really much more. And so, while the feet serve to ground us in space, we perceive through our *entire* bodies not only the eyes, all contributing to the haptic perception of place. Walking, therefore, integrates the senses of sight, sound, smell, and touch (Lund 2006). Both Ingold (2002) and Tilley (2004) refer to the synaesthesia of the experience; in other words, the fusion of the senses, as for example, the idea of the hearing ear, the seeing ear, etc. Nevertheless, both scholars acknowledge there to be a biased ocular-centrism of the Western tradition that privileges sight over other senses as sources of human knowledge.

Not only is the walking experience multi-sensorial (Ingold 2002, 2011; Wunderlich 2008), it is also multi-conceptual in that is fosters peripatetic ponderings (Middleton 2010). For as O'Neill and Roberts (2020: 216) describe, 'a walk in a garden is a phenomenological sensual experience and a physical activity, but one also set within memory experience – with invocations and resonance that draw us towards, or that we search out through, our senses and our emotions. A garden walk is also a joint, relational experience – a 'conversation' not merely with oneself, but also a 'dialogue' with oneself and the environment.' Landscapes exist in the mind as much as they do on the ground; in other words, they are cultural just as much as natural objects (Creswell 2004), being places of memory (Schama 1996). Being cultural phenomena, landscapes are built from subjective experience, à la Heidegger's *Dasein* or 'being there,' facilitated by the concordant flows of bodies, perceptions, and conceptions along 'emotional pathways' (Viik 2011). And it is for this reason that the sensory experiencing and perceptual memory of the form of regenerative urban designs is best facilitated through the act of walking (Degen and Rose 2012).

References

Adams, T.E., H. Jones and C. Elles. 2015. Autoethnography: understanding qualitative research. Oxford University Press, Oxford, UK.

Anonymous. 2012. Tips for writing thick descriptions for ethnographies and case studies. Available at: www.vanderbilt.edu/writing.

Aoki, J. and A. Yashimizu. 2015. Walking histories, un/making places: walking tours as ethnography of place. Spacea dn Vulture 18: 273-284.

Bannon, B.E. 2016. Nature and experience: phenomenology and the environment. Rowman and Littlefield, Lanham, MD, USA.

Bassett, K. 2004. Walking as an aesthetic practice and a critical tool: some psychogeographic experiments. Journal of Geography in Higher Education 28: 397-410.

Bates, C. and A. Rhys-Taylor (eds.) 2017. Walking through social research. Routledge, London, UK.

Brown, C.S. and T. Toadvine. 2003. Eco-phenomenology: back to the Earth itself. State University of New York Press, Albany, NY, USA.

Careri, F. 2017. Walkscapes: walking as aesthetic practice. Culicidae Architectural Press, Ames, IA, USA.

Clarke, P. and O. Jones. 2001. Dwelling, place and landscape: an orchard in Somerset. Environment and Planning A 33: 649-666.

Coles, B.F. 2014. Making the market place: a topography of Borough Market, London. Cultural Geographies in Practice 21: 515-523.

Coverley, M. 2012. The art of wandering: the writer as walker. Oldcastle Books, Harpenden, UK.

Creswell, T. 2004. Place: a short introduction. Blackwell, Hoboken, NY, USA.

Degen, M.M. 2008. Sensing cities: regenerating public life in Barcelona and Manchester. Routledge, London, UK.

Degen, M. and C. Lewis. 2020. The changing feel of place: the temporal modalities of atmosphere in Smithfield Market, London. Cultural Geographies 27: 509-526.

Degen, M.M. and G. Rose. 2012. The sensory experiencing of urban design: the role of walking and perceptual memory. Urban Studies 49: 3271-3287.

Doring, M. and B. Ratter. 2021. 'I'll show you *my* coast...' – A relational study of coastscapes in the North Frisian Wadden Sea. Maritime Studies 20: 317-327.

Edenson, T. 2000. Walking in the countryside: reflexivity, embodied practices and ways to escape. Body and Society 6: 81-116.

France, R.L. (ed.) 2001. Reflecting heaven: Thoreau on water. Houghton Mifflin, Boston, MA, USA.

France, R.L. 2003a. Deep immersion: the experience of water. Green Frigate Books/Libri, Farringdon, UK.

France, R.L. (ed.) 2003b. Profitably soaked: Thoreau's engagement with water. Green Frigate Books/Libri, Farringdon, UK.

France, R.L. 2013. Exploring the bonds and boundaries of water management in a global context. Journal of Cleaner Production 60: 1-3.

France, R.L. and J.B. Campbell. 2015. Payment for agro-ecosystem services: developmental case-history descriptions of Canada's grassroots 'ALUS' programs. Research Journal Agriculture and Environmental Management 4: 405-431.

Francis, M. 2001. A case study method for landscape architecture. Landscape Journal 20: 15-29.

Geertz, C. 1973. Thick description: toward and interpretive theory of culture. In: Interpretations of cultures. Basic Books, New York, NY, USA.

Gibson, J.J. 1979. The ecological approach to visual perception. Houghton Mifflin, Boston, MA, USA.

Harper, D. 1987. The visual ethnographic narrative. Visual Anthropology 1: 1-19.

Hermann, A. 2012. Criteria against ourselves? Embracing the opportunities of qualitative research. International Review of Qualitative Research 5: 135-152.

Howard, P., I. Thompson and E. Waterton. 2013. The Routledge companion to landscape studies. Routledge, London, UK.

Humphries, M. 2005. Getting personal: Reflexivity and autoethnographic vignettes. Qualitative Inquiry 11: 840-860.

Ingold, T. 2002. The perception of the environment: essays on livelihood, dwelling and skill. Taylor and Francis, Abingdon, UK.

Ingold, T. 2004. Culture on the ground. The world perceived through the feet. Journal of Material Culture 9: 315-340.

Ingold, T. 2011. Being alive: essays on movement, knowledge and description. Routledge, London, UK.

Jarvis, R. 1997. Romantic writing and pedestrian travel. Macmillan, New York, NY, USA.

Jenks, C. and T. Neves. 2000. A walk on the wild side: urban ethnography meets the flaneur. Cultural Values 4: 1-17.

Johnsen, S., J. May and P. Cloke. 2008. Imag(in)ing 'homeless places': using auto-photography to (re)examine the geographies of homelessness. Area 40: 197-207.

Kaplan, R. and S. Kaplan. 1989. The experience of nature: a psychological perspective. Cambridge University Press, Cambridge, UK.

Kharel, D. 2015. Visual ethnography: Thick description and cultural representation. Dhaulagiri Journal of Sociology and Anthropology 9: 147-160.

Kusenbach, M. 2003. Street phenomenology: the go-along as ethnographic research tool. Ethnogeography 4: 55-485.

Lund, K. 2006. Seeing in motion and the touching eye: walking over Scotland's mountains. Anthropological Journal 18: 27-42.

Mason, J. and K. Davies. 2009. Coming to our senses? A critical approach to sensory methodology. Qualitative Research 9: 587-603.
Middleton, J. 2010. Sense and the city: exploring the embodied geographies of urban walking. Social & Cultural Geography 11: 575-595.
Middleton, J. 2011. Walking in the city: the geographies of everyday pedestrian practices. Geography Compass 5: 90-105.
O'Neill, M. and B. Roberts. 2020. Walking methods: research on the move. Routledge, London, UK.
Ponterotto, J.G. 2006. Brief note on the origins, evolution, and meaning of the qualitative research concept thick description. The Qualitative Report 11: 538-549.
Rendell, J. 2006. Art and architecture: a place between. I.B. Tauris, London, UK.
Richardson, T. 2015. Walking inside out: contemporary British pychogeography. Rowman & Littlefield, Lanham, MD, USA.
Schama, S. 1996. Landscape and memory. Vintage Books, New York, NY, USA.
Seamon, D. 1980. Body-subject, time-space routines, and place-ballets. In: Buttimer, A. and D. Seamon (eds.) The human experience of space and place. Croom Helm, London, UK.
Sidaway, J. 2009. Shadows on the path: negotiating geopolitics on urban sections of Britain's South West Coast Path. Environment and Planning D: Society and Space 26: 1091-1116.
Solnit, R. 1999. Wanderlust: a history of walking. Viking Press, New York, NY, USA.
Springgay, S. and S.E. Truman. 2018. Walking methodologies in a more-than-human world: Walking Lab. Routledge, London, UK.
Thibaud, J.P. 2015. The backstage of urban ambiance: when atmospheres pervade everyday experience. Emotion, Space and Society 15: 39-46.
Tilley, C. 1994. A phenomenology of landscape: places, paths and movements. Berg, Oxford, UK.
Tilley, C. 2004. The materiality of stone: explorations in landscape phenomenology. Berg, Oxford, UK.
Tuan, Y-F. 1990. Topophilia: a study of environmental perceptions, attitudes, and values. Columbia University Press, New York, NY, USA.
Unwin, B.C. 2008. Phenomenology and landscape experience: a critical appraisal for contemporary art practice. Ph.D. Thesis. University of Hertfordshire, UK.
Vergunst, J. 2010. Rhythms of walking: history and presence in a city street. Space and Culture 13: 376-388.
Vilk, T. 2011. Human spatiality: a cultural phenomenology of landscapes and places. Problemos 79: 103-114.
Wallace, A.D. 1993. Walking, literature, and English culture: the origins and uses of the peripatetic in the nineteenth century. Clarendon House Publications, Sheffield, UK.
Wilson, B.M. 1980. Social space and symbolic interaction. In: Buttimer, A. and D. Seamon (eds.) The human experience of space and place. Croom Helm, London, UK.
Wunderlich, F.M. 2008. Walking and rhythmicity: sensing urban space. Journal of Urban Design 13: 1215-139.
Wylie, J. 2002. An essay on ascending Glastonbury Tor. Geoforum 33: 441-454.
Wylie, J. 2005. A single day's walking: narrating self and landscape on the South West Coast Path. Transactions of the Institute of British Geographers 30: 234-247.
Wylie, J. 2012. Landscape and phenomenology. Routledge, London, UK.

Part 2
Visits

Chapter 3
Healing economies: business and tourism

3.1 Commercial enterprises

Anecdote

Personal experiences of a young garden stand entrepreneur. The first income I ever earned, notwithstanding receiving a weekly allowance for doing household chores, was when, as a ten-year-old and following the advice of my maternal grandfather who had himself been an immigrant farmer, I set up a small garden stand. I grew up on an unpaved suburban street alongside a public golf course. At a short distance from our house, the course was bisected by a paved road across which golfers would have to move back and forth in order to complete their round of nine holes. This chokepoint became the targeted location for my first entrepreneurial venture. For a period of a week – until such time as the produce was depleted and my youthful interests diverted elsewhere – I harvested pods of peas and tufts of carrots that I and my grandfather had planted in the backyard garden. With these, a little foldable card-playing table, and a marketing sign extolling the virtues of the vegetables as a quick source of energy that could boost their chances of scoring a good round, I set up my little market stand, eager for the revenue to start rolling in. In my grandfather's advisory role, he suggested the benefit of offering added value in order to attract business. And so I also sold glasses of ice-cold lemonade, with the understanding that its purchase would facilitate a discount on the garden produce. Lastly, I branched out and sold golf-balls that had ended up on our front lawn or which I had searched for and found in the street-side strip of trees, both resulting from those with bad hook-shots off the second tee (the fact that I was asking golfers to buy back their own misguided balls was a source of good humour). Lemonade consumed, veggies in their hands, the golfers would continue on with their round. Some, if they were heading out on the course, would be return customers an hour later when they re-crossed the road on their return back to the clubhouse. I still remember cycling to the neighbourhood pharmacy and, to the slight disappointment of my grandfather who had lived a hard life with few luxuries, buying several comic books and chocolate bars with my first earnings, amazed at the ability to transform what had been simple seeds into products of personal pleasure.

DOI 10.3920/978-90-8686-938-1_3, © Wageningen Academic Publishers 2022

Book

Uncommon cents: Thoreau and the nature of business, by Robert M. Abbott, is based on the idea that Henry David Thoreau's opinions on business, wealth and wellbeing, simplicity and self-reliance, and leadership are not only quaint, parochial nineteenth-century postulations and pronouncements, but offer contemporary guidance in how to navigate a path to a more just, humane, and caring world through a lens of what is now called sustainability and corporate social responsibility. Each section begins with commentary about the shape of current thought, followed by quotations from Thoreau's copious writings, and concluding with provision of resources where readers can 'Learn More, Do More.' The bottom-line, as Abbott writes (2008: 18), is that 'Thoreau understood that our capacity as individuals, firms and societies to create social and economic wealth, or well-being, is inextricably linked to the natural world. Human-made capital can offset some decreases in natural capital, but the two are not perfect substitutes.'

Thoreau was writing at a time when new industry and rapid economic growth were contributing to what would become America's dominance in the world. He worked as a surveyor and a pencil-maker and was critical of dams, factories, railroads, clearcut forests, and ice-harvesting that despoiled the landscape through irresponsible industry. Farmers, too, were a target of his criticism when they used practices that treated the soil as non-living property, robbing its ability to support livelihood and thereby leading to what he referred to as the deformation of the land. Farmers who rake riparian meadows for cranberries and marsh grasses based on a desire for monetary profit are likewise to be worthy of criticism. Blinkered focus on financial gain is wrong, as is judging the prosperity of a farm simply by the size of its barn. In *Walden*, Thoreau famously quantified the expenses and revenues of his tiny kitchen garden, reminding readers that at one time farming was regarded as a sacred act in which processes and festivals were celebrated. He was critical of the incessant bustle of work and instead championed what might be called 'slow gardening:' 'As I had little aid from horses or cattle, or hired men or boys, or improved implements of husbandry, I was much slower, and became much more intimate with my beans than usual' (Thoreau *in* Abbott 2008: 47). According to Abbott, the great lesson that Thoreau offers the modern businessperson is to position the purpose of business in the wider frame of society that gives value to what is now referred to as 'natural capital' or 'natural assets.'

Thoreau's great lesson is to distinguish between monetary versus personal, societal, and environmental wealth: 'He saw that too many of his contemporaries were preoccupied with the accumulation of material wealth – without regard to the cost to the community of which they are part, to say nothing of the cost to their soul' (Abbott 2008: 83). Thoreau is almost Marxist in his criticism of the acquisition of monetary gain, believing that men decay in proportion to the money they accumulate, and warning that of those whom have become tools of their tools. Instead, he stated that the richest men are those whose pleasures are the cheapest. All this contributes to what Abbott highlights is, based on contemporary research on the pursuit and worth of happiness, Thoreau's belief that true wealth is that which collectively contributes to the quality of one's life experiences. 'Give me the poverty that enjoys true wealth,' the scribe intones (Abbott 2008: 88). In terms of agriculture, Thoreau insisted that the value of fruits is not in mere possession or consumption but 'in the sight and enjoyment of them,' and that true real estate is our stock in life, not just houses and land, the value of the ground functioning as 'a pasturage for our thoughts' (Thoreau *in* Abbott 2008: 77). Riparian agricultural landscapes offer much more than merely being locations where food if produced; they can provide recreational use:

> Now I see gentlemen and ladies sitting at anchor on the lakes in the calm afternoons, under parasols, making use of nature, not always accumulating money. The farmer hoeing is wont to look with scorn and pride on a man sitting in a motionless boat a whole half-day, but does he not realise that the object of his own labour is perhaps merely to add another dollar to his heap, nor through what coarseness and inhumanity to his family and servants he often accomplishes this (Thoreau *in* Abbott 2008: 72).

Working man, Thoreau believed, has much to learn from watching plants in that 'simplicity is the law of nature as well as for flowers' (Thoreau *in* Abbott 2008:94). It is wrong to confuse simplicity with poverty. Loosening one's grip on materiality and replacing it with emphasis on life's necessities is essential, for in the end 'more work is not better, it is simply more' (Abbott 2008:101). And in a quote bound to have resonance with today's urban gardeners, Thoreau championed the importance of doing things for oneself: 'I take some satisfaction,' he wrote, 'in eating my food, as well as in being nourished by it (Thoreau *in* Abbott 2008: 95).

Sites

3.1.1 London Cru Urban Winery (London, UK)

London's only winery opened in 2013 and is found in a densely urbanised neighbourhood a short walk from the popular tourist accommodation area of Earl's Court. The winery is located down a small alleyway and operates in an old Victorian warehouse which had once been a gin distillery. The winery champions being steeped in the London traditions of trade, innovation, and eccentricity, and has a stated goal of demystifying winemaking while bringing excitement to the local wine scene (www.londoncru.co.uk). It accomplishes this through a modus operandi that, in the opinion of one wine journalist and critic, is 'both perfectly logical and maverick in its originality' [note: all unreferenced quotes throughout derive from material obtained from the various case study websites]. Simply put, the winery is based on the geographical separation of the growing and the processing of grapes. In other words, grapes are hand selected from traditional vineyards in southern Europe and 'carefully delivered to London where they are nurtured into great wine.' Refrigerated trucks are used and grapes arrive at the London winery within a day-and-a-half of being picked from the vines (since the time of my visit and due to the 2016 Brexit vote, the winery has adapted to using only grapes sourced from English vineyards).

Operating like a small boutique winery of the sort found on the Continent, small batches are produced of about half a dozen varieties which are sold at medium prices. Most tasks from crushing to fermentation to bottling are undertaken or guided by hand. Several wines have received prestigious rewards after only their third vintage, and are now available in Michelin-starred restaurants in London. Wines are described with the lyrical flourish of purple prose that is typical to the profession, as for example:

> Full of citrus and stone fruits, with just a touch of gentle spice. The grapes were grown within a stone's throw from the Atlantic Ocean in the verdant hills of Pontevedra, Spain giving the wine a lovely salty tang. Vinified in London using concrete tanks and lees ageing for an elegant, refreshing finish. Try it with chargrilled octopus doused in a zippy lemon and parsley dressing.

And:

> Our very first rosé has a delicate copper hue and a fresh, aromatic nose with subtle flavours of grapefruit, lychee and strawberry. The grapes were grown on the outskirts of a traditional hamlet in Piedmont, Italy after which they were made into a limited-edition wine in London. Try it with an unctuous risotto Milanese and peppery rocket salad.

Other sources of revenue accrue through tours and wine tasting events, as well as hosting private parties, all occurring within the small space surrounded by dozens of French oak barrels and steel fermentation tanks, and accompanied by London-produced cheeses and charcuterie. Images of fancily-dressed crowds milling about or seating at tables are shown on the website.

A small sign on a busy London street directs visitors into a somewhat dodgy alleyway [e3.1.1.1] [note: square brackets here and throughout denote e-photos found on the affiliated website https://doi.org/10.3920/978-90-8686-938-1] where, at the end, a warehouse is reached [e3.1.1.2], otherwise nondescript save for the large, colourful painting of a wine bottle (Figure 3.1). Inside, the cramped space of the winery consists of a series of open-doored rooms with whitewashed brick walls and grey concrete floors (Figure 3.2). Despite a few nods given to the use of the space for tasting tours, such as the presence of a serving table upon which the winery's products are

Figure 3.1.

Figure 3.2.

displayed [e3.1.1.3], this is very much a place of work. The process of winemaking can be followed from the observed equipment: from grapes – towers of plastic trays stacked in the corner [e3.1.1.4]; through fermentation – arrays of steel tanks filling alcoves [e3.1.1.5]; to aging – clusters of oak barrels tucked into corners [e3.1.1.6]; all guided by oenology, as shown by the tiny laboratory [e3.1.1.7]. A map of London in the form of a stylised leaf, with the course of the Thames as the twisty stem, is used on all bottles and for advertising [e3.1.1.8].

When I first read that there is an 'urban winery' in the middle of London near where I customarily stay when visiting the city, it took a while to wrap my mind around what at first blush seemed to be an oxymoron. I knew the area well, and tried to think of greenspaces in the neighbourhood where grapes might be grown. But upon learning that the urban winery is based on separating the production of wine (oenology) from the production of grapes (viticulture), it made sense, and indeed, seemed rather brilliant. When during my private tour of the facilities, I learn details about the logistics and how with rapid air flights and ground transportation that people and grapes can effortlessly move from one location to another, it makes me reconsider concepts of *terroir* and 'locavorism'. The highlight of my visit comes through being allowed to taste the wine. As it is the late afternoon and but a half hour walk back to my hotel at Earls' Court, I select a recommended variety and imbibe generously while engaging in pleasant banter with the winemaker.

3.1.2 Brother Nature Produce (Detroit, MI, USA)

When it comes to opportunities for urban agriculture to contribute toward the regeneration of cities, there is simply nowhere else like Detroit. Thousands of properties lie vacant in the urban core, having been cleared of their dilapidated houses and buildings following the exodus of people and businesses (see Chapter 7.2.5 Detroit Dreams). It is no exaggeration to state that the ensuing ramifications have been profound, the only debate being whether to refer to the situation as a crisis or a catastrophe. Many approaches have been implemented to take advantage of the available space through urban farming. One such, 'Brother Nature Produce' (www.brothernatureproduce.com), is a modest, husband-and-wife enterprise that began in 2007 and operates on the idea that 'healthy food helps regenerate the economy.' Located in the North Corktown neighbourhood, the farm grows salad mix and herbs for sale at the Eastern Market (see Section 3.2.6) and other markets, as well as several nearby restaurants. The owners herald the merits of their produce as being extremely fresh (sold within half a day of being picked), inexpensive, diverse in flavour, and packed with nutrients. Also heralded is that it is grown locally to reduce food miles and the ecological footprint, through use of 'better than organic' growing practices. Finally, the point is made that purchasing from them supports small local business which goes toward helping to create a more stable economy as well as adding to food security.

Sadly, as is the case for much of downtown Detroit, the neighbourhood is comprised of broken-surfaced roads and driveways that lead to abandoned lots (Figure 3.3) [e3.1.2.1], in addition to scattered homes occupied by those whom bravely struggle on [e3.1.2.2]. The urban farm is situated along a largely empty street [e3.1.2.3] where are found a densely planted garden, plastic hoop-house, large compost pile, and handful of beehives (Figure 3.4) [e3.1.2.4].

Figure 3.3.

Figure 3.4.

While visiting the farm, I keep in mind a judgmental comment made to me by the manager of Detroit's Eastern Market in regard to the plethora of urban agriculture sites cropping up throughout the city. He suggested that it is often easy to visually distinguish between those operations established with the purpose of being serious farms in which food is grown in commercial enterprises, and those operations set up by well-meaning dilettantes to address any number of ancillary social needs through the boutique gardening of food. Certainly, as I walk about the no-nonsense site and look at the rich bounty being extracted from the soil, there is no confusion in regard as to which category the present farm positions itself.

3.1.3 Hope Blooms (Halifax, NS, Canada)

'Hope Blooms' was established in the North End of Halifax in 2008 as a youth-led, community-building initiative to create young leaders through hands-on mentoring and a fostering of hope that arises from entrepreneurism. The city donated a piece of abandoned land that had been filled with trash and weeds, and it is this that was transformed into dozens of family allotments and a youth garden. It is the latter which is of interest as the garden is used to enable 'youth to learn how to grow food, produce and successfully market value-added products, grow a small social enterprise from the ground up, and give back to their community' (www.hopeblooms.ca). It is through the growing and marketing of food that the at-risk youth are empowered to contribute to their own futures and wellbeing by becoming engaged in actions that contribute toward improving the social-environmental health of their community. The project is based on the importance of establishing community relationships as well as changing the inherent attitudes the youth may have about what it means to be 'rich' or 'poor.'

The youth garden is managed to provide a sense of ownership while teaching skills and increasing food literacy. Herbs grown on the property are used to create a series of salad dressings (Fresh Basil Pesto, Maple Sage Balsamic, Orange Rosemary Dijon, Fire Roasted Oregano) which are sold at the Halifax Seaport Farmers' Market (see Section 3.2.5). Entrepreneurial and leadership skills are learned as youth partake in the complete supply chain from organic growing, to production, to marketing and sales. Doing so, they receive training in sustainable farming, food science, nutrition, business management skills, and public relations. In response to comments made about the 'cute idea' and the nice 'fundraising' effort while at the farmers' market, the youths selling their fresh herbal salad dressing retorted that 'We're not fundraising. This is our company!' (Brunet 2012). Revenue is recycled back into farm operations, as well as helping fund the establishment of a scholarship and supporting a homeless shelter and school lunch programme.

The initiative has received considerable press coverage. This is especially true after the youth were successful in obtaining investment from the millionaire entrepreneurs on the nationally televised reality show *Dragons' Den* (Henderson 2014; Spur 2013). With the ensuing wide publicity has come strong sales in the amount of more than six thousand bottles a year, such that it is customary for the salad dressing to be sold out every week at the market. Hope Blooms has also attracted the attention of scholars with interests in food economics, viable communities, nutritional health, and landscape design (O'Reilly 2017; Raine 2014; Seth *et al.* 2019).

Hand-painted signs announce arrival to the community and youth gardens [e3.1.3.1]. Nearby apartment blocks and high-rise offices look down upon the site [e3.1.3.2], which spreads out in verdant plots along the edge of the sidewalk and across the street from an eighteenth-century church [e3.1.3.2], one of the oldest buildings in the city [e3.1.3.3]. A sequestered area exists for production [e3.1.3.4] and storage [e3.1.3.5] of compost. The no-nonsense utilitarian area includes storage bins for compost, assorted tools, a wood working table, and seating stumps [e3.1.3.6]. A few notable attempts have been made to enliven the gardens through attention to aesthetic details. A service box attached to a telephone pole is painted in a scene reminiscent of an English country garden (Figure 3.5) [e3.1.3.7]. The hoop-house, tucked away in the corner of a parking lot [e3.1.3.8], is decorated with a brightly coloured illustration (Figure 3.6) [e3.1.3.9]. In addition to germinating seeds, the structure also functions as a meeting place and demonstration area, with one supervisor remarking that 'some of the most magic moments have been running as a large group of laughing children into the greenhouse to take shelter from the rain and sitting quietly under its protective canopy listening to the rain drops in our green-filled amphitheatre.' Other embellishments include a portion of the chain-link fencing being shaped into an archway festooned with colourful diamonds of plastic [e3.1.3.10], and a seating area which features raised wooden planters and a stone slab table [e3.1.3.11].

Figure 3.5.

Figure 3.6.

It seems as if I have the site entirely to myself... or so I think. It is when I am sitting down and staring dreamingly out across the garden plots while admiring the artfully designed fence pergola that I witness the capture. Suddenly my attention is grabbed by a blurred streak that arrives as if out of nowhere. I watch, transfixed, as a hawk plunges into the foliage only to reascend but a moment later and fly away with a vole or mouse in his/her talons. That this little food-web drama played out in the very heart of the province's capital city rather than in the rural countryside is remarkable, and demonstrates the multifaceted role of urban agriculture.

3.1.4 Dill's Pumpkin Farm (Windsor, NS, Canada)

Windsor, a town of about four thousand located on the Bay of Fundy, has a rich agricultural heritage. Its agricultural fair, begun in 1765, is the oldest in North America. A short walk from the town centre and bordered on two sides by suburban development, is 'Dill's Pumpkin Farm' (run by Howard Dill Enterprises), an example of how specialising in heirloom varieties can produce a lucrative business (Radovich 2011). Through generations of careful cross-pollination, the farm produces the world's largest pumpkins (some exceeding a thousand kilograms). These gargantuan pumpkins are purchased for use in regional fall fairs, or, after being hollowed out, to create boats to be used in a renowned regatta that takes place in a nearby lake. Seeds are sold around the world for international pumpkin competitions. Visitors to the suburban farm select and use wagons to transport their own modestly-sized Halloween pumpkins as well as various squashes and decorative gourds. A shop within an onsite barn functions as a mini-farmers' market that sells local products in addition to gift items such as pumpkin-scented candles and other souvenirs. The farm is immensely popular among tourists, people from the region, and sports 'pilgrims.' The latter come to visit Long Pond, a small body of water situated in a forested glen on the property that is touted as being the location where the first game of hockey was played in Canada. The farm caters to bus tour groups who arrive to marvel at the 'orange giants,' and to wedding parties whom, somewhat perplexingly, arrange to have their photographs taken with such (in hope of fertility, perhaps?).

Due to limited space for parking, visitors often have to walk several blocks through Windsor to arrive at the peri-urban farm, thereby experiencing the transition from retail core to residential suburbs to agrarian landscape. A long driveway lined with small pumpkins leads to the entrance shop. Inside, admix the pumpkin-themed souvenirs and local jams and honeys, are bins filled with decorative gourds of all manner of shape and size. Outside, a bucolic scene unfolds with a rustic, cedar-shingled barn behind which rolling fields are filled with thousands of orange pumpkins still attached to their stems. Pumpkins are displayed everywhere: strategically balanced on rusted pieces of antique farm machinery, situated in old wooden wheelbarrows, hollowed out and used as flower planters, or stacked in pyramids atop wood or straw-bales (Figure 3.7). People mill about selecting pumpkins arranged in vast arrays raised on wooden pallets to keep them above the moist ground. These are then loaded onto children's wagons to be pulled back to the shop for purchase. Here and there, orange giants approaching the size of kitchen tables lie atop pallets ready for transport to the back of customers' trucks through use of a tractor. Signs inform about the ethnobotany of pumpkins: how they were used by Mayans for treating burns and by other North American Indigenous peoples for kidney infections, that they are the most important ingredient in Chinese plum sauce, and that they get their name from the Greek word 'pepon,' which means 'cooked in the sun,' such that if someone was thought to be soft in the head, s/he would be called a 'pompion.' A trail leads past ornamental flower gardens and the expansive fields of pumpkins to enter a grove of trees and then arrive at the small pond in which the original game of hockey occurred (Figure 3.8). Elsewhere on the property, a small barn functions as a museum of idiosyncratic hockey memorabilia.

Figure 3.7.

Figure 3.8.

That the town of Windsor takes its history seriously is evident as one approaches on the highway and sees the roadside sign claiming it to be the 'birthplace of hockey.' The day of my visit is during a weekend in mid-October when the farm is at its busiest. Dozens of families are enjoying the outing, fathers and mothers pulling wagons in which their young children share space with pumpkins which have been selected for purchase. A festive spirit of happy harvest pervades the scene. Carved jack-o'-lanterns abound. Parents scramble after children who cry out at spotting pumpkins, each of an increasing Brobdingnagian proportion. Here, people jostle for selfies, while would-be participants examine the suitability of the giant gourds for their use as boats in the upcoming regatta. Surrounding maple trees proudly boast their presence by exhibiting the peak of their fall colours. Later, at Long Pond, I watch as a father explains to his two sons the significance of the site to the history of Canada's national winter sport (and some would say, obsessive secular 'religion'). It can be no significance, I think, that both generations are wearing sports jerseys bearing the insignia of their favourite NHL team (several years after my visit I learn that the site of the pond is up for sale, with the mayor expressing hope that it would not become a retail strip-mall or hotel).

3.1.5 Fresh City Farms (Toronto, ON, Canada)

Established in northwest Toronto on the site of a former military base that had closed in 1995, 'Downsview Park' is being designed to become Canada's first national urban park. The development of the 230-hectare site will take decades and become a hallmark example of the potential for landscape and ecological urbanism to create dynamic, evolving places (Czerniak 2002; North 2012). The intention of the park is to be 'a place for exploration, innovation, and learning for grass roots groups, neighbourhood organisations, educational institutions, businesses and citizens' with urban agriculture playing 'a significant role in its programming' (Gorgolewski *et al.* 2011: 46). Part of the concept plan included a 'Cultivation Campus,' with community gardens, a nursery, an orchard, and space for commercial enterprises involving urban agriculture, whose 'internal pathways and farmyard layout are subject to change at the discretion of the operator' (www.downsviewpark.ca). One organisation established here is the 'Toronto Beekeepers Co-operative,' and another is 'Fresh City Farms,' which is billed as the nation's largest commercial city farm.

In operation since 2011, Fresh City Farms is an organic growing and delivery retailer that provides direct farm-to-table food (www.freshcityfarms.com). Working under the motto of 'better life through food,' their mandate is to provide meticulously sourced food that is well curated with overt transparency. From its 4-hectare farm and hoop-houses, weekly or biweekly produce bags are delivered throughout the greater Toronto region (although, according to some, there is need to expand the service into hard-to-reach neighbourhoods – Romero *et al.* 2014). Much is made of the fact that their 'food chain' is not a chain but a direct link of customers to organic and sustainably produced food. Fresh City Farms partners with dozens of other regional farmers and owns several stores in the city where organically raised dairy and meat, as well as artisanal baked goods are sold alongside produce from the organisation's farm in Downsview Park. Goods ordered from their online shop are delivered in reusable bags and occurs via vans, or in the downtown core, with electric peddle-car bikes. Touting themselves as 'Toronto's organic grocery store,' the organisation markets organic food that is 20-25% cheaper than that available at large chain stores, even with the charged delivery fees. Like Hope Blooms (see above), Fresh City Farms was also a participant on the national television show, *Dragon's Den*. In this case, however, they were unable to attract investment due to confusion as to whether they were competing based on either the price or quality of their food. Benefitting from advice rather than capital received, Fresh City Farms decided to focus on quality. Doing so has enabled their expansion to more than two thousand deliveries per week. They now also offer ready-to-eat meals as well as fruit beverages and mixed salads. Staffing has increased to more than fifty, including farmers, drivers, cooks, and an in-house nutritionist.

Trails and roads wind their way past sculpted hills and plantings in the newly landscaped Downsview Park [e3.1.5.1] to lead to the utilitarian agricultural area. Here, accompanied by a warning sign [e3.1.5.2], is the apiary attended by the professional beekeepers (Figure 3.9) [e3.1.5.3]. Nearby is a phalanx of hoop-houses (Figure 3.10) [e3.1.5.4] used for seed germination and potting [e3.1.5.5]. Signage informs that this is part of Fresh City Farms' program of 'good food, grown right' [e3.1.5.6]. Large growing fields spread from the hoop-houses to the edge of the busy urban street, across which is a line of apartment blocks, single-family homes, and assorted retail development (Figure 3.11) [e3.1.5.7]. Many of the vegetables being grown are easily recognisable [e3.1.5.8] as this is not a boutique operation catering to exotic appetites. What is obvious is that this is a no-frills, sleeves-rolled-up, commercial farm that is being expertly tended by a dedicated cadre of young agriculturists (Figure 3.12) [e3.1.5.9].

Figure 3.9.

Figure 3.10.

Figure 3.11.

Figure 3.12.

My experience in visiting the urban farm is one of displaced reference. On the early summer day of my visit I am able to observe half-a-dozen people diligently at work in the vegetable plots. In one instance, a young woman is wearing a conical straw hat of the variety common in Asia, and so, for a brief moment, I find myself reminiscing about previous visits made to farms in China, Japan, Cambodia, and Thailand. Later, when another young woman is observed striding through a cereal patch with buildings in the background, I am struck by the semblance of the scene to the famous 1982 landscape art project, *Wheatfield - A confrontation,* in lower Manhattan that I used in the introductory chapter of my book about integrated urban agriculture.

3.1.6 Very Local Greens (Dartmouth, NS, Canada)

Situated across the harbour from Halifax, and connected to it by several bridges and the oldest saltwater ferry service in North America, the municipality of Dartmouth is a bedroom community for the regional capital as well as a working class community with many small businesses in its own right. King's Wharf is a new development of waterfront condos beside a former shipping wharf upon which the commercial operations of 'Very Local Greens' are located on a small parcel of land whose rental is based on its size footprint of 6 parking-lot stalls. Tucked into the side of the sole building on the onetime industrial site is a 12-metre-long shipping container of the type used for overseas transport, but which here is used to grow local produce. The use of shipping containers in environmentally controlled urban farms is a trend that is gaining in popularity, and that used by Very Local Greens was purchased, ready-to-run, from a Boston-based outfitter specialising in such. By growing plants in a system of recirculated water rather than soil – i.e. through 'hydroponics' – and by using vertical growing columns, the small container is able to grow up to five thousand leafy plants, the equivalent to that produced in a traditional, one-hectare field. In consequence, no soil depletion occurs and pesticide use is avoided. Another advantage is the year-round rather than seasonal provision of a high-quality supply of produce, until recently, a novelty in Canada. Also, as stated in an interview about the farm (France *in* Smith 2018), because a major hurdle faced by urban agriculture can often be a lack of space, 'the advantage of having various containers are that you can pick up your production system and move it somewhere else should another use come in to dictate the previous site you were in.'

Very Local Greens advertises itself as producing 'ultra-fresh, hyper-local boutique greens' (www.verylocalgreens.ca). It is presently being run by a single individual with no prior experience in farming. A broad range of microgreens, herbs, edible flowers, and lettuces are produced (e.g. cilantro, basil, dill, Kale, cornflowers, borage, radish, curly cress, chard, beet leaves, sorrel, mustard fronds, and six rare varieties of lettuce). This diversification of plants is important given that some restaurants desire to order from a single supplier and that they wish for more than just standard lettuce. Produce is sold to high-end restaurants, to select markets, and directly to online customers. During the pandemic, a small-scale CSA (community supported agriculture) was established to continue to provide greens to private citizens. Produce derives from non-GMO seeds and through use of the fewest resources by reducing reliance on fossil fuel-heavy imports, and in so doing, increasing food security through year-round production. Achieving high quality rather than quantity is the operating mandate. Heads of lettuce that are transported to Nova Scotia by truck from California will experience loses of up to half their mass due to waste. In contrast, the urban farm owner states that his local, container-grown produce is regarded by fine restaurateurs as having a stronger flavour and a substantially longer shelf-life compared to the normal fare that is imported.

The container farm has received considerable press reportage. In an online post by the federal government representative for the area, the politician heralded Very Local Greens as being an innovative example of how bright entrepreneurs can make a difference in the local community: 'Small businesses are critical to our economy, and they play a huge role in growing the middle class.

We've been working hard to make it easier to do business here by lowering the small business tax to 9%, supporting innovation and we're cutting red tape.' Future plans are to transfer management of the container farm to a nearby community garden, making it one of the few examples anywhere wherein a hydroponics system with an established record of commercial viability is formally integrated into a social urban agriculture enterprise.

The container farm is easy to spot on the abandoned wharf (Figure 3.13), with the view of Halifax on the other side of harbour, one of the world's largest deep-water ports. Inside the container, on two sides down a long corridor [e3.1.6.1], hydroponic towers of plastic [e3.1.6.2] sport tufts of leafy greens and edible flowers (Figure 3.14) [e3.1.6.3]. Climatic control panels and liquid chemical mixtures (Figure 3.15) can be operated by computer from a remote location. This means that the only time a person needs to be physically present is when transferring seedlings from their sprouting beds (Figure 3.16) or harvesting produce from the easily disassembled plastic towers (Figure 3.17) laid out on the metal work bench.

Figure 3.13.

Figure 3.14.

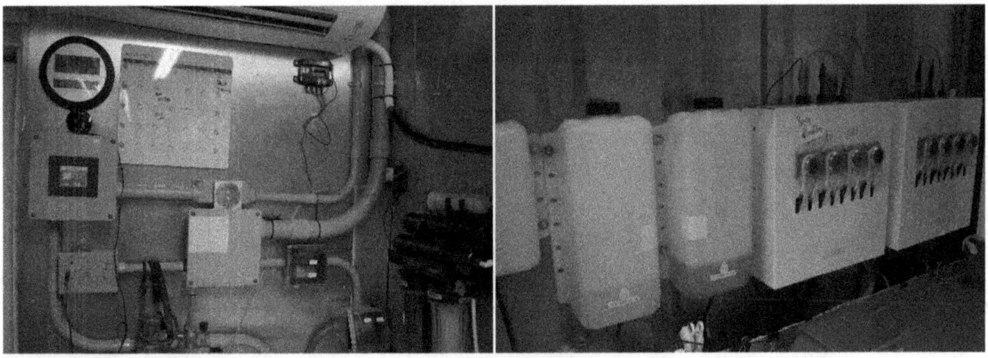

Figure 3.15.

Figure 3.16.

Figure 3.17.

I am struck by how much King's Wharf on the Dartmouth waterfront is a place of sharp contrasts. Emerging from my car, I see that one side of the street is occupied by a row of new and obviously expensive condos, their glass fronts staring out across the water toward Halifax. On other side of the street, a vast expanse of concrete characterises the former commercial wharf. This mostly barren landscape juts out into the harbour towards where, in 1917, the Halifax Explosion occurred, the largest and most costly in terms of human life, man-made single detonation before the atomic bombs. I enter the forlorn site and gaze at the shipping container propped up on wooden blocks beside the nondescript building. The sense of pervading contrast is emphasised by knowing that what is going on inside the shipping container is so remarkably different from the lifelessness of the parking lot in which it is situated. The contrast between the bright natural sunshine outside and the purplish glow from the LED growing lights that bathes the inside of the container is also jarring. As I look about at the climate control panels and vertical arrays of growing plants, it takes but a moment for me to appreciate the aptness of the moniker given to the site by the owner's child: the 'spaceship farm.' That all this vegetative bounty can be produced from recirculating water in the absence of soil seems somehow magical.

During my visit, the owner relays, in the form of a cautionary story, his frustration in dealing with city bureaucrats. Before deciding upon the present site for his container farm – chosen due to its high visibility location – the owner researched and asked numerous municipal authorities about needed permissions. Allowed to proceed, it was after operating for more than a year, during which time numerous politicians had visited, that he received a cease-and-desist order from the planning department which cited bylaws existing against the presence of old shipping containers in the city. Even support from the wharf redeveloper stating that they were allowed up to four shipping containers on their site during construction and that no additional permits were needed did not budge the bureaucrats who insisted on a site plan. A lengthy exchange ensued concerning what exactly constituted 'a shipping container', with much discussion being about the physical nature of the outside material from which such were constructed. Moreover, the owner argued that if the bylaw specifically stated that containers previously used for shipping were not to be allowed, how then did this pertain to the present situation given that the specific container under review was bought brand new and fitted to its use as an urban farm from the very get-go. Moreover, what, the owner asked, was needed in order to make the structure not be considered a shipping container: perhaps adding a window or tiny office? The city planners countered that as this would constitute a building, there were all kinds of safety issues which must be addressed, etc. Finally, the owner took advantage of the fact that the farm had been visited by politicians. Senior city administrators from the mayor's office stepped in to end the 'administrivia' after the owner argued about the bad optics of shutting down a facility that had been providing local produce for more than a year in line with the government's own regional sustainability platform, as well as the absence of any complaints from the public. The net result of the saga is that now and henceforth shipping containers are a formally recognised part of the city's urban agriculture plan. I depart, shaking my head and thinking that the battles faced by urban agriculture innovators can indeed be legion.

3.1.7 Food from the Sky (London, UK)

Located in north London, 'Thornton's Budgens' refers to itself as 'the community supermarket that really cares'. This is demonstrated by their in-store removal of plastics and placing of doors on chillers, as well as their partnering with a local group to transform the flat expanse of their roof into a permaculture community garden (Barnett 2011). 'Food from the Sky' believes in using food as a fertile and healthy vehicle for change, and operates with a vision 'to grow Life, Food and Communities on our most cemented places and to plant seeds in people's hearts' (https://foodfromthesky.wordpress.com). The non-profit does this by growing vegetables on the roof and selling them in the supermarket located directly below. In so doing, it takes the concept of 'food miles' and reduces it to the extreme, in this case, there being just 8 metres in vertical distance from soil to shelf. Begun in 2010, and operated with the help of several dozen volunteers, the farm is composed of hundreds of plastic recycling boxes filled with compost and a rich diversity of plants (peas, potatoes, kale, broccoli, strawberries, raspberries, lettuce, spinach, edible flowers, tomatoes, carrots, beans, herbs, radishes, artichokes, cauliflower, melons, chard, onions, squash, and others) with attention paid to seed provenance. Also on the roof, is a composting area where unsold fruits and vegetables from the supermarket below are mulched before being added to the potting soil. The microclimate of the roof is favourable for cultivation due to the rising warmth from the supermarket's heating and lighting systems. Surveys in coordination the Natural History Museum have been conducted on the diversity of insects, birds, and bats (Cockrall-King (2012) noted the abundance of bees when she visited the site).

Seen from the street, there seems to be nothing particularly special about the supermarket [e3.1.7.1]. It is only once inside, past the signage announcing the company having won a prestigious retail award and that it cares for its community [e3.1.7.2], that the uniqueness of the site becomes apparent. There, a Food from the Sky blackboard lists the availability and prices of its produce [e3.1.7.3], and a sign states that the vegetables were 'Made in London' and are '0 in food miles, but miles ahead in taste' through being grown directly above [e3.1.7.4]. The rooftop farm's produce is displayed in its own set of wicker baskets (Figure 3.18) [e3.1.7.5], presumably due to capitalising on an imagined concordance in the minds of shoppers of natural foods and natural material containers.

Returning outside and going around to the rear of the building reveals a ladder scaffold [e3.1.7.6] that is identified as the way to access the rooftop farm [e3.1.7.7]. There, spread out over a 450-square metre area, vegetables are growing in imported soil and compost while being sustained by London's frequent rains and its 'heat island' microclimate. Some are contained in elongated wooden planting beds [e3.1.7.8]. But most are embedded in hundreds of plastic recycling tubs (Figure 3.19) [e3.1.7.9] that are veritably over-brimming with a wide variety of plants [e3.1.7.10]. Picnic tables serve as work benches (Figure 3.20) [e3.1.7.11]. A hoop-house for early germination has been ingeniously assembled through use of recycled plastic bottles strung together to form the walls (Figure 3.21) [e3.1.7.12]. A banner on the perimeter fence signifies the location to be

Figure 3.18.

Figure 3.19.

Figure 3.20.

Figure 3.21.

part of a government campaign to develop urban agriculture in London [e3.1.7.13]. On the back wall a motivational quote from Goethe instructs that 'Whatever you can dream of begin it now. Boldness has genius, power, and magic in it' [e3.1.7.14].

The site is visited by groups of schoolchildren and lectures are offered on topics such as permaculture, natural beekeeping, biodynamic growing, food preservation techniques, and the importance of linking biodiversity with food security. Events have also been hosted on the roof, such as performances, poetry readings, and book launches. Professionals and the lay public alike have lauded Food from the Sky:

> It is about people taking responsibility and taking action.
> It is a beacon of change for how a project can inspire people and their community.
> It's a premium example of how innovation and creativity can help provide a framework for new ways of doing things in our outdated food system.
> It typifies the kind of entrepreneurial spirit that is needed to shake people out of the pessimism they hold in relation to the environmental challenges ahead.
> It has the potential to transform London's skyline, bring living life back to the city, and connect more Londoners with where their food comes from.
> I can hardly believe that I am so lucky to have this beautiful vegetable garden on the roof of my local supermarket.
> Grassroots action. Linking business and communities with clean low carbon food. Spot on! This is exactly what we need to revolutionise our food system and transition us to low carbon Britain.

In addition to this anecdotal praise, the project has received considerable press coverage, has been visited by the then-mayor Boris Johnson, and won an environmental achievement award. It is therefore unfortunate that the project closed in 2014 due to the store being sold and the new owners not wishing to continue to have a farm on a roof that they considered was in need of repair. Although some of the Food from the Sky's staff relocated to a small organic farm in south London, the innovation and pedagogy has been lost of having food being grown atop of the building in which it is sold.

As I ascend the scaffolding staircase, I'm struck by the pioneering aspect of the project. Upon gaining the roof and being greeted by the sight of the vast array of potted plants, I immediately stare out at the blank expanses of the neighbouring rooftops. It is as if they are all there, just silently lying in wait for being transformed into similar productive spaces. I find it hard not to reflect on the missed opportunities that abound with respect to the greening of our cities. Although I have visited green eco-roofs designed for purposes of stormwater management and microclimate, these have been vast carpets of sedums or grasses. Nothing like the edibles that I now see about me. And so I spend considerable time circulating about the recycling boxes, examining their plant contents. Nowhere is this more enjoyable than when I pinch off sprigs of herbs and crush them in my fingers to inhale their aroma. With permission, I pick a young carrot to munch upon while walking around, not because I'm in the least bit hungry, but simply for the novelty of being able to state that I've eaten a vegetable grown on a roof in the middle of London. The innovative use of the environmental bane of plastic water bottles to form the walls of the hoop-house is simply brilliant. I realise at the moment that never again will I look at a hoop-house made of vast sheets of new plastic without being reminded that it could have been constructed with such recycled material. Visit over, I make my way down the stairs to re-enter the supermarket and purchase salad greens that had been grown on the roof, sold, I'm happy to determine, in a bag that is 100% biodegradable. Together with a small loaf of artisanal bread, this will be my lunch for the morrow.

3.1.8 GrowUp Box Urban Farms (London, UK)

Through use of the techniques of controlled environment agriculture, 'GrowUp Urban Farms' has become part of the ballooning urban bio-economy (Buck 2018; McChrystal 2015). The company's mandate is to expand the way in which food can be grown in cities while educating people in the story of production from farm to fork (www.growup.org.uk). By directly repurposing unused urban space, their projects lower the environmental impact of farming. GrowUp is a commercial enterprise rooted in, and guided by, a set of core values:

> We care about People, Profit and the Planet.
> We believe we can Do Good and Make Money.
> We search for the best environmental solution available at the time – and then revisit those decisions regularly to see if a better option is available.
> We think learning about and eating food is something everyone should get the chance to do.
> We should be producing food in cities at a commercial scale in an ecologically sustainable way.
> We can improve the level of food knowledge, education and enthusiasm in cities.
> We should reduce the environmental impact of food productions for cities.
> Our systems and processes can lead innovations in urban food production.

The company runs several operations, the most interesting of which is the 'GrowUp Box,' a one-time shipping container that has been 'up-cycled' with construction of a glass greenhouse on top. Of equal interest and imagination is the container's location on the upper level of a multi-storey carpark in east London that overlooks the site of the 2012 Olympics. Established in 2014, the mini-farm shares space with 'Roof East Stratford,' a rooftop urban park with an outdoor bar and cinema (www.roofeast.com).

The mini-farm produces fresh salads, herbs, and fish through a combination of 'aquaponics' and vertical growing techniques. Aquaponics are recirculating systems that combine hydroponics (growing plants in water without soil) and aquaculture (fish farming) to create an efficient closed-loop system. The only inputs are fish food and energy for the pumps and heaters. Following feeding, the nutrient-laden water containing fish excretions is used to fertilise the plants, which, through taking up the nutrients, purify the water, which is then pumped back into the fish tanks. By reusing the water back and forth between plants and fish, the process is especially suited to high density, limited space situations, such as exists in urban cores (including on rooftops). Plants are grown in the upper greenhouse through combined natural and LED lighting and via vertical farming. Year-round production is achieved through strict environmental control of temperature, humidity, lighting, airflow, and nutrient concentrations. Tilapia are kept at low-stress densities to ensure maintenance of quality taste. The vertical columns allow for 400 individual salad, herb, and micro-green plants (e.g. watercress, caraway, lettuce varieties, rocket, mustard, basil, dill, parsley) to be grown at a time.

Shiny modern buildings define this area of East London [e3.1.8.1] as one approaches the Stratford Centre. The entrance plaza is marked by a large art installation of metal lozenges [e3.1.8.2], behind which can be seen the railings on the roof of the carpark [e3.1.8.3]. Past the jostling crowds in the shopping mall, a staircase leads up to the sign announcing arrival at the Roof East Bar, Cinema, and Urban Park [e3.1.8.4]. Outside, what had once been a parking lot is now occupied by an attractive wooden service bar, seating tables, and planter boxes (Figure 3.22) [e6.1.8.5], providing views over the 2012 Olympics' site and the 115 metre-high ArcelorMittal Orbit Tower, Britain's largest piece of public art [e3.1.8.6]. A series of long benches provide additional seating [e3.1.8.7], and a large screen and rolled up matting indicates where cinema shows take place [e3.1.8.8]. Most intriguing is the use of fancy cars that recall the site's original function and which now have been repurposed as decorative planters [e3.1.8.9]. Behind one such can be seen the urban mini-farm (Figure 3.23).

The GrowUp container stands in proud isolation in a corner of the roof (Figure 3.24) [e3.1.8.10], a counterpoint of functionality to the frivolity of the rest of the parking lot, its emblazoned logo declaring its mission to provide 'sustainable food for a local market' [e3.1.8.11]. It is immediately apparent that this is like no other shipping container one has seen. The upper level of the glasshouse showcases vertical columns of leafy greens (Figure 3.25) [e3.1.8.12], which upon closer inspection provide a wonderful example of hydroponic growing [e3.1.8.13]. Inside the lower level of the container are the aquaculture components of the project: interconnected tubs of cascading water and tanks holding fish (Figure 3.26) [e3.1.8.14]. A large information panel that is obviously used for tours educates about aquaponics [e3.1.8.15]. Rainwater from the roof is collected in a rain barrel and an attempt has been made to establish a bit of a green wall through installation of a hanging burlap planting bag [e3.1.8.16].

Figure 3.22.

Figure 3.23.

Figure 3.24.

Figure 3.25.

Figure 3.26.

Since my visit to the rooftop site during its first year of operation, several changes have occurred. The urban park event space has grown in popularity to become London's 'rooftop playground for grown-ups' through expanding its offerings of social activities, which now include game tables, mini-golf, batting cages, lawn bowling, as well as offering food. GrowUp continued to expand through several iterations of large warehouse facilities before being acquired by another company. In this respect, the role of the original GrowUp Box container was critical in the scaling up of urban agriculture. Originally funded through a Kickstarter campaign, the mini-farm situated on the popular rooftop site functioned to attract interest from the press, restaurants, and London investors. It is in this role as a catalyst rather than a commercial endeavour in itself that, according to the company co-founder, the 'small but visually exciting' container mini-farm proved its worth in terms of engaging people and fostering investment.

Roundabout public transport routing has taken me to a location in east London that I have not previously visited. The sight of the expansive art installation in the entrance plaza sets the bar high for my visit to the shipping container mini-farm. I emerge onto the roof and immediately break out into a loud laugh upon spotting the spiffy cars that have been repurposed as glorified planters. I find the visual absurdity delightful as it calls to mind the famous 'Cadillac Ranch' public art installation. It is the morning of an overcast autumn day in the middle of the week when I visit. In consequence, the only person about is the individual tending the mini-farm whom proceeds to give me a private tour. As I look about at the other amenities of the event space, I am struck by the brilliance behind the obvious strategy of transforming the upper level of carparks into popular meeting spaces. Staring past the outdoor movie screen toward the spectacular Orbit Tower, I wonder what urbanite would not relish hanging out on such a rooftop bar? I walk over to the aquaponics system while imagining that its presence in the high-visibility rooftop urban park has done more to educate the public about the possibilities of urban agriculture than any dozen news articles or television interviews ever could. Indeed, as I soon discover, there is simply nothing like seeing with one's own eyes in order to appreciate the mutualism between the fish and flora in such aquaponics systems.

3.1.9 FARM:shop (London, UK)

In preparation for hosting the 2012 Olympics, London focused attention and poured considerable money into refurbishing areas in the East End near the sporting venue. One such program was a challenge issued to local artists to imagine different ways to occupy empty shops in the Dalston neighbourhood. An eco-social boutique design firm was selected to regenerate a derelict and dilapidated, four-storey, Victorian terrace building into an urban food factory. Located on a busy street across from the 'Dalston Eastern Curve Garden' (see Chapter 5.1), the 'FARM;shop' opened in 2011 with the bold (and contestable) statements that it is the world's 'first urban farming hub' and 'first farm in a shop,' in addition to being London's 'first and largest aquaponics system.' What is certain is that, in the words of several journalists, it is 'a compelling combination of public space and agricultural experiment' that 'is strikingly original and demonstrates a rare ingenuity.' The farm shop operates as an imaginative experiment to discover how many different types of food can be grown in the limited space – 14 square metres indoor and 12 square metres outdoor – occupied by a small urban shop (www.farmlondon.weebly.com).

Each room in the building contains something growing in it. The premier operation utilises aquaponics, based on water and waste from fish being used to provide nutrients for lettuce, which in turn cleans the water before it is recirculated back to the fish. An early challenge was to find the ideal temperature for the fish in order to make sure that the tilapia didn't grow too rapidly and produce too much nitrate for the plants. The system is augmented by an aeroponic unit that sprays the roots of tomatoes, peppers, and basil with water. A chicken coop is housed on the roof, and attempts have been made to grow mushrooms in the basement.

With a motto of 'eat, drink, grow, work, play,' FARM:shop advertises itself as a workplace, café, and events venue that functions as the meeting place of choice for London's food lovers; a place where people can be reconnected to their food, something which has become severed due to the physical and conceptual separation of city and countryside. The operation is principally financed by the café, which serves sandwiches and salads made from produce grown onsite in addition to that sourced from partnered, independent farms around London. Given that only a few minutes exist between the picking of the greens and their incorporation into dishes, there is justification to the farm shop's claim to serve the freshest food in all of London. The shop is working with a university in an attempt to create a market for freshwater fish to be a replacement for marine fish as the traditional staple of British diet.

The farm shop champions that it has zero food miles and that aquaponics is environmentally better than hydroponics since the latter's nutrients are often fossil fuel based and imported. At the same time, they note that it is not enough for urban agriculture to be ecologically sound; it must also be economically sound. Toward this end, the operation broke even in its very first year of operation and now employs four people. A local business and development newsletter referred to the FARM:shop in a teasing and attention-grabbing title as a 'fishy business,' and another noted it to offer promise as a potentially long-term, financially viable model of business. The FARM:shop has received national and international coverage (Rawsthorn 2011; Webster and Dawson 2012), including being referred to as a 'business, an experiment, [and] a new idea' (Hawkins 2011). The success of the operation has led the FARM:shop to become the anchor for a wider organisation called FARM:London, which provides consultation on other design/build projects. Fund raising efforts are underway to support plans to build both a large urban rooftop farm and a micro-mushroom farm and training centre in London.

It is but a short walk along a busy street from the Underground Tube station to reach the FARM:shop (Figure 3.27). Inside the front room a large table is surrounded by fish tanks and growing beds (Figure 3.28). The glass-fronted tanks [e3.1.9.1] are filled with tilapia which, when not actively swimming about [e3.1.9.2], seem to take great interest in the goings on in the room

(Figure 3.29). Along one side of the room, leaves of lettuce emerge from their growing containers [e3.1.9.3], whereas, on the other side, water recirculates through several troughs [e3.1.9.4] on its way back to the fish tanks. Tomatoes and other plants grow under artificial light in small hydroponic systems on the next floor [e3.1.9.5]. And several chickens are busy pecking at grains scattered in their rooftop coop (Figure 3.30) [e3.1.9.5].

Since my visit, the chickens have been removed from the roof to the backyard 'due to the health and safety concerns of our landlord.' A green growing wall with plants in stacked pockets now exists in the front room. Different hydroponic growing systems have been experimented with, and a shift has occurred to LED lighting to reduce the carbon footprint. The backyard contains an herb garden and hoop-house, the latter available for rental for corporate meetings or events. In fact, the entire building can be rented for private functions, with the opportunity for catering with food that is produced onsite. Following the pandemic, the FARM:shop has been taken over by new management.

Figure 3.27.

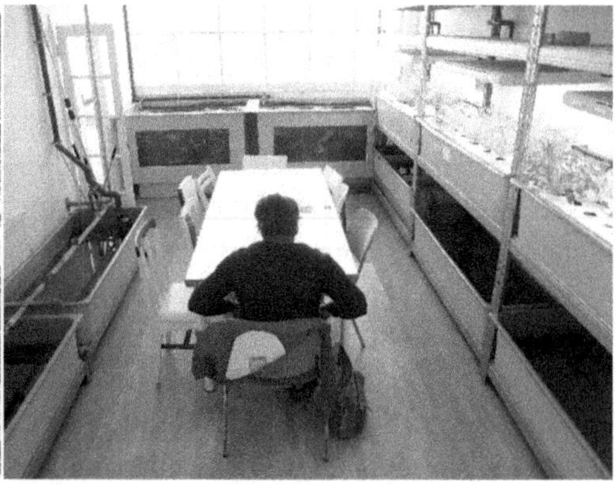

Figure 3.28.

Figure 3.29.

Figure 3.30.

There is nothing particularly notable about the frontage of the FARM:shop to suggest that there is something special going on inside. It is merely one of a string of small shops arranged cheek-by-jowl along the busy street. Even the identification sign itself offers little help, for it could pertain to one of those idiosyncratic charitable organisations commonly encountered on municipal High Streets throughout Britain. Upon entering the building, however, comes the immediate realisation that this is indeed a different kettle of fish altogether. Even before walking into the large front room, I can hear it. The noise turns out to be emanating from the recirculating aquaponics operations that are busily gurgling away. I spend half an hour studying the intriguing system before wandering through the rest of the building during which I become increasingly impressed with the manner in which the different growing systems are crammed into small spaces. Explorations of the farm shop over, I move to the kitchen and place an order for a green salad and chopped egg sandwich after being told that there is no fish available today. It is while sitting at the table and waiting for the arrival of my lunch that any feelings of disappointment about food choice availability are forgotten. For, as I look up and at the staring eyes of the dozens of fish whom are closely watching me, I think it would be somewhat unsettling to be under their lidless gaze while in the process of eating one of their fellows.

3.1.10 Valley Grown Salads (Harlow, UK)

The Lea valley is a major transportation and industrial corridor that runs north-south through East London. The area underwent extensive regeneration in the years leading up to the 2012 London Olympics (Lewis 2011a), including increased attention to urban agriculture (as, for example, Grow Up Urban Farms, above, and Dalston Eastern Curve Garden, see Chapter 5.1). For more than a half century, the northern, peri-urban part of the valley has been the centre for the greenhouse growing of vegetables in the United Kingdom (Langton and Cockshull 2013; Lewis 2011b), although this declined following the Second World War (Petts 2008). 'Valley Grown Salads' is one of a dozen 'glasshouse' vegetable wholesalers that have set up in the area. Established in 1985, and building on a family tradition of decades spent in greenhouse farming, the company is still owned and managed by the family. Today, the extensive operations include a workforce of 120 people farming more than a hectare under glass. The first crops to be produced were cucumbers (the major vegetable still grown in the Lea valley), followed by lettuce and sweet peppers, and more recently by aubergines and varieties of tomatoes. The company runs its own fleet of trucks and operates a packhouse where produce is graded and size-sorted through an automated production line prior to distribution. Produce is grown through hydroponics and in rock wool bags in order to prevent root disease and to obtain a more controlled growing regime. Insect pests are treated through biological control without the use of chemicals. Special machinery has been installed for waste reduction, shelf-life extension, and energy conservation, all directed toward a goal of reducing the carbon footprint. Heating is generated from modern gas burners, said to be 'more environmentally friendly.'

In recent years (since my visit), a 92,000-square metre glasshouse extension has been built following a five-year battle with local planning authorities. These new high-tech growing facilities, referred to as a turning point toward achieving a British self-sufficiency in food production, merited a Royal visit. The company was the 2004 winner of the coveted UK Salad Grower of the Year award, and its produce has passed the United Kingdom's food safety audit with highest achievable grade 'AA.' In his overview of all the growers in the Lea valley, Lewis (2011b) particularly singles out Valley Grown Salads for continuing to commit time and resources to research and development programmes. In his mind, they 'represent a microcosm of best practice and demonstrates how other growers in the Lea Valley region are successfully adapting their businesses to meet the challenges of the modern world, driven by consumers who are demanding fresh and chemical-free produce, reduced transport miles, and above all, value for money' (p. 77). It is for these reasons, that it is this particular company that I have selected to visit for the present survey.

The extent of the growing operations is immediately apparent the moment one arrives on site. Massive glass greenhouses, each the size of a soccer pitch, abut one another, their windows steamed with humidity as it is a cool day early in the spring [e3.1.10.1]. Trucks [e3.1.10.2] are arranged near loading docks [e3.1.10.3] to receive pallets of stacked plastic crates filled with peppers [e3.1.10.4]. Inside the administration building, a large stuffed mascot of a red bell pepper belle greats visitors to the main office [e3.1.10.5]. The greenhouses are filled with row after row of green bushes suspended from stringed trellises (Figure 3.31) [e3.1.10.6] that rise up from their basal hydroponic system composed of piping, values, and tubing (Figure 3.32) [e3.1.10.7]. Trackways run in between the rows of plants [e3.1.10.8] to enable movement of small dollies [e3.1.10.8]. Upon these, small carriages allow pickers to effortlessly slid along as they fill up their tubs (Figure 3.33) while also allowing them to be jacked up in height [e3.1.10.9]. The bounty of different types of peppers still hanging from the small bushes [e3.1.10.10] or as harvested produce filling crates [e3.1.10.11] is indeed impressive.

Figure 3.31.

Commercial operations on such a massive scale necessitate careful attention paid to growing conditions. An extensive array of high-tech climate control, plumbing, mixing, and circulation equipment is needed (Figure 3.34) [e3.1.10.12], as are containers, tubs, and bags of various fertilisers (Figure 3.35) [e3.1.10.13]. Infestations of plantivorous insects and fungi are major problems in greenhouse horticulture (Figure 3.36) [e3.1.10.14]. Sachets are therefore deployed from which predatory mites emerge to attack thrips larvae feeding on the plant tissue (Figure 3.37).

Figure 3.32.

Figure 3.33.

Figure 3.34.

Figure 3.35.

Figure 3.36.

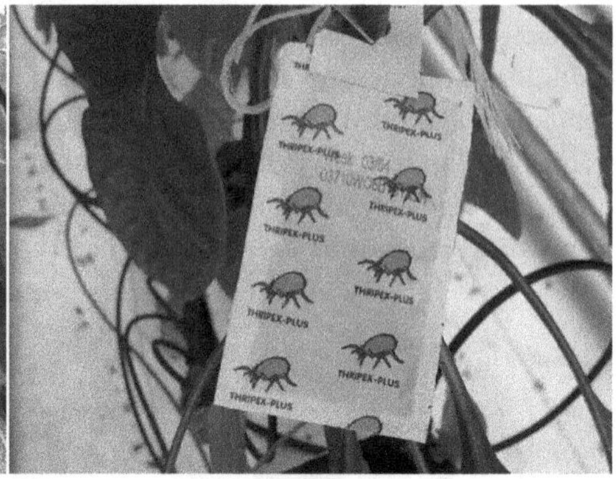

Figure 3.37.

As I stand inside the greenhouse and stare out at the vast array of climbing plants it is impossible not to be impressed by the sheer magnitude and professionalism of Valley Grown Salads. This is peri-urban agriculture on an industrial scale far removed from the one- or two-person entrepreneurial efforts of, say, Brother Nature Produce or Very Local Greens (see above). Gone too is the prioritisation of sustainability as the paramount concern that often characterises many urban agriculture ventures. Here, in a bedroom community of London, is a possible glimpse to what a future might look like with increased recognition and widespread acceptance of climate-controlled agriculture undertaken in the shadow of great cities.

It is fascinating to watch a worker, dressed in shorts and a T-shirt, make his way along the track to harvest the peppers. The elevated temperature has meant that I have had to wait until my camera lens ceases to fog up. Examining the infested leaves in one section makes me realise the challenges faced by greenhouse farming. Later, upon seeing the jugs and mixing tanks of various fertilisers, my initial judgmental reaction as an environmentalist is quickly supplanted when I realise that the amount of land needed to achieve a comparable level of productivity of the year-round hydroponics system would have been immense and would have generated concerns related to soil fatigue and chemical use (notwithstanding increased transportation miles related to distributing produce to the nation's largest metropolis). For urban agriculture just as for traditional rural agriculture, it can be a delicate and sometimes difficult balance between industrial-scale food production and environmental sustainability concerns.

3.1.11 Zero Carbon Food (London, UK)

'Zero Carbon Food' and its subsidiary 'Growing Underground' are an agri-tech company that builds and operates controlled environment farms for the sustainable growing of food in dense urban areas. With the twin mottos of 'shaping the future of the growing industry' and 'from farm to fork in under four hours,' this is a commercial venture that uses abandoned underground spaces to produce leafy green, herbs, and microgreens via hydroponics and LED lighting. Like all such systems, productivity occurs year-round (as the website intones 'every day is the perfect summer's day' – www.zerocarbonfarms.co.uk), with substantially higher yields due to more harvests per year (60) compared to either greenhouses (25 to 30) or outdoor growing (5 to 6). An overarching goal is to reduce the carbon footprint of the food chain by producing 'hyper-local food for the city from within the city' at the same time as offering employment opportunities to inner city residents. The company has received considerable coverage in the international press as well as being invited to give a TED talk and to be included in forum run by the Gates Foundation. The website contains a quotation from noted environmentalist Sir David Attenborough that perfectly sums up what the innovative firm is doing: 'It's entirely possible for us to apply both low-tech and high-tech solutions to produce much more food from much less land. We can start to produce food in new spaces, indoors, within cities, even in places where there's no land at all.'

In 2015, Zero Carbon Food established the world's first subterranean urban farm in the South London borough of Clapham. Remarkably, the farm is located 33 metres *below* ground in a disused air raid shelter from the Second World War. Here, where once eight thousand people had sought shelter during the Blitz, two tunnels of about a half a kilometre in length and with an area of more than six thousand square metres are used for the growing operations. A mezzanine floor divides the tunnel, with the upper level dedicated to growing space and the lower level containing the operations equipment. As LED spectrums greatly influence the taste of plants in terms of phenols, sugars, and starches, much attention has been devoted to finding the most appropriate 'light recipe'. The company has partnered with a university undertaking research on growing technology in tight spaces, which has led to the publication of several studies (Jans-Singh *et al.* 2019, 2020).

The website contains an entire section about sustainability goals. All of its energy use comes from renewable sources; specifically, electricity is created from waste sent offsite to a convertor to offset the farm's carbon footprint by purchasing wind and solar power from the grid. An outside sustainability auditor was hired and in 2021 a milestone was reached when the company became a 'Certified Carbon Neutral Organization' through various onsite and offsite (e.g. sponsoring tree planting) activities such that more greenhouse gases are now offset than are emitted. The closed-loop hydroponic system makes use of filtration technologies to clean water without needing to use chemicals. Harvested greens are packaged in plastic containers which are made from 70% recycled and 30% virgin material (this is the highest percentage legally possible due to concerns about food safety). Investigations about using compostable packaging were explored until it was determined that the material was actually not accepted in many garden waste collections throughout the country, in contrast to the presently used plastic containers which are 100% home recyclable.

Necessary improvements in technical, biosecurity, and human resources were made to achieve the certification required for the nation's largest retailers to stock their produce. Growing Underground's salad mixes are distributed directly to high-end restaurants and caterers as well as to specialty grocers and major supermarket chains through a nearby fresh produce market, the largest such in the greater London area. Varieties include Asian Salad Mix ('A punch blend of sweet pea shoots, spicy purple radish and fragrant coriander'), Italian Salad Mix ('Pea shoots blended with peppery salad rocket and spicy garlic chives'), and English Salad Mix ('Bright pea shoots provide a tasty based for these spicy English mustard leaves mixed with broccoli shoots').

The above-ground operations of the Zero Carbon Food at the time of my visit are conducted from a temporary office trailer set up on the site [e3.1.11.1]. Inside, the cluttered space is filled with books, technical papers, data sheets, and schematics of how the farming activities are carried out below [e3.1.11.2]. Access to the farm is through a locked door in the original air raid shelter building. Here, one descends underground through the shaft of the elevator and spiral staircase [e3.1.11.2]. Once the bottom is reached, empty tunnels disappear into the darkness (Figure 3.38) [e3.1.11.3]. One section, however, is encased in white protective lining resembling parachute material (Figure 3.39). Here, growing tables contain a series of seed density and yield trials that were used to determine the optimum growing conditions in the hydroponics system as well as to attract investment (Figure 3.40) [e3.1.11.4]. Air circulation fans and climate control equipment and sampling machines occupy the rest of the laboratory (Figure 3.41).

Since my visit, growing tables have expanded such that the tunnel is lined with five tiers of hydroponic beds that stretch into the distance under a square roof with no traces left of original cylindrical shape of the tunnel (Figure 3.42). The above-ground offices have moved from the temporary trailer to the building. Today, the company is comprised of more than thirty employees. Those who work as the troglodyte farmers dress in lab coats and surgical boots, with hairnets and face masks, all done in order to maintain the pest-free status of the environment so as to avoid the use of pesticides to maintain the organic status.

Figure 3.38.

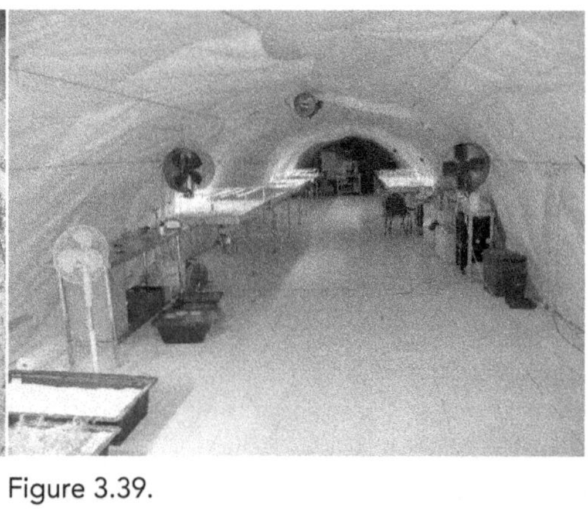

Figure 3.39.

Figure 3.40.

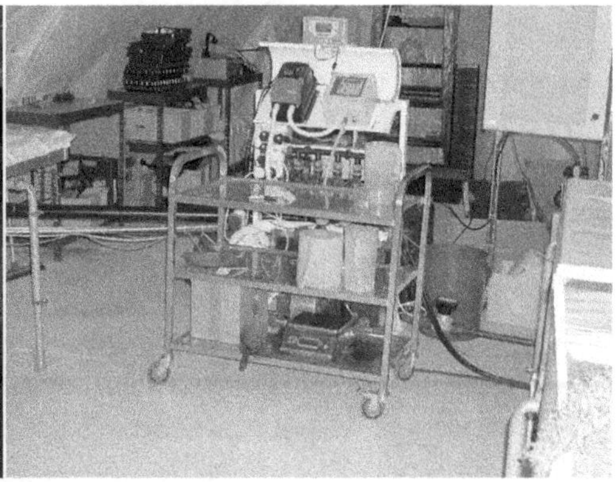

Figure 3.41.

Figure 3.42.

Controlled environment agriculture has traditionally been focused on producing simple plants such as lettuce and microgreens. Growing Underground plan to expand the planting palette to include more exotic varieties from around the world. They also have a desire to develop beyond the fresh produce industry. Toward this end they have partnered with a high-end beverage company to produce limited edition, plant-infused gin. The importance of natural ingredients (flavours and fragrances) from plants is also being explored for application within the cosmetics and healthcare industries. For example, a certified organic coriander face cream and a coriander and mint shampoo are now available. The possibility of growing medicinal cannabis is being considered. Long-term plans are to establish or build a fully-automated urban farm somewhere on the outskirts of London.

My excitement builds as the subway train approaches the Clapham Common Tube Station. For, unbelievable as it may seem, there is actually a farm located beneath where I am now sitting in the carriage! I disembark, exit the station, and after a brief walk along the edge of the urban park, arrive at the trailer that serves as Zero Carbon Food's office. Once inside and sitting around the cluttered space, the enthusiasm of the group is contagious as they explain – each interrupting the other – their current operations and describe their long-term production and marketing plans. And then the otherworldly tour begins. The door to the surface building is unlocked and we descend into the bowels of the earth. Upon reaching the bottom, 33 metres below, a series of tunnels lead off into ominous darkness. An eerie silence pervades. It is an effort to repress memories of the film *An American Werewolf in London*. I shudder and wonder what it must be like to be down here working 'alone.' We walk along, the sound of our footsteps echoing off the concrete walls. And then in a flash, it is as if we've suddenly left one sound stage on a film lot and entered into another right next door. The horror movie set is switched for one for a science fiction film: the white sheeting, the jumble of monitoring equipment, the growing beds, all bathed in the weird pink glow of the LED lights. It is certainly one of the most remarkable urban agriculture sites that I've ever visited. The ingenuity is marvellous, and the mind races in thinking of all the out-of-the-way places in cities that could be dedicated to growing food along the same lines, the only limitation being that of our imagination.

3.1.12 Lufa Farms (Montreal, PQ, Canada)

The island of Montreal, today a metropolis of more than one-and-a-half million, has a rich agricultural heritage. Notable in this regard is the Montreal melon, a nutmeg-flavoured melon that was once extremely popular, but which gradually disappeared as the city expanded after World War II. Thought to be lost forever, seeds were discovered in a United States seedbank in 1996 and the gourd is now experiencing a revival. Since the early 1970s, community gardens have become increasingly popular in the densely urbanised city (Bhatt and Farah 2016). Today, there is high potential for urban agriculture throughout the city (Haberman *et al.* 2014). There is much interest in reconfiguring traditional food networks (Cheveley 2021), including repurposing urban roofs as sites for food production (Hoogstraten 2019). In regard to the latter, it is 'Lufa Farms' that has by far become the runaway success story (www.montreal.lufa.com).

Lufa Farms' motto, proudly proclaimed, is 'Eat fresh, local, and responsible. Join us making cities more sustainable'. Their vision is to change 'the way we feed cities [by] creating an ecologically and economically sustainable model for urban farming'. The company accomplishes this through their mandate to build an improved food system based on reconnecting people to where their food comes from through partnering with hundreds of local farmers and food makers, and providing food to their customers via an online, one-stop farmers' market and delivery service. Customers, whom they refer to as 'Lufavores,' subscribe to receiving weekly food baskets for $30. These can be customised for dietary flexibility and contain rooftop-grown, picked-to-order vegetables in addition to thousands of other products from their network of partners, including eggs, dairy, pasta, bread, meats, and vegan alternatives. Baskets are delivered to a suite of pick-up locations around the city, or for another $15, directly to one's door. And though Lufa Farms are loath to admit the parallel, if this sort of online retail and distribution model sounds at all remotely familiar, think of the retail giant Amazon.

Following several years of planning, the firm claims to have built the world's first commercial rooftop greenhouse in 2011 in the Ahuntsic district of Montreal. Fifteen types of vegetables were grown to fill two hundred food baskets a week, with no opportunity for customising orders. Since then, the growth of the company has been nothing short of remarkable. In 2013, a second rooftop greenhouse was built in a different suburb of Montreal, and the online market was developed to allow subscribers to customise their orders. By 2016, there were thousands of customers who received home delivery of food baskets by electric car, with the company becoming cash-flow positive. The following year, a third and more-technically advanced greenhouse became operational. In 2019, corporate offices were moved to a new location, their packing system was redesigned, and they began construction of the world's largest rooftop greenhouse in another neighbourhood of Montreal. When this facility opened the following year, Lufa Farms doubled its total growing capacity. This has enabled the company to go from feeding 1 to now 2% of the entire city's families, a truly remarkable feat for a single entity in operation for less than a decade. Today, hundreds of different types of vegetables are grown on more than one hundred thousand square metres of rooftops. More than three hundred food partner farmers and food producers are part of the online market. And about twenty thousand food baskets are delivered each week to four hundred pick-up points and private dwellings.

All greenhouses grow vegetables and herbs hydroponically based on their five principles of 'responsible agriculture': using no land, recirculating water, conserving energy, reducing waste, and using bio-controls instead of synthetic pesticides. The original 2011 farm is the most diverse, with different types of growing systems producing dozens of herbs, microgreens, cucumbers, and peppers. Their later, higher-tech farms are much more specialised, with one focusing on tomatoes and eggplants, another on lettuces, herbs, and leafy greens, and the newest, which they dubiously claim to be the largest urban farm in the world, on eggplants and tomatoes. The rooftop farms are extremely efficient, to such a degree that they claim to only need to convert the roofs of nineteen average-sized shopping centres to be able to grow enough vegetables for much, if not all, of the city of Montreal.

Lufa Farms recognises that since everything cannot be produced on a city roof, their facilities are not meant to replace local farms and food makers. For this reason, much attention has gone into building a diverse network of partners who are selected not only for their products but also for the transparency of their production processes. The company states that 'It's essential for people to know their farmers, know their food, know where it's from and how it's grown'. Lufa Farms goes on to acknowledge that reinventing a broken food system without consideration of the community is impossible. To this end, tours are given and open houses take place in their original facility, in addition to social offerings such as yoga sessions and workshops on caramel apple making, pumpkin carving, and ceramic painting. Finally, cursory mention is made on their website that they work with local non-profits to sponsor individuals and families-in-need to become customers through contributions from their well-off clients. What is obvious is that this is very much a for-profit enterprise that is far removed from the socially responsible idealism that imbues the urban agricultural efforts of many of the case studies examined in the present book.

Lufa Farms' original roof-top greenhouse is located close-by a busy thoroughfare in the north-central garment neighbourhood of the city [e.3.1.12.1]. Approaching the building, it is easy to recognise that this is a serious enterprise in that, looking up from the ground, it seems that nearly the entire roof surface is occupied by the jagged, multiple peaks of a single, contiguous glass structure (Figure 3.43) [e3.1.12.2]. Inside, up a staircase, and past a company sign [e3.1.12.3], brings one to a suite of company offices and a staging area in which crates filled with plastic bags of freshly harvested leafy greens [e3.1.12.4] await being taken downstairs for loading onto several large trucks in the parking lot. Visitors taking one of the regularly run weekly tours are given a brief introductory talk on the genesis, ethics, and mission of Lufa Farms. Here, we are told that the building had been in industrial use, although specifically what that might have been is unknown by the guide (today the building is dedicated to office space for an assortment of companies). Interestingly, we learn that the reason that this particular building was selected was that it had been originally planned to have another storey which was never actually built. This meant that the building was structurally sound to be able to accommodate the addition of the rooftop greenhouse without the need for expensive engineering work. Another criteria was the presence of a good freight elevator in which to move material back and forth from ground to roof.

We ascend to the rooftop growing operations to view a section of the farm devoted to vertical hydroponics (Figure 3.44). Here, long allées of bell pepper plants [e3.1.12.5] can be closely observed along with their complicated network of plastic bases and tubes through which sustaining water and nutrients circulate [e3.1.12.6]. The harvest vehicle that moves along the tracks between the towers of plants is also present [e3.1.12.7]. Also here can be seen a hive for pollinating bumblebees [e3.1.12.8] in addition to suspended sachets that release aphid-eating mites.

After a few minutes of observation, the tour group is shepherded through a door into another region of the enormous greenhouse. This is where a seating area of attractive wooden benches and worktables (Figure 3.45) [e3.1.12.9] is arranged facing a single-step-up, small wooden stage [e3.1.12.10] upon which the cicerone delivers a lengthy lecture about the operations of the farm. As this is also the area where yoga classes and workshops are held, attention has been paid to aesthetics in the form of a canopy of passion fruit [e3.1.12.11] and several box planters painted – in a cute node of homage to the surrounding rooftop greenhouse – to resemble buildings atop (i.e. out of) which decorative plants emerge (Figure 3.46) [e3.1.12.12]. Immediately behind and beside this social space are vast hydroponics beds. Unlike the typical vertical arrays characteristic of many, small-scale urban farms, such as Very Local Greens and Grow Up Urban Farms housed in shipping containers or even the large operations of the Lea Valley Green Salads (see above), the novelty here in that the considerable space of Lufa Farms allows for the systems to be horizontal.

The beds display several species in various stages of growth, from the first sprouts emerging hesitantly from their plastic embryonic solutions to mature plants overflowing and completely obscuring their plastic bases in an incredible display of productivity that almost beggars belief (Figure 3.47) [e3.1.12.13]. Climate regulating equipment is pointed out and explained to the visitors: the lighting system to augment the natural daylight at certain times of the year and the heat-retaining and sun-blocking screens which can be lowered from the ceiling [e3.1.12.14], the humidifier matrix wall of cooling water and circulated air [e3.1.12.15], as well as various electrical control devices.

Lufa Farms has been internationally singled out as an urban agriculture model in which to emulate. For example, 'What Lufa is doing in Montreal could be an integral part of any urban centre' (CITIES 2016:149). Others, however, are not so certain, at least in terms of the implications for local food networks. Due to their growing popularity among the food-conscious, 'locavore' public, alternative food networks originally initiated by small-scale farmers, such as community supported agriculture (CSA), are now being adopted by corporations. A critical paper from McGill

Figure 3.43.

Figure 3.44.

Figure 3.45.

Figure 3.46.

Figure 3.47.

University in Montreal (Allaby *et al.* 2021) is important in that it challenges the myth that all urban agriculture is benevolent and unproblematic. The study focused on the online marketplace enterprise run by Lufa Farms and its consequences for regional farmers. Allaby *et al.*'s study was admittedly constrained by the company refusing to talk to the scholars, instead responding to a handful of their questions in an email that did not provide any information beyond the few snippets of such available on their website. Clearly this is a firm which has no time or interest in discussing their operations with academics.

The problem the authors found is that many customers are being misled that the socially conscious values traditionally associated with CSA are being co-opted by Lufa Farms. Instead, these values, argue disgruntled small-scale farmers who were interviewed, have been conflated and mimicked by the profit-based model of Lufa Farms. The corporation's rooftop greenhouses only grow about one quarter of the food that is sold through its marketplace. A key tension relates to the confusion about the degree of local food sourcing in such a distribution model as used by Lufa Farms. The authors note, for example, that several partnering farms are 'local' only if one considers Ontario, British Columbia, Georgia, and even Florida to be so in relation to Quebec. Moreover, some of these partner farms are large-scale, single-product operations far removed from diversified family-run farms. Peri-urban organic farmers are frustrated that the public do not understand the differences between the terms 'organic' and 'sustainable,' erroneously believing that Lufa Farms' hydroponic greenhouses practice the former form of agriculture. Overall, a David-and-Goliath confrontation has developed in Montreal with some regional family farms falling by the wayside if they either fail to specialise in one or two crops or if they choose not to become part of Lufa Farms' distribution partnership. It is impossible not to draw analogy to the ramifications that the online giant Amazon has had on the marketplace of independent bookshops.

My initial interaction with Lufa Farms is not a positive one. I had written, explaining that I was a university professor who had hosted and produced a book from the first international conference held at a Canadian university on urban agriculture, and that I was hoping to be able to visit several of their farms accompanied by someone with technical knowledge as it was my intention to include Lufa Farms in the present book. The curt reply was that I could simply sign up for one of the existing tours they put on for the general public, which, furthermore, are limited to only their original location. In short, and unlike the other sites visited for this research endeavour, this particular company was completely unwelcoming to scholars. Perhaps with the publication of Allaby *et al.*'s paper as well as the slight dis here, they might learn the lesson for the future not to annoy academics who then go on to write about them.

I am made slightly uncomfortable from the very start of the tour, feeling that I may have entered a cult when, in response to the question from the cicerone, everyone else in the group enthusiastically state that they are 'Lufavore' customers. I try to keep my prejudices at bay, formed from having read Allaby *et al.*'s paper for the first time just several days before. The bragging, in terms of Lufa Farms being the 'first,' the 'biggest,' etc. may be true, but seems unneeded. The guide is highly capable and engaging, although it takes but a few questions to realise that she does not have the technical expertise to answer anything out of the norm expected from the members of the general public. Given the dearth of technical details on their website – something also alluded to by Allaby *et al.* – means that I reconcile myself to having incomplete information about the site. My inquiry as to whether this or any of their other rooftop facilities grows the fabled heirloom Montreal Melon is met with a puzzled frown and an 'I don't know.' What a missed opportunity, I think, to generate some good PR; something which might be needed in the months ahead should the press learn of the competitive casualties of Lufa Farm's success from Allaby *et al.*'s incisive article. Discussion by the cicerone of the recent expansion of the company as well as its future plans to increase its customer base in Montreal seems to go over well with the group of visitors. Afterwards, a colleague who was accompanying me on the tour, remarked that she could certainly recognise the analogy made that Lufa Farms is positioning itself to become the Amazon.com of urban agriculture. A thought flashes through my head while the tour guide is talking; it is that of Dr. Suess' Once-ler from *The Lorax*, and his goal of 'biggering an' biggering'.

Despite some reservations about the competitive ramifications ensuing from the Lufa Farms' success, it is impossible not to be duly impressed by the physical operations of, and the imagination behind, the enterprise. Walking about while looking at the luxuriant bounty in the foreground, framed by a view of city rooftops seen in the distance through the glass ceiling and walls, is a memorable experience... and perhaps, should expansion continue on its current trajectory, a glimpse of what the future may hold, in Montreal and beyond. As we leave the greenhouse, another group of keen visitors is already mustering for the next tour, this one in French. I reflect that there is no doubt that Lufa Farms is well on its way to becoming the worldwide standard in which to compare for commercial rooftop urban agriculture enterprises combining both elevated ingenuity and lofty ambition.

3.1.13 Case study lessons

- Importance of out-of-the-box thinking; i.e. not everything in urban agriculture has to be entirely 'urban,' as for example, the separation of production and processing in viticulture and oenology (London Cru Urban Wintery).
- Small-scale, no frills operations can be viable additions to commercial urban agriculture (Brother Nature Produce).
- Youth can be engaged in successful commercial ventures (Hope Blooms).
- Specialising in heirloom crops can carve out a profitable niche (Dill's Pumpkin Farm).
- Capitalising on embedded local history helps attract people to sites as shopping tourists (Dill's Pumpkin Farm).
- Urban agriculture can be integrated into world-class, landscape-scale, recreation parks (Fresh City Farms), just as it can in small, dynamic, and fun social spaces (Grow Up Urban Farms).
- Focusing on the higher quality of produce being grown compared to less expensive competitors can be the key to marketing success (Fresh City Farms, Very Local Greens).
- Who said that urban agriculture has to be conducted outside and dependent on the vagaries of the weather? i.e. innovative, controlled environment systems utilising hydroponics (Very Local Greens, Valley Green Salads, Zero Carbon Food, Lufa Farms) and aquaponics (Grow Up Urban Farms, FARM:shop) can achieve high productivity in small indoor spaces throughout the year.
- Bringing local political leaders on board early in the planning process can be critical for successfully circumventing roadblocks from narrow-minded bureaucrats that are crippling to innovation (Very Local Greens).
- Use of vacant roofs is an obvious solution to reanimate the urban framework while simultaneously reducing food miles (Grow Up Urban Farms, Lufa Farms), especially if situated immediately above supermarkets (Food from the Sky).
- There may be no need to use fancy, high-tech growing systems when something as simple as plastic tubs can suffice (Food from the Sky).
- Cradle-to-cradle recycling of previously-used materials can be imaginatively utilised, such as former shipping containers to house growing operations (Grow Up Urban Farms) and disposable plastic bottles to construct hoop-house walls (Food from the Sky).
- Food production can be combined with food consumption through onsite cafes (FARM:shop).
- Crop specialisation can be the key to scaling-up production for large-scale commercial enterprises, as can consideration of some but not all issues related to sustainability (Valley Green Salads, Lufa Farms).
- Movement toward sustainability can be achieved through using waste to offset energy use, planting trees to obtain carbon neutral status, and changing packaging materials (Zero Carbon Food).
- Expanding operations from the production of food to include its distribution through an online marketplace is an approach to maintain a base of returning clients (Fresh City Farms, Lufa Farms).
- The imaginative repurposing of old and abandoned spaces, including underground locations (Zero Carbon Food), proves that urban agriculture can thrive without the need to construct fancy new buildings as envisioned by some architects.

- Plants can be grown and marketed for uses other than their consumption, as for example, for cosmetics, beverages, and health purposes (Zero Carbon Food).
- Providing onsite access through site-tours is important for fostering a community of engaged and happy consumers (Lufa Farms).
- Attention to site aesthetics in the form of art installations and attractive furniture go a long way toward contributing to memorable experiences of visitors (Hope Blooms, Grow Up Urban Farms, Lufa Farms).

3.2 Food market tourism

Anecdote

Personal experiences of a gastronomic tourist. I have fond memories of being a food tourist. Ethnic restaurants are often visited by tourists just as much for an experience of exotic ambience as for a gustatory experience of culinary offerings. Sometimes it is possible to travel and sample the gastronomic world without leaving one's city. I grew up in Winnipeg (Manitoba, Canada) at a time when there were few ethnic restaurants despite the cultural diversity of the city, and thus had little opportunity to explore exotic cuisine. That was all to change with the advent of 'Folklorama' in the 1970s, which has since gone on to become the province's largest festival and most important annual tourist attraction (with visits numbered in the hundreds of thousands and contributions to the local economy in the millions of dollars – Selwood 2003). Indeed, the festival's success has been so marked that it has been copied by other cities around the world. Folklorama was an event eagerly awaited each year. With my mother – who was always an adventurous cook – we would visit, over a two-week period, more than a dozen pavilions scattered around the city, each representing a cultural group or nationality. There we would savour the distinctive food, watch dances, examine crafts, and of course get our mock 'passports' stamped. Years later, now old enough for independent foreign travel myself, I visited family friends in Frankfurt. There, in addition to attending the well-known local apple wine festival, I marvelled at the intriguing idea of one city street being dedicated to pedestrians. There, stretching for several blocks, were a number of food kiosks. Each of these was topped by a different national flag and offered the characteristic food of one of Frankfurt's sister cities from around the world, all sold by actual restaurateurs who were residents from those distant locations and present in Germany on limited work visas.

And who does not remember that first visit to a large and dynamic food market when travelling abroad? For me, it was a during a backpacking trip to Sicily, when on a Saturday morning I spent hours strolling about the local market in Messina, trying to recognise the genus of the mysterious sea-life displayed in front of me. I left with a large bag filled with dried sausages and impossibly juicy blood oranges, both of which would sustain me for days of travel and sight-seeing ahead.

Book

Food tourism around the world: development, management and markets, edited by C.M. Hall, L. Sharples, R. Mitchell, N. Macionis and B. Cambourne, contains 18 chapters by contributors from around the world. Most striking in this collection is the remarkable range of tourism experiences related to food, as for example, halal food for Muslims, cookery-school holidays, tea museums, aboriginal foods, nouvelle cuisine, food inspired festivals, wine-food linkages, development planning, and food markets.

Food is often a recognised characteristic of a region's identity and thus is an important component of its culture and heritage. Food tourism focuses on consuming the production of a particular place, in other words, it's *terroir* in the French tradition of idiosyncratic, regional wines, or it's 'foodscape' in professional parlance. In England this takes the form of the imaginatively titled 'Eat the View' campaign, aimed at promoting sustainable local products within specific regions. Tourists crave authenticity and a sense place, both of which can often be found at the end of one's fork.

Therefore, given that food is integral to the experience of travel, it is no surprise that it is now considered as a key element in destination marketing in that tourism planners are increasingly recognising interests in artisanal foods and food markets. Such place-marketing depends on long-term competitive advantages derived from differentiating locations on the basis of their unique food products. Numbers shown in one chapter support adopting such a strategy. For example, a survey of tourist attitudes and actions towards regional and local food in the United Kingdom found that almost three-quarters of tourists took an interest in local fare during their visit, with 39% making at least one purchase, and 26% expressing a strong interest in having restaurant menus identify food that has been produced locally. In the introductory chapter to this volume, the authors believe that food tourism reinforces local food economies, encourages the conservation of food and biodiversity, and helps sustain local identities. Furthermore, as another contributor points out in his chapter, the contribution of food to the tourism economy can be considerable due to the intensive use of labour in food preparation and services.

The dramatic recent upsurge in farmers' markets is one of the great success stories of food tourism. Such markets not only make substantial contributions to farmers' revenue, allowing them to maximise their returns, but they also help bring life into towns and cities by aiding regeneration. In this respect, farmers' markets, in which the financial expenditure of visiting tourists is becoming increasingly significant, stimulate local economic development by supporting businesses and keeping the money within the local community. They also attract new retailers to the vicinity and employ locals. Case studies in the book highlight the significant economic impacts from farmers' markets on communities, including one such investigation which showed that 60-70% of market-goers in Ontario visited neighbouring businesses outside of the physical bounds of the market itself. Two strategies that are becoming increasingly used by farmers' markets as a way to collect tourist dollars and funnel it into the local economy are the presence of onsite food kiosks or restaurants and the regional branding of products to emphasise their local distinctiveness. When tourists purchase food at a farmers' market they are not only provided with an interesting and enjoyable shopping experience, but also an opportunity to learn about local food culture while directly supporting that culture. Furthermore, this aspect of food tourism, by facilitating and shortening the distance between producers and consumers, also contributes to education about environmental sustainability.

Sites

3.2.1 Plaza de Armas (Ancud, Isla Chiloé, Chile)

Farmers' markets, as well as being recognised as instigators of economic revitalisation and spatial regeneration, are also public places characterised with a dynamic, socio-behavioral ecology (Sommer 1989; Sommer *et al.* 1981). Landscape architects and urban planners work to integrate these concepts into the public realm (Corum *et al.* 2001), the physical manifestations of which – the promenade, the working area, the market landscape, and the abutting neighbourhood – can be spatially experienced and examined (Francis and Griffith 2011).

Ancud is a coastal fishing town of forty thousand inhabitants on the southern Chilean island of Chiloé. On Saturdays, a small food market is set up in front of the church in the central plaza (Figure 3.48). Here, as is the case in many sleepy towns around the world, life for a time becomes focused in the plaza. Musicians play, dogs laze and cats prowl, the elderly sit and chat while the young mingle and flirt, both groups periodically interrupting their socialising to sample items from the food trucks (Figure 3.49) parked along one side of the town square in front of quaint buildings [e3.2.1.1]. Tables display honey, jams, fruit, and vegetables from local farms, in addition to traditional handicrafts from regional artisans. Apart from the fascinating statues of figures from the pantheon of Chilote mythology that are arranged about the plaza [e3.2.1.2], and the information booths set up by regional organic growers associations [e3.2.1.3] that display bags of gourmet heritage potatoes, there is nothing particularly unique about this market to distinguish it from the thousands of such that exist throughout the Global South. It is simply an unpretentious and genuine facet of urban life which, on Saturdays, is focused around food.

Figure 3.48.

Figure 3.49.

From the room window in my quirky boutique hotel, I look out to see the plaza bustling with activity. The scene is strikingly different from when I walked about the town following my arrival the day before. Then, Ancud had struck me as a somewhat sad and lifeless backwater. The contrast to that ill-formed first impression could not be clearer than on this present Saturday market-day. Shaking off my jetlag, I hurry down to immerse myself in Chilean urban culture. Several times I break my perambulations to return to the food trucks to sample the various types of fresh juice and *carne asada* street food. The island archipelago of Chiloé is one of two geographical nuclei of potatoes, with more types existing here than anywhere else. It is with fascination that I spend time at the information booths, where I strike up conversation about the challenges and rewards of organic farming in the similar temperate climes of southern Chile and southern Canada. My visit is made perfect when I am offered several varieties of potatoes that have been slowly simmering away in a large pot. I breathe in the aroma as a blast of steam arises when the lid is lifted. Even this olfactory prelude underestimates the sumptuous flavour of the assortment of potatoes that I soon sample.

3.2.2 The Goods Shed (Canterbury, UK)

Mentioned in internationally published guidebooks, 'The Goods Shed' in the historic cathedral city of Canterbury is a major attraction. Here, as the prominent sign beside the train station proclaims (Figure 3.50), one can feast on a variety of soups and sandwiches. Begun in 2002 as a farmers' market, it has expanded to include both a food hall and restaurant, all contained within an old Victorian brick building that was once used to store coal and to service train engines [e3.2.2.1]. A steady growth in popularity has led to the food being available for sale six days of the week, leading to questions as to whether such a permanent mercantile situation still counts as being a true, ephemeral farmers' market per se (Hall and Sharples 2008).

Though modest in size, this loose coupling of market and restaurant is historically significant, as explained in the 'Where to Eat' section of a guidebook for pilgrims, walkers, and tourists arriving to their destination city (Curtin 2006: 166): 'If you like fresh quality produce the chances are you'll want to eat at The Goods Shed ... England's first full-time farmers' market restaurant is housed in a revamped Victorian engine shed with high oak-beamed ceilings and huge windows. Bringing the kitchen to the market, the menu constantly changes using the produce available on the day so you get the likes of pan-fired skate and pumpkin risotto.'

The market restaurant has garnered praise in national newspapers (www.thegoodsshed.co.uk), including from *The Guardian*: 'Ingredients are 'sourced' from roughly 2 ft. away in the market: finest Kentish produce, British traditions and techniques, Spanish sensibilities. The result is balm to the soul, not to mention the kind of grub that sends you off with a giant grin on your well-fed chops;' and from *The Telegraph*: 'Enter, and you are confronted by the kind of scene that makes

Figure 3.50.

you wonder if you've died and been whisked off to gastronomic heaven: the finest Kentish fruit and veg, glorious, doughy pillows of fresh bread – and the said dead animals, lots of them, all heaped under a rustic wood-beamed roof, flooded with light from grand Victorian windows. It's enough to make a Londoner burst into tears and decamp to the country.'

Operating for several decades, Canterbury's The Goods Shed has long served as a model in how to boost revenue by selling farmers' food for immediate onsite consumption. Also, being situated right beside the train station on the major line to London means that for many tourists this is the first eatery they spot upon disembarking. A testimonial on the website reads:

> I went to the market ... and bought fresh stocks from Baldlesmere, some traditionally reared pork chops, local mackerel, bread pulled straight from the oven and vegetables collected that morning from Jonnie Sandwich with a glass of wine from Clive and pondered Sunday lunch. The restaurant menu inspired local venison, celeriac, black pudding and cobnuts. The Butcher gave me cooking advice for the haunch of Stour Valley venison and I grabbed the black pudding from Murrays General Stores. The Cheesemakers of Canterbury furnished me with their own Ashmore and other fine cheeses for a board, local juices and homemade pickles. At Wild Goose, folk were enjoying hedgerow-inspired cocktails and little dishes. Next time, I thought...

Each of the handful of merchants present herald their local fare. For the butcher, this means stating that animals come from local Kentish farmers and were raised and slaughtered humanely; for the fishmonger, this means noting that fresh fish arrive each morning after being caught in nearby waters; and for the produce stalls, this means mentioning that a significant proportion of the vegetables are grown in the fields of East Kent.

For several days I have been riding the train out from Canterbury towards London, hopping off at a handful of locations in order to walk back toward the 'Cathedral City' along the medieval pilgrim way made famous by Chaucer. Each morning I pass the still quiet environs of The Goods Shed, pausing to purchase a tea and fresh muffin and to watch the merchants set up for the day while I wait for the train. With its old wooden cross-beams and aged pink, red, and white bricks, the scene looks like a giant food pantry in an old country manor house.

Sitting in a carriage on the return ride back to Canterbury, it is always the first sight of 'Bell Harry,' the cathedral tower, that brings a quick rush of anticipation for knowing I will soon be able to pick up some fresh bread, local cheese, and bottles of craft ale and cider, all for my evening pilgrim repast after a long day spent walking through the English countryside. I also always think back to my first pedestrian arrival to the city, at the conclusion of having walked hundreds of kilometres from Glastonbury along the so-called Pilgrims' Way. At that time I resisted the temptation to immediately enter the city through its imposing medieval Westgate. Instead, I diverted a few steps to The Goods Shed in order to celebrate having completed England's most iconic pilgrimage route. It was a truly memorable time, well-spent engaged in savouring local delicacies and consuming several martinis made with infusions of herb spirits and flower syrups inspired from nearby English country gardens. My justification for imbibing the latter was that it was part of my agriculture research, since, after all, Kent is referred to the 'Garden of England.' Later that evening, I returned to The Goods Shed restaurant for a sumptuous meal, all the time admiring the way the multi-coloured bricks and arched windows of the repurposed building provided such lovely ambiance for the dining experience.

3.2.3 Kensington Market (Toronto, ON, Canada)

Canada is one of the world's most culturally diverse countries, and nowhere is that better displayed than in its vibrant farmers' and culinary markets located in its major cities. Anyone who has visited, either as a tourist voyeur or a resident consumer, markets such as Atwater in Montreal, The Forks in Winnipeg, or Granville Island in Vancouver, well knows the importance of these institutions to the Canadian urban experience. Toronto boasts two such markets, striking in their physical dissimilarity.

Toronto's 'Kensington Market' is a very different shopping and tourism experience. Instead of being clustered under a single building, the open-air market stretches out over a series of city blocks that span 27 hectares. The area is known for its multi-cultural diversity and was the location for a much-loved television sitcom of many decades ago. Due to being an evolving cultural landscape of immigrant occupation – Irish and Scottish, then eastern European Jews and Italians, followed by those from East Asia, the Caribbean, Africa, and Central America – the area has been designated as a Canadian national historic site. Because if this, the Kensington Market is a collective 'memoryscape' (Li 2015) that radiates a strong sense of place. It is referred as 'Toronto's cultural mecca' and is a noted tourist destination, leading to it being the most photographed neighbourhood in the city. The narrow streets – free of cars on Sundays – are bordered by densely-packed Victorian-era houses. The whole area was saved from being converted into high-rise apartments during the 1960's plague of urban renewal, and is now protected through strict construction and renovation ordnances.

Today, in addition to shops prominently displaying fresh produce on the sidewalk [e3.2.3.1] and other specialty food locations [e.3.2.3.2], a suite of venues such as bars and café patios (Figure 3.51) and street-side cooking operations [e3.2.3.3] from restaurants exist to enable urban hipsters to linger and support the local economy while rubbing shoulders with their neighbours in Canada's richest cultural melting pot. Furthermore, the market champions its artsy, counter-culture funkiness with numerous building facades given over to brightly coloured murals [e3.2.3.4]. The bohemian neighbourhood abounds with indie shops, vintage boutiques, and art spaces. And, as a further incentive to attract food tourism to the area, the Kensington Market abuts one of Canada's largest Chinatowns (Figure 3.52) [e.3.2.3.5].

Figure 3.51.

Figure 3.52.

It is worth noting that the flip side of popularity is often competitive shopping. The increasing size of crowds at farmers' markets in Toronto has produced critics who rankle at the bad manners often being displayed. As, for example, an article entitled 'Food fights: Jockeying for fresh fruit and vegetables reveals middle-class Torontonians at their competitive, grasping worst' (Richler 2010). Also, with popularity comes the threat of gentrification. Working-class immigrants live beside artists and writers, both working together to thwart colonising incursions by international chain stores (Shaw 2005). It remains uncertain for how long these successes against globalisation can last (Johnson 2010b; McLean and Rahder 2013).

It is a Saturday afternoon in the early summer when I visit the Kensington Market. All is a-bustle with shoppers jockeying one another on sidewalks as they pull along small carts, each, minus the spilling goat horn, a brimming cornucopia on wheels. The babel of languages overheard recalls feelings of walking through European or Asian villages. Almost all the outdoor seating of more than a dozen establishments is already occupied, the voices of happy patrons spilling out into the neighbourhood. And so, I opt for some Jamaican jerk chicken purchased from a street-side vendor. I find a curb upon which to sit down and munch away while reminiscing about the joy of eating ethnic street food in previous travels.

More than anything, it is the colourful murals adorning a handful of buildings that are most interesting feature about this particular market. I continue my explorations of the endemic artistry of the area, searching out, first, the statue of the actor who figured prominently in the television show about the neighbourhood, and then the 'Garden Car', a vehicle that has been repurposed as a public art piece while also doubling as a micro-community garden. Alas, although successful in locating the former, I strike out with regard to the latter. Above all, while strolling about the area, I find myself awash in nostalgia for the time when I lived in the city and with my fellow graduate students would frequent the shops and market stalls of this bohemian gem, Toronto's answer to the Latin Quarter, Soho, or Greenwich

Village. Later, I wander over to the adjacent Chinatown to examine the diverse array of specialty fruits offered for sale in the sidewalk market stalls, before entering a restaurant also remembered from decades before. There, I cap off the experience with a delicious meal of Chinese dumplings.

3.2.4 Haymarket (Boston, MA, USA)

The 'Haymarket' is *the* place to be on a Saturday morning in downtown Boston (Figure 3.53). Located nearby the Old State House [e3.2.4.1] and situated beside the famous and historic Quincy Market and Faneuil Hall [e3.2.4.2], the former now given over to shops and restaurants catering to tourists, the outdoor market has itself become part of an itinerary marketed in tourism guidebooks and by tour companies. The latter pitch the visiting experience as being an opportunity to interact with the cultural diversity of colourful Bostonians (Figure 3.54) [e3.2.4.3], while smiling at the over-the-top antics of the vendors who are known for their affected abrasiveness (e.g. one dare not handle their produce lest you become the object of vehement chastisement) and booming sales voices. Not without reason has the location been featured in a handful of big-budget films.

Given that some form of market has existed at this location for hundreds of years, the 'Haymarket Pushcart Association,' established in 1820, refers to it as being one of the oldest such in the United States. Although in the early days, vendors were primarily of Italian heritage, today more than twenty nationalities are represented among the forty or so independent sellers. This being Boston – a place that goes to great lengths to champion its role in creating the American identity – much is made of the history of the market in a book (Goodstein and Turino 2015) and various websites. To preserve the living history of the area, the 'Haymarket Project' presents interviews of long-time vendors and recent immigrants along with a photographic record of sellers and customers filmed over the span of a year (www.historicnewengland.org). On its own website, the

Figure 3.53.

Figure 3.54.

market association provides a montage of old photographs along with contemporary interviews of vendors and a time-lapse video showing the set-up and bustle of the market over a two-day period, accompanied by the sound of buzzing bees (www.haymarketboston.org). Of particular interest are the interviewed vendors who are recent immigrants, and their expressions of enjoyment at being able to meet people from all over the world for the first time in their lives.

Abutting the many fruit and vegetable stalls which snake around several city blocks [e3.2.4.4], are shops selling seafood and specialty meat [e3.2.4.5], and of course, this being Celtic-infused Boston, a pub dispensing Guinness on tap [e3.2.4.6]. The spill-over economic boon to the surrounding businesses is enormous. All this contributes to making a visit to the Haymarket an outing that is equal parts functional (picking up the next week's supply of food) and fun (people watching and lingering to enjoy the many street buskers). Due to a history of conflicts between vendors and city administrators brought about by piles of trash left behind after the market closes, refuse is now collected and processed right away in disposal systems that are front-and-centre rather than being hidden behind the scenes [e3.2.4.7]. Moreover, as a nice artistic detail, several gilded brass installations have been set into the pavement to represent dropped containers from which produce has spilled (Figure 3.55). The artists, well-known for their environmental pieces (Harries and Heder 2005), describe the project as representing both a living experience that occurs when actual debris covers the sculptures when the market is in full swing, and as a memory from when the waste is cleared away to the next Saturday morning.

Figure 3.55

As I walk from the subway station, I know I am nearing the Haymarket even before the crowds are spotted. This is due the boisterous shouts that echo along the streets. I follow in the wake of what is obviously a tour group, possibly either from a cruise ship docked in the nearby harbour or from several companies that specifically offer guided tours of the market. All are wearing identical nylon jackets and trundling along behind a woman holding a yellow umbrella aloft despite it being a sunny morning. Their presence provokes a flurry of loud outbursts from the market stall sellers, each vying to outdo his neighbour in encouraging the tourists to stop and sample his produce. The voices are a precious slice of multi-culturalism, the distinctive dropped 'r's' of the Boston Brahmin booming alongside the variously accented shouts of immigrants, all in a cacophony of competition. I move away from the immediate hustle and bustle of the stalls to find a location where I can stand still for a few moments. With eyes closed, I concentrate on the aural soundscape of the market that bustles around me: the bellowing sales pitches, the rustle of grocery bags being filled, the crunch of wooden crates of vegetables being dragged across the pavement and ripped open, the babel of inquiring customers, the squawk of seagulls, the apologies offered for inevitable collisions in the throng...

Later, while strolling about taking photos, I become amused at the venial sins being committed by those whom are in obvious ignorance of the rules of etiquette posted on the market association's website. Several potential customers are screamed at for reaching to touch the displayed vegetables themselves; others who attempt to haggle are barked at to go away and not to return until after 3 pm when such negotiation is permitted; and several people are chastised for being daft and bringing their baby-strollers or dogs, neither of which can be tolerated in such teeming crowds.

3.2.5 Farmers' Markets (Halifax, NS, Canada)

North America's oldest continually operational farmers' market is found in Halifax, Nova Scotia. Since its inception in 1750, the location of the market has changed several times (Kilcup 1998). The most recent incarnation is the 'Halifax Seaport Farmers' Market' which is strategically situated on the waterfront beside the pier where immigrants once entered the country (now operated as an Ellis Island-like museum) and where today tourists disembark from moored cruise ships. A case can be made that the most interesting feature of this market is its innovative architecture. The beautifully renovated building (Figure 3.56) [e3.2.5.1] is LEED Platinum-credited for sustainability, including a pump heating and cooling system, natural lighting, an air-purifying bio-wall, and cisterns for harvesting rain for use as greywater. There is also a rooftop deck where visitors can take a break from food shopping down below [e3.2.5.2] to admire the panoramic views of the harbour and city while being surrounded by small wind energy turbines and a green eco-roof composed of sedum plants for rainfall absorption and evapotranspiration cooling (Figure 3.57) [e3.2.5.3]. Repurposing of the semi-abandoned warehouse into the new 5,200-square-metre market has been referred to as 'a stellar example of adaptive reuse and incisive architectural intervention' (Carroll 2012). Because it addresses the province's sustainability priorities (Bibeau 2020), the site was a finalist in the Great Places in Canada competition run by the national organisation of planners, who referred to it as a model to be emulated elsewhere (Buckland and Azab 2020). All this is by way of demonstrating that farmers' markets are now recognised as urban places of such paramount significance as to warrant the investing of the considerable capital needed to implement exceptional designs.

Figure 3.56.

Figure 3.57.

Another interesting feature of the Seaport Market is its waterfront location on Pier 20 adjacent to the Canadian immigration museum. Halifax itself is not generally recognised for its cultural diversity compared to, for example, larger centres in the country such as Toronto or Vancouver. A decision was made to open an international market in the building on Fridays, with the specific purpose of enabling recent arrivals to the country to be able to mingle among one another while selling foods and crafts imported from their home countries (Froese-Stoddard 2012), thereby disregarding, for the day, the normal requirement of the market for only local products to be sold.

But all is not well at this market, in that no amount of well-meaning green building design can make up for a poor management scheme. Since opening in 2010, the market has suffered from a string of different managers, an inability to attract shoppers to an ill-advised attempt at a mid-week market, as well as drowning under a $12-million debt due to expensive overruns during construction. In 2012, facing a bevy of lawsuits for unpaid bills, the Farmers' Market Development Co-operative and the City Market of Halifax Co-operative surrendered their lease on the property to its landlord, the Halifax Port Authority (Stephenson 2012). The result in all this was an exodus of vendors (Lambie 2013; D. Burton, Dalhousie Univ., pers. comm.) who left the co-operative to set up shop (and in some cases return) to other markets. One of these is the nearby and much smaller 'Brewery Market,' located in an old brick warehouse which had been the previous site of the market. This new/old competing market advertises itself as being the true, 'historic' farmers' market of small-scale producers [e3.2.5.4], something emphasised through wonderful murals portraying scenes of bygone days (Figure 3.58) [e3.2.5.5]. A book containing stories and recipes pertaining to this market is available, harkening to when it was more of a bustling place than it is today (Kilcup 1998). Another market that houses vendors who have relocated from the Seaport Market is found across the harbour at Alderny Landing in the city of Dartmouth. Notwithstanding these sibling rivalries among competing farmers' markets, people, including many tourists, crowd into the new Seaport Market on Saturday mornings, making it difficult to negotiate one's way along the aisles [e3.2.5.6].

Figure 3.58.

My visit to the Seaport Market leaves me slightly downhearted. I've been coming here since its opening and it seems that over time more and more of the stalls are selling new-age health concoctions and craftwork by artisans rather than actual locally-produced food. I wonder if this is a trend of diversifying offerings such as occurs in bookshops that now seem to give increasing floor space to non-literary products as a strategy to attract a wider shopping clientele. If this is the case, I wonder if this is based on catering to tourists over locals, given that the former are possibly more likely to purchase a bar of expensive lavender soap than they are a head of organic lettuce. I am also alarmed at the disappearance of the touted interior bio-wall that had been filled with plants, as well as of all the interpretive signage that had once provided environmental education about the building's meritorious eco-friendly design. Worried that the green eco-roof might have suffered the same fate, I ascend the stairs to the roof deck and am relieved to see that it is still there. However, I have always wondered at the obvious missed opportunity of growing only sedum plants instead of edibles on the roof of a building whose primary purpose is related to local and quality food.

I walk a few blocks to the original Brewery Market and find it to be a somewhat sad, under-visited vestige of what it had been in its heyday. Although the physical structure of the repurposed building is still a charming feature, little of market's original, claustrophobic vibrancy exists since its competitor, the new Seaport Market, lured away many of the vendors. To complete my tour, I take the ferry across the harbour to visit the Alderny Landing Market. Situated in its own green building, this particular market seems to have found the right balance between food and non-food related products.

3.2.6 Eastern Market (Detroit, MI, USA)

Opened in 1841 and relocated to its present location fifty years later, Detroit's 'Eastern Market' (Figure 3.59) [e.3.2.6.1] has been designated a Michigan State Historic Site. It has its own visitor centre (Figure 3.60) and has been used for several well-known film shoots. At 17 hectares in size, it is the largest historic public market district in the United States (Deeb 2012). Up to 45,000 people, including many out-of-state tourists, visit on Saturdays to purchase food from vendors based from throughout Michigan as well as Ohio and even southern Ontario. In addition to locally produced farm fresh produce, the market is also known for its meat and seafood products, herbs and spices, nuts and candies, and a variety of condiments, some international in origin (Terek 2011). As well as hosting the Flower Days festival each year, the Eastern Market calls itself as the world's largest open-air bedding flower market [e.3.2.6.2]. And of course the market on a Saturday morning offers an experience that is so much more than simply being a location in which to buy food (Deeb 2012):

> The Eastern Market means 'family', as several generations have shopped together, bringing their children to see the market, its wares and the farm animals when available. It is also a people watcher's delight. A melting pot of rich and poor, old and young browse and actually bargain old-world style for what suites their interests, absorbing the sights and sounds, colour and excitement of the Eastern Market area. Many couples and families make an 'outing' of shopping here, arriving before 7 a.m., having breakfast at one of the area's restaurants [e.3.2.6.3], then spending the morning shopping for their weekly needs.

The market, which has been identified as being important for Detroit's foodscape (Hesterman 2016) as well as for its 'inner city regeneration' (Bohn and Viljoen 2016b: 47), is run by the Eastern Market Corporation. The mandate is to mobilise leadership and resources to achieve the stakeholders' vision to 'capitalise on the attractions and legacy of the Eastern Market in order to create a vibrant and diverse urban district in Central Detroit' (Anonymous 2012a). Serving as a centre for urban agriculture through its adjacent market garden [e3.2.6.4] and offering urban gardening classes, and billing itself as 'Detroit's kitchen,' the Eastern Market's development plan

Figure 3.59.

Figure 3.60.

celebrates its 'gritty environment' and 'authenticity (aka 'funkiness')' as a means for attracting economic development and creative people 'to live, work, visit, and invest in the district.' Nearby shops, for example, sell gardening supplies (Figure 3.61). In other words, Eastern Market is much more than a simple farmers' market where revenue leaves the area. Rather, it is a vehicle for urban revitalisation of an entire neighbourhood, which, this being Detroit, is in sore need of such (Figure 3.62). Plans exist for improving businesses, enlivening streetscapes, renovating buildings, and improving the urban design of the local area. With the operating concept of a full integration of food, housing, and access to work, the Eastern Market District is dedicated to creating a robust food system to grow jobs (Anonymous 2012b).

Figure 3.61.

Figure 3.62.

My visit to the site is on a non-market weekday morning. As such, footsteps echo in the empty great hall, birds nibble undisturbed in the untended market garden, and the only people seen are those watering the enormous array of bedding plants. It therefore takes a bit of an effort to envision the scene on a busy Saturday morning and to empathise with the concerns raised by my guide, the market manager. Of course, part of the success of such a place, he tells me as we wander about the site, derives through the popularity of the market and the amount of dollars coming in from Detroit residents as well as out-of-city visitors and tourists. But he fears that with upwards of two million people visiting each year, the market may now be at the cusp of becoming a victim of its own success. Over the last several days I've learned to appreciate that there are few cities where issues of race, regeneration, and food justice and accessibility are of such paramount importance as Detroit. The problem I'm told, is that the influx of tourists visiting to absorb the colourful 'experience' of the market may be crowding out locals who are there to simply procure food. As we cross a street to have lunch in one of the restaurants, the manager mentions that one strategy being deliberated for not alienating the local and loyal shopping clientele is to work toward establishing a weekday market for those people who are now intimidated or in some cases even afraid of the enormous Saturday crowds. But all such future concerns are put aside as we pause on the sidewalk to deliberate on which of the enticing delis or diners should be visited.

3.2.7 Feria Alcalde José Sandoval Gomez (Castro, Isla Chiloé, Chile)

Castro, with a population of forty-two thousand, is located on the island of Chiloé and is the third oldest city in Chile. Devastated by an earthquake in 1960, it has been rebuilt as a modern urban centre while at the same time preserving the atmospheric charm of its colourful *palafitas*, the shingled houses of fishermen precariously perched on stilts over the estuary [e3.2.7.1]. Away from the water and in a working-class neighbourhood with several shopping markets, can be found 'Feria Alcalde José Sandoval Gomez', the city's fresh food market. It is housed in a remarkable building that was opened in 2017 (Figure 3.63) [e3.2.7.2]. A large atrium (Figure 3.64) is bathed in natural light entering through the glazed glass roof and is surrounded by upper level walkways [e.3.2.7.3]. The sharply-angled structure celebrates the island's famous, UNESCO-acclaimed, architectural heritage of wood construction, including the massive crossbeams, skeleton of complicated struts, and floors, walls, and railings [e3.2.7.4].

The central atrium is filled with dozens of local farmers, each having a small, folding table upon which bunches or bags of produce are laid out (Figure 3.65) [e3.2.7.5], including heritage potatoes for which the island is known (Figure 3.66) [e3.2.7.6]. The surrounding walkways are wide aisles that house the permanent booths of larger sellers, where produce is arranged in overflowing plastic baskets (Figure 3.67). Here, too, and concentrated in a corner on the main floor, is the fish market (Figure 3.68). Elsewhere, an area is given over to a handicrafts market in which traditional woolen apparel and woven natural-fibre baskets are sold along with an assortment of wood and plastic knickknacks (Figure 3.69) [e3.2.7.7]. Upstairs, a half-dozen *cocinerias* cook meals, such as the traditional Chiloé specialty of *curanto* [e3.2.7.8], which can be eaten at tables in the food court (Figure 3.70) or purchased for take-away. Outside, people load their purchased foodstuffs into cars before continuing their shopping at the nearby dry-goods market.

Figure 3.63.

Figure 3.64.

Figure 3.65.

Figure 3.66.

Figure 3.67.

Figure 3.68.

Figure 3.69.

Figure 3.70.

It is a long walk to reach the food market from the location where tourists normally congregate in Castro. The central plaza in front of the UNESCO World Heritage-designated cathedral is so vibrant that I am a bit apprehensive about what to expect as I leave to traverse through the nondescript streetscapes to reach the market. But it turns out that the distance is a blessing since it dissuades the type of touristic affectations that characterise many of the other markets I've visited for this research project. Here, instead, is a market that radiates authenticity, the sort that can be described as truly 'public produce: food by the people, for the people' (Nordhal and Fisher 2016). I wander about, agog at the lovely architecture, my head and camera pointed every which way to admire the wood construction. The crafts market does not attract, since I've encountered better products elsewhere during the last few weeks spent on the island. It is the central atrium that captures my attention: the limited produce placed on the tables – no artsy 'food porn' artistic arrangements here, the only nod to marketing being some surfaces draped in a simple sheet. Although for some so-called farmers' markets, one should rightly be a little suspicious concerning provenance, this is not the case here. I spend several minutes watching the sellers, most of whom are women. Although I am conscious that I am being overly romantic, I can imagine being able to see traces of black earth upon their fingers. Mostly, it is the varied shapes and colours of heritage potatoes, Chiloé being the Globe's biodiversity hotspot for such, which I find fascinating. Visiting the fishmongers a few minutes later, the varied forms of shellfish catch my 'stomach's eye'. I take my bowl of fresh steamed mussels outside and sit on a playground bench, eating lunch while watching trucks unload bags of produce and containers of fish to the market's loading dock. It is a perfect way to spend my last day on this trip, and is something that will be fondly looked back upon as the calm before the storm, in this case, that of the pandemic which would become a global concern a few weeks later.

3.2.8 St. Lawrence Market (Toronto, ON, Canada)

Toronto is one of the most multicultural cities in the world, something that is reflected in its diverse food culture (Bond and Feagan 2013). And nowhere is this better displayed that in the vibrant 'St. Lawrence Market' located in the downtown core near the lakefront. Replacing temporary structures which had existed on the original site of the food market since 1803, a brick building served as the seat of municipal government during the middle of the nineteenth century. The old official chair of the mayor from this period is displayed in the exhibition hall of the present St. Lawrence Hall and Market. This iconic building (Figure 3.71), completed in 1902, was patterned on London train stations and was at the time the largest civic building in Canada. It is referred to on its bas-relief, street-side sign as being 'an important source of revenue ... [and] part of Ontario's commercial history.' Its imposing architecture enshrined concepts of Victorian social hierarchy by physically separating butchers, cheese mongers, and fruit and vegetable wholesalers, who occupy the South Market, from small-scale farmers who were relegated a less prestigious building across the street in the North Market (Dickau *et al.* 2011). Today, more than one hundred and twenty permanent vendors have been brought together under the roof of the South Market, a structure which sprawls over three city blocks and is open 6 days a week, 'offering visitors the unique and lively atmosphere of an authentic farmers market,' according to the official website (www.stlawrencemarket.com). The North Market (undergoing reconstruction at the time of my visit) typically hosts a traditional farmers' market on Saturdays, which on Sundays is turned over to an antique market where one can purchase items such as antiquarian books, vintage jewellery, and pieces of architectural salvage. In a nice homage to the historical continuity of the site, present-day vendors' tables are situated outdoors directly in front of a well-executed mural that illustrates such tables being set up here in bygone days (Figure 3.72). On the other side of the street, the commanding bulk of the South Market signals its embedded importance to Toronto. It is therefore shocking to learn that the building was lucky to escape the urban renewal wrecking-ball of short-sighted city planners who wished it to be demolished in the early 1970s. It is in this impressive fin de siècle structure where the interest lies and my tour takes me.

Figure 3.71.

Figure 3.72.

Upon entrance, a welcome sign claims that in the 'daily sensorial tapestry of the Market' one can purchase 600 types of cheese, 300 types of olive oil, 600 varieties of rice, 38 types of mustard, up to 400 varieties of fruit, and 100 kinds of seafood. Expectations are raised by other signage: a quote from an enthusiastic visitor from a Zagat survey, a 4.5-star certificate of excellence award from Tripadvisor, and a large banner that claims that *National Geographic Magazine* voted it as the 'No 1 food market in the world'. Permanent food stalls fill the hall (Figure 3.73), selling the usual fare: seafood, meat (including entire pig carcasses, presumably for roasts), and fruits and vegetables [e3.2.8.1]. The diversity of imported exotic meat is truly remarkable [e3.2.8.2], as is the number of shops specialising in nuts, chocolates, mustards, deserts, maple syrup, caviar, sausages, honey, mushrooms, bagels, rice, and melons. One can visit other specialty shops selling ethnic foods (e.g. Indian, Greek, Ukrainian) or healthy alternatives of grass-fed meats and organic produce. Restaurants, florists, and boutiques with jewellery, clothing, and crafts are scattered throughout the hall, including at least one shop flogging tourist kitsch (Figure 3.74).

Figure 3.73.

Figure 3.74.

Situated on the second floor [e3.2.8.3] and part of the last surviving section of Toronto's original City Hall – it had been the Council Chambers – is the Market Gallery. Here, more than a hundred public photo and archival exhibitions have been staged about the city's rich social and physical history (Figure 3.75) [e3.2.8.4]. The area is also used for talks, educational programmes, and 'special presentations for tourists'. The Market Kitchen occupies the mezzanine floor and functions as a cooking school 'where culinary inspiration and education meet'. It is also available for rental as a venue for social gatherings. The St. Lawrence Market publishes a glossy newsletter and hosts food and beverage sampling soirees, referred to as 'Mmm Nights', that usually sell out within a few days of being advertised. The market is so popular that it has an official historian who gives guided walking tours. A small eating court, lined with historic photos is tucked away in a corner [e3.2.8.5], near what used to be the basement of the municipal jail (Figure 3.76).

Figure 3.75.

Figure 3.76.

I pause partway through the entrance vestibule into the market and find myself wrenched back in time to when I was a resident in the city three decades ago. It is the fond memory of delaying grocery shopping when arriving to this precise spot in order to listen to the buskers who were always there on Saturday mornings, one of whom would go onto become the country's best-known World-Celtic musician. I wonder about whether my meagre graduate student coinage dropped into their various collection hats contributed to their economic survival during the lean times that characterise the start of music careers.

Inside the market on this new visit I circulate about the shops, admiring the artfully arranged piles of produce. I am startled by the offerings of imported exotic meat. Buffalo is not uncommon to see these days in such markets, nor is ostrich unheard of, but this is the first time I have ever seen turtle, camel, or Australian crocodile and even kangaroo! I linger for several minutes in the hope to see the sort of person who might buy such; but to no avail, and so I move on. Upstairs, in the Market Gallery, I examine the historic photographs, finding one to be particularly atmospheric of an earlier time. Vendors are shown standing in front of their stall in which the carcasses of hundreds of poultry carcasses are hung. It is hard to determine if the overt pride exhibited by the pictured vendors is more to do with the extent of their displayed fowl or that of the size of their prodigious Victorian moustaches. I head back down into the market to purchase a peameal bacon sandwich. Touted as the 'culinary icon' of the market, it was popularised as a tourist attraction in the 1970s and is still sold in the thousands every week. Lunch over, I venture into the old dungeon of the nineteenth-century jail and chuckle over the ball and chains arranged like Christmas decorations over the bricked-in fireplace. A quick check of my watch reveals that it is time to rendezvous with my guide for a prearranged tour of the market and surrounding buildings. And so I ascend back up to the food hall for another walkabout.

3.2.9 Mercado do Bolhão (Porto, Portugal)

Porto's historic centre, particularly its picturesque riverfront, has been designated as a World Heritage Site by UNESCO. As is the case for Venice (France 2011), this has come at a cost in that the city has lost a significant proportion of its population since the middle of the last century and now seems poised on the cusp of gentrification designed for the benefit of tourists over locals (Queiros 2019). Until recently, one exception in the rush toward urban renewal has been the 'Bolhão Market'.

A market has been held on this location since the middle of the nineteenth century, with the neoclassic wrought-iron building constructed in 1914. The attractive and admittedly decrepit structure sprawls over an immense city block that is ringed by busy streets (Figure 3.77) [e3.2.9.1]. It is a highly valued resource for the city, having been classified as a 'Property of Patrimonial Interest' in 2006 and as a 'Monument of Public Interest' in 2013. The food market, visited by both locals and tourists, is the most famous one in the country. Although some decry it as a pricey tourist trap, other travel websites sing its praises as a 'must see' that holds the essence of the city, where one can discover 'the real Porto' by getting 'an idea of the daily life and the citizens of this enchanting city'. One gastro-tourism website (www.partaste.com) heralds the market's sensorial epicurean delights:

> It is one of the most authentic markets you are likely to come across in all of Europe. With the smell of freshly baked breads and the sounds of merchants bartering across pridefully arranged stalls, a stroll along the aisles of Mercado do Bolhão, is a feast for all the senses. At the same time, there's a sweet nostalgia that underlies the market's activity. People interact in a manner of days gone by; greeting each other with a smile, gossiping and exchanging pleasantries.

Farmers' stalls line two floors of wide aisles [e3.2.9.1] that surround and look down on the open-air atrium that is chock-a-bloc full of shops and stalls (Figure 3.77). As in most markets, foodstuffs are segregated: fishmongers [e3.2.9.2] and butchers in one area; green grocers and florists [e3.2.9.3] in another. Vegetables are for the most part arranged simply in tiers of plastic trays [e3.2.9.4], although several vendors have given some attention to the presentation of their displays [e3.2.9.5]. An air of nostalgia is provided by an organ grinder playing a hole-punched strip of music [e3.2.9.6]. A broad staircase leads down and out to the ground-floor atrium [e3.2.9.7]. Here, in addition to bars and shops selling wine [e3.2.9.9], are located stalls specialising in Porto's famous blue tiles and linens [e3.2.9.8], thereby catering to the numerous tourists who mill about [e3.2.9.9]. Most interesting are the dozen or so buildings with steeply-pitched roofs that resemble tiny country cottages. Each of these is a restaurant accompanied by outside seating (Figure 3.78), many with a menu focused on regional seafood for which Portugal is famous.

Figure 3.77.

Figure 3.78.

Travelers comment on the Bolhão Market's worn-down appearance as being part of its charm, one going as far as to state that it is 'beautifully ugly, like traveling back in time'. Such a rosy picture, however, contrasts with the wish of the municipal government to improve the site's appearance and, they argue, also its functionality. As such, the building has been undergoing extensive renovations since 2017 as part of the city's larger urban renewal plan (Coeho 2018), despite a fear voiced by some (Queiros 2019) that it may lead to the consequent gentrification of the neighbourhood.

Following several weeks spent sailing aboard a tall ship from Rotterdam to Porto, my legs are still wobbly as I leave the historic river port and make my way up into the city to see its famous food market. My walk is interrupted by detours to visit several blue-tiled churches, an interesting building that sports a green eco-roof, and what is claimed to be the world's most beautiful bookstore. Once I arrive at my destination, it takes but a moment to become enamoured by the time-worn charm of the massive building. The visual contrast is striking between the brightly coloured flower shops bathed in sunlight and the organ grinder sequestered in a dark alcove as if he has just snuck off the page from a Dickens' novel. Hanging strings of garlic and red peppers catch my eye, and partially opened bags of nuts and dates attract my business. The present trip now being over, I buy some traditional Portuguese crafts to take back home as gifts. Then, after comparing menus, I settle at a table to enjoy a lazy lunch of expertly grilled and spiced octopus and a carafe of white wine while watching the crowds and listening to the cries of vendors extolling the virtues of their particular fare.

3.2.10 El Mercadeo Central (Valencia, Spain)

With its futuristic 'City of Arts and Sciences' development, Valencia has reinvented itself as a leader in modern architecture (Prytherch 2003). That said, one of the most attractive, visited, and loved buildings in the city is the old 'Central Market,' a structure that many consider to be an apogee of Art Nouveau through its homage to the Gothic style (Delgado 2012). An outdoor market has existed in this area of the old quarter since the first half of the nineteenth century. The present building, replacing an earlier Victorian one, was completed in 1928 and is now designated as a 'Heritage of Cultural Interest' by the Spanish Government. With a surface area of over eight thousand square metres, this is the largest fresh food market in Europe. As well as being an acknowledged architectural masterpiece that attracts visitors in its own right, the experience of the market has become so popular among general tourists that companies offer guided tours and more than half-a-dozen videos have been posted online by enthusiastic travellers. One tourism website proclaims the market as being 'an explosion for the senses' due to its floods of smells, tastes, and sounds. Not without reason then, is the site used prominently in Valencian tourism marketing, identifying it as one of the city's 'must see attractions.'

Located in a neighbourhood of historic buildings [e3.2.10.1], the beautiful Central Market, even when viewed from the outside (Figure 3.79) [e3.2.10.2], makes a strong statement representing a 'commitment to progress and agricultural production' as well as a 'cathedral of the senses' (www.mercadocentralvalencia.es). Passing through an entranceway decorated with glazed tiles [e3.2.10.3], one enters a contender for what might be the most beautiful farmers' market in the world. Inside, and on first impression, it is easy to ignore the street-like grid of more than a thousand food stalls by being distracted by the glass, iron, and ceramic construction of the building itself; especially the remarkable dome, thirty metres in the air, and the sloping roof that radiates out at different heights and orientations (Figure 3.80) [e3.2.10.4], everything designed to enable natural light to diffuse into the vast interior space [e3.2.10.5]. The website extols, with a flourish of hyperbole, the virtues of the market as 'a sacred temple where the professors of fresh produce receive and transmit the most substantial wisdom, our food. The magical light that enters from its dome and its stained-glass windows, the permanent whisper that caresses the ears, the explosion of colours and aromas, the taste at the end of the classic and external flavours,' and concludes with calling it 'an authentic gastronomy theme park.'

The visual character of the market is celebrated on its website with encouragements for customers to bring their cameras. Included here are also photo-archives of the market's history, architecture, activities, hosting of special functions, as well as visitations by dignitaries. Open every day except Sunday, the market conducts school visits and hosts artistic performances, both of which are attended by a costumed figure who is the market mascot. Books, DVDs, and assorted souvenirs are sold on the market website.

The market stalls are really permanent shops (Figure 3.80), over ninety percent of which sell the sort of food one expects to find at such locations, such as fresh produce, dairy, bread, olive oil, nuts, and herbs [e3.2.10.6], with, however, a strong emphasis placed on those particular products for which Spain is famous, namely *marisco* in the form of mussels, crustaceans, and fish (Figure 3.81) [e3.2.10.7], and cured Iberian ham (Figure 3.82). Small tapas bars and other restaurants are centrally situated in the Latin cross-shaped hall.

Figure 3.79.

Figure 3.80.

Figure 3.81.

Figure 3.82.

As well as a location for shopping, the market is used by locals as a socialising place and is considered part of their lifestyle identity. It has been likened to being an organism of living history where one does not just buy and sell, but where one tastes and talks. This has raised concerns about the influx of tourists and their potential influence on transforming cultural heritage and local customer habits (Esteve and Arnandis-Agramunt 2020). However, statistics on the market's website indicate the following breakdown of clientele: 20% from the neighbourhood, 60% from the rest of the city, 18% from the surrounding province, and only 2% from those from elsewhere in Spain or foreigners. But differences no doubt exist between the numbers of tourists who simply visit the market and the smaller number of those who actually purchase food and can be regarded as clients. As such, the threat of 'touristification' may indeed be real. Certainly, given the disproportional importance that has recently been placed on tourism by the municipal government, the potential for losing local culture remains a meritorious issue of concern.

From the very start I am captivated by the marked contrast between the glass, iron, and ceramic of the modernist market building and that of the stone and brick of the older buildings in the neighbourhood. Approaching the front entrance, it is easy to witness how the commercial influence of the market overflows into the tiny piazza. Here, the outdoor seating of a handful of tapas bars is completely occupied, patrons being serenaded by buskers while they imbibe their snacks and drinks, many with shopping bags brimming with market-acquired groceries at their feet. I pause for a moment to admire the parrot weathervane atop the roof before entering the building.

Once inside, the first thing I do is to laugh out loud, startling some of the nearby shoppers. For the initial impression is more of being in a cathedral than a food market, something that is reinforced when standing directly under the central dome. I am simply gob-smacked by the sight. It is as if I've been magically transported from southern Spain to northern Italy and am in the famous Galleria Vittorio Emanuele II shopping arcade in Milan. The next sensorial experience is an aural one, for the loftiness of the place causes the typical market sounds to resonate in a way that I have never experienced before. It takes an effort to break free of the architectural splendour of the site in order to focus on the food. Here, it is the fishmongers whom I find most interesting given that, in several cases, the displayed animals are still alive. I spend a few moments watching a lobster crawling over a bed of ice in a vain attempt to escape its fate. And I cringe when watching eels slither over each other in disregard to the hand that periodically removes one in order to chop it up into little pieces but a short distance from its unsuspecting fellows.

3.2.11 Borough Market (London, UK)

Since the Middle Ages, the right to be able to stage a weekly market has been avidly sought by European towns as a means to ensure their prosperity. The granting of a royal charter to become a market town was a guarantee for long-term civic wellbeing. Today, farmers' markets in the United Kingdom are experiencing a resurgence in popularity among locals and tourists alike. One reason is that they are the 'original street theatre', as Barton (2013) elucidates in an article in a popular tourist magazine:

> 'Juicy pine-ap-ell!' 'Who likes pies and pasties?' 'Last of the season English straw-ber-ries!' A chorus of calls rings outs from different corners of a typical British market. Colourful stalls, selling every variety of fruit and vegetable and provisions galore, stand in the shadow of enormous department stores. A garishly dressed woman tells the fortune of anyone proffering a five-pound note and a queue forms at a kiosk selling a local specialty: baps filled with generous servings of hot pork, herb stuffing and apple sauce. All human life is here.

London has a suite of notable food markets which collectively contribute millions of pounds sterling to the city's economy. Several of the most famous markets, such as Convent Garden and Spitalfields – known respectively among tourists for their association with the Victorian characters Eliza Doolittle and Saucy Jack – are today given over to high-end shops or merchandise stalls not associated with foodstuffs. However, the grandfather of urban food markets in the English-speaking world remains the 'Borough Market'.

Goods have been traded and sold at this location since the eleventh century and possibly as far back as Roman Londinium. It is the dramatic location of Borough Market, nestled as it is beside the towering bulk of a cathedral [e3.2.11.1] that adds to its enjoyment for visitors. With the people milling about the hundreds of vendor stalls in a chaotic streetscape in the shadow of Southwark Cathedral [e3.2.11.2], itself strategically situated at the end of the original London Bridge, one can step out of the present and time-trip back to medieval times. Or if you prefer, the Victorian glassed arcades and rattling railway viaducts that run overhead (Figure 3.83) [e3.2.11.3] might foster Dickensian-period reveries (after all, the Marshalsea prison that featured in *Little Dorrit* is just around the corner). But all one has to do is look upward to be wrenched back to the present. For towering overall, and symbolic of London's affluence, is The Shard, one of the tallest buildings in Europe.

With crowds of thousands jostling inside the covered arcade it is easy to become disoriented, necessitating reliance on signage to negotiate the three abutting markets should one have a particular destination stall in mind (Figure 3.84). Most individuals, however, are more than happy to simply wander about, pausing by happenstance here and there to shop when something catches their eye. The market's website refers to the place as being 'a riot of colours, smells and human engagement' (www.boroughmarket.org.uk). Coles and Crang (2011), in turn, describe the market as a place where activity engages the senses, where 'textures and sensations are experienced closer to the body than the other senses.' They offer the following piece of atmospheric place-writing: 'The air is thick with food smells, filled with the sounds of people, sticky with sweat as bodies collide. I can almost taste the market and feel its excitement'.

Figure 3.83.

Figure 3.84.

London is among the most popular tourist destinations in the world, and visiting the Borough Market feels like a trip to the United Nations. It is easy to overhear a babel of languages from tourists as they excitedly point at the various foodstuffs artfully arranged as if in a photo shoot for a foodie magazine (Figure 3.85). Coles and Crang (2011) describe the artistry involved in setting up the stalls: 'A fishmonger and game seller is building the stall's display of fish and seafood, propping open a shark's mouth, wrapping an octopus around a piece of driftwood and attaching salmon heads to the sides of the cage. Some vendors are pulling tarps off the wooden boxes and wine casks that make up the 'architecture' of their pitches'. In short, the Borough Market sells experience as much as it does food (Coles and Crang 2011). Vendors' use of old wheelbarrows and push carts, wicker baskets and wooden tables, all contribute to a sense of nostalgia, a harkening back to a romanticised past when, in the minds of many, more attention, if not love, was involved in the growing and marketing of foodstuffs. All this performance theatre is not lost on the many tourists who put their own cameras to frequent use, an act which, interestingly, is technically not permitted according to the market management's website. Accolades by happy customers about the market are emblazoned in huge letters on several large banners that are mounted throughout the arcade [e3.2.11.4].

Figure 3.85.

The market emphasises that its food is of higher quality and has a more ethical provenance than anything available in supermarkets. Consumers are reconnected to producers through the latter's storytelling of how their products came to be (Coles and Crang 2011). In addition to an almost unbelievable plethora of produce and gastronomic delights to take home [e3.2.11.5], it is the many opportunities to purchase and eat food onsite that attracts the crowds of visitors (Figure 3.86). Indeed, for some vendors, the income generated from food sold for immediate consumption can be a substantial part of their total earnings. Dozens of visitors buy their lunches and stroll a short distance to the pubs that skirt the edge of the market where they imbibe an accompanying pint while standing at tiny outside tables and watching the parade of people.

Figure 3.86.

There is no particular emphasis on locally produced fare. Instead, at the Borough Market one finds a 'gastronomic wonderland' of sumptuous traditional and exotic food gathered from across the whole of England [e3.2.11.6], as described by Barton (2013):

> Aromas of Vietnamese-style curries bubbling in cauldrons merge with those of mulled wine and giant cheeses from every corner of the land, stacked ceiling-high ... Customers fill every passageway, grazing on smoked wild boar sausages and spicy kedgeree, often washed down with something from the organic juice bar. Many pause to purchase ostrich, reindeer burgers or exotic casserole mix (kangaroo, springbok and water buffalo).

In reality, for every hipster-foodie buying exotic ostrich meat and artisan bread [e3.2.11.7], there are dozens of others buying more traditional British fare such as fresh eggs and meat pies [e3.2.11.8].

In addition to directly supporting the livelihoods of its vendors, many of whom are from across the country, the Borough Market also contributes to the local community. The market is owned by a charitable trust and is run by a board of trustees committed to operating it as a public amenity. The market hosts a lecture series, a digital book club, and cookery demonstrations. Financial surpluses are reinvested into the neighbourhood. These take the form of funding for community events, providing health education, and supporting local projects and schemes.

It is a cool Saturday afternoon in early spring for this particular visit to the Borough Market. Over the years, I've made a point of coming here whenever I have found myself in London for more than a day. This time I want to test Coles and Crang's contention that 'Borough Market is a multi-sensory consumptive experience where place is sold alongside the market's foods'. One element that I always find attractive is how this market, with its arcade of impressive glass and metal roof and lack of sidewalls such that it is open to the elements, blurs spatial impressions of what constitutes 'inside' and 'outside'. For several minutes, I watch the boisterous crowds move back and forth, in and out, grocery bags empty or filled. Mustering the courage, I then immerse myself in the tidal flow of cosmopolitan humanity. Diffuse light pours in through the market hall's glass roof, casting a warm glow on all. Steam rises from counters and pans where take-away food is being cooked. I bring my nose close to a pile of bread fresh from the oven and cannot resist buying a round loaf from which I tear off a warm portion to immediately eat while placing the rest in my daypack. A few moments later I allow myself to be lured into a faux wine cellar from which I purchase a bottle. I continue to circulate about, succumbing to buying several pork pies to have for lunch. I locate a small alcove in which to stand to eavesdrop in on the conversation of the perambulatory crowd, detecting what seems to be at least half a dozen difference languages in the time it takes me to consume one of the sumptuous meat pies. Periodically, from overhead, comes a loud rattling that signals the passage of another train. Only one thing remains to complete the experience of my visit.

I exit the hall and make my way across a pedestrian lane to enter a pub and buy a pint of real ale, taking it back outside to a tiny stand-up height table. There, I sip the beer and munch my other pie, while watching the milling masses. I start up a conversation with a middle-aged couple with whom I am sharing the table. They too are nibbling away on items that they've just purchased from the market while sipping their ciders. This, it turns out, is part of their established weekend ritual. They inform me that they are of the opinion that as many people visit the market now in order to sample the food *in situ* as to take it home. And surprisingly, this is by no means limited to only tourists (who of course have little need or desire to purchase quantities of perishable foodstuffs) but also pertains to local Londoners like them.

Pint and pie finished, I return the plastic glass to the barkeep. With one last look at my favourite food market in the world, I start walking to the nearby Tube station, already looking forward to a future visit when I am next in London. And of course, I still have that bottle of wine and fresh loaf of bread to look forward to in my hotel room this evening. Which reminds me that I need to buy some cheese and dried sausage. So with a quick pirouette, it is back to the market. 'Once more unto the breach,' I voice with a smile, as I rejoin the crowd.

3.2.12 Mercado de La Boqueria (Barcelona, Spain)

The 'Boqueria Market,' situated in the bohemian quarter of Barcelona along its famous pedestrian artery, Les Ramblas, is the quintessential urban food shopping experience in many people's eyes. As a public agora where culture is shared and celebrated among citizens and sought after by visiting tourists (Crespi-Vallbona and Dimitrovski 2017; Crespi-Vallbona *et al.* 2017; Dimitrovski and Crespi-Vallbona 2017a,b), the Boqueria reigns supreme, and has become a singular model for the ability of food markets to be transformative agents in urban regeneration (Fuertes and Gomez-Escoda 2020).

After centuries of itinerant existence, the location of the present market was established in the middle of the nineteenth century. With its Modernist style metal façade and roof (Figure 3.87), La Boqueria operates on the belief that gastronomy is a lifestyle infused with emotion and delicateness, and something that needs to be experienced through the senses of sight, hearing, and smell, in addition to taste (Mas and Ubide 2011). Due to its global fame, the market runs its own tourist kiosk from where one can purchase tote bags, books, and DVDs (Figure 3.88). There are more than two hundred stalls, each created as a special reflection of its owner. Some emphasise their modernity through use of boldly coloured ceramic and glass, whereas others purvey their tradition with wrought-iron or wood. In one of the very few books that have produced about an individual food market, Mas and Ubide (2011) provide descriptions of many of the stalls, some having been in the same families for generations. Each is introduced through way of the history of the stall, its unique architecture, the goods provided (to which are added a side-bar recipe and a tip on food selection and consumption), all accompanied by a rich catalogue of 'food porn' shots from professional photographers.

Figure 3.87.

Figure 3.88.

Once through the congested entrance off Les Ramblas (Figure 3.87) and inside the market, one navigates through a typical assortment of food stalls, each a permanent shop. In addition to bakeries with crusty bread, fishmongers (some of whom specialise in salt-preserved fishes), greengrocers with vegetables artfully arranged in tiers and draped with hanging fruit, and butchers with full carcasses of suckling pigs for roasting in addition to rabbits, goat or lamb heads, piles of sausages, and hanging cured hams [e3.2.12.1], there are stalls specialising in an incredible diversity of nuts (Figure 3.89), artisanal salt [e3.2.12.2], candies, dried fruits, mushrooms, and international fare, as well as organic growers and those from ecological farms (Mas and Ubide 2011). In some cases, the attention to display is remarkable, with one stall owner assembling tiny hills of eggs upon which cute models of chickens are precariously balanced (Figure 3.90).

Figure 3.89.

Figure 3.90.

Customers perch atop stools sidled up to the counters of restaurants while watching staff cook tapas, seafood, casseroles, omelettes, and Catalan stew (Figure 3.91). Some of the bars have become famous following their appearance on magazine covers around the world (Mas and Ubide 2011). The outflowing of commerce and confraternity from the market radiates into the surrounding neighbourhood wherein locals enjoy glasses of wine from a shop-cellar whose outside tables are made from repurposed casks (Figure 3.92).

The 'Food Markets in the Belly of the City' project has produced documentaries about five of Europe's leading markets, each designed to be 'a journey of the senses ... with lively characters [where one] uncovers their gastronomic secrets' (www.stefil.it). One of these was the 'deliciously seductive, queen of all food markets', La Boqueria (Ferrando and Tealdi 2013). The video provides glimpses into the fascinating inner workings of the market through showcasing the diverse activities of a handful of representative food producers and vendors. Much emphasis is placed on the multi-sensorial experience of food.

The first person featured on the film is a chef of a Michelin-starred restaurant who is observed shopping at the market in the early morning. He is shown feeling and smelling mushrooms, and selecting and examining, by squeezing, a hare carcass, while expressing his desire for 'the excitement of seasonality' that comes from visiting the market. Next, a vendor is shown interacting with patrons about her calçots. She is also shown on her own farm harvesting, barbequing, and eating these special Catalan green onions. Next, the manager of a tapas bar that is famous for its prawns is shown interacting with fishmongers in the market. He handles the food while explaining how to recognise its freshness: for shrimp, this is based on physical traits and colour; for cuttlefish, this is based on rubbing a finger to see the colour change; and for fish, this is based on the presence of protruded eyes and bright red gills. A fishmonger holds up a handful of razor clams, some poking out, while she smiles and encourages the other clams to not be so shy. Next, the owner of a meat stall in the market is filmed at a local abattoir where he patiently examines each carcass by touch and smell before selecting those he deems to be the best ones. He is then

Figure 3.91.

Figure 3.92.

shown back at his market stall descending into a refrigerated storage room below the ground level where the butchered meat is kept. Next, a vendor is shown collecting different types of eggs from his nearby farm, and then delivering some to restaurant managers in the market, before arranging the bulk in his own stall. A final merchant is shown arriving at 2 am and cooking and delivering breakfast omelettes to her usual customers, the stall owners, before the market opens to the public.

The film continues by addressing the delicate balancing act between catering to tourists and to locals, each now representing about half of the daily total of 45,000 visitors. One stall owner who now concentrates on selling prepared fresh juices and salads to tourists expresses sadness at losing the local clientele who had been regular customers in her father's day when the family sold produce. The managing director of the 'Boqueria Market Traders Association' is shown walking around the market and working in his behind-the-scenes office. He notes that the balance is presently at 200 stalls selling fresh food with 36 catering to take-away or in-place consumption. Locals continue to visit the market since, with all the displays, they regard the experience as somewhat like going to Disneyland. The presence of restaurants is a challenge to vendors, but they add magic to the market, with some people from the neighbourhood regularly visiting for breakfast and lunch. The manager goes on to state his belief that the market must fix the total number of restaurants and that the primary purpose of the market needs to remain one of providing healthy food to locals, with tourism being a bonus. Presently, there is a temporal segregation wherein locals shop in the mornings, tourists in the afternoon, and on Saturdays people arrive from the wider region. If the Boqueria Market relies too much on tourism it will suffer in the long run, as tourists will begin to avoid it in their search for more meaningful interactions with locals elsewhere. The market, according to the managing director, needs to evolve into becoming a modern gastronomic hub and meeting point, and the only way to accomplish this is through extending its opening hours into the evening in order to attract local cliental that otherwise will continue to stop off at supermarkets rather than here on their way home.

Rather than recounting my own experiences, I will refer to those from journalist Gerry Dawes, taken from his Foreword (pp. 14-16) in Mas and Ubide's (2011) book about the market. His place-writing provides an atmospheric portrait of La Boqueria, a 'gastronomer's paradise,' where he admits to losing himself for hours. He too is struck by the beauty of the artistically arranged tiers of fresh produce, as well as by the diversity of Spanish hams hanging from stall ceiling hooks, and of eggs from all manner of fowl. Dawes refers to the 'stunning' selection of fresh fish arranged on beds of ice, and mentions years of photographing one fishmonger friend while he chops up the fish, wraps up the slices, and exchanges them for currency from customers, all from a stall that bustles with constant movement. Dawes states that he always finishes his tour of La Boqueria by visiting his favourite market bar wherein he knows 'I will be fed like a king. He may make me a bowl of steaming *berberechos* (cockles); a plate of grilled asparagus or deep-fried artichoke hearts; an exquisite dish of shrimp with the heads still on; my beloved *mongetes* with *botifarra* and *allioli*; then a perfectly cooked slab of foie gras with crispy fried leeks, all accompanied by glasses of cava *rosat* (rosé), Catalan Champagne.'

Inspired by this description, I finish my glass of wine which I've been drinking while reading the book at one of the wine bars situated adjacent to the Boqueria Market. And so, I follow the journalist's lead and renter the market to complete my visit by consuming some of the delicacies from the very stall he mentions.

3.2.13 Markthal (Rotterdam, the Netherlands)

Just as landscape architects have embraced designing food gardens (see Chapter 5) so too have architects done the same for food buildings. As J. Johnson (2010a) shows, beautiful farmers' markets are now being constructed around the world. For example, the replacement of a decrepit fish stall in Istanbul's Besiktas neighbourhood took the form of a roof structure designed to resemble a giant seashell. Touching down at three points, the pillar-less and wall-less building permits access from all sides and is a true open-air market. While waiting for the final market to be built, vendors at the Barceló complex in Madrid clustered underneath a temporary structure composed of six giant polycarbonate pentagons arranged in a floral configuration. The apogee of this trend is the food market in Rotterdam. Occupying a full city block, the new food hall houses one hundred vendors beneath a 34-metre archway building that contains more than two hundred apartments from which residents can look down upon the bustling market and its associated restaurants, bars, and cafés in what has been termed a 'culinary cathedral'.

With the intent of bringing vitality back to the central area of the city that was destroyed in a firestorm following bombing in 1940, Rotterdam's 'Markthal' is perhaps the most spectacularly designed food market in the world. Due to rumours that new European Union restrictions would come into play to restrict the selling fish, meat, and cheese in the open (something which actually was never to be legislated), Rotterdam moved to create the first covered market in the Netherlands. Already known for its edgy modern architecture, their goal was to build a completely new concept of a food market that was like nothing else in the world. Opened by the Queen in 2014, and costing 175 million euros, the resulting hybrid market hall and housing complex is an architectural marvel which makes a strong contribution to the local economy.

Shaped like a giant horseshoe or triumphant arch (Figure 3.93) [e3.2.13.1], the 40-metre high by 170-metre long by 70-metre wide structure contains 228 luxury residential apartments [e3.2.13.2], half being rented and half owned, which span the 4,600 square metres of retail space in the market hall. Beneath the structure is a four-storey underground parking garage. As part of the Rotterdam's branding itself as the most sustainable harbour city in the world, the building is an embodied political act (Cretela and Bvenger 2016). Sustainability concerns were factored in the building's construction and operations. For example, an urban ecologist helped with the design of outside perches for bats and nests for swifts (www.archdaily.com/553933/markthal).

Figure 3.93.

As eye-catching as the building is when viewed from the outside, its true magnificence is revealed upon entrance. There, towering up on both sides and encompassing the ceiling vault, is a 11,000-square metre artwork which is referred to the 'Sistine Chapel of Rotterdam' and touted as being the largest painting in the world (Brooks 2017). With a size of about two soccer pitches, the *Hoorn des Overvloeds*, or Horn of Plenty, depicts fruits, vegetables, seeds, fish, flowers, and insects, in addition to the nearby medieval church, all shown in vibrant colour (Figure 3.94) [e3.2.13.3]. Production of the massive artwork entailed a complicated production process that required assistance from Pixar Studios and involved printing on four thousand aluminium panels. The artist describes the subject matter as showcasing the spiritual miracle of food. Rising up and poking out from the painting are the tiered windows of the apartments (Figure 3.94) [e3.2.13.4]. Each interior window facing the market hall is triple-glazed to block out sounds and smell, with their angles decreasing with height until they become light-wells in floors of the 24 penthouses that occupy the ceiling. Mid-level walkways provide views down upon the market hall [e3.2.13.5].

Twelve thousand people an hour pass through at weekends to visit and shop at the hundred vendors and restaurants that are listed and briefly described on the official website (www.markthal.nl/en). Under the grand arch [e3.2.13.6] permanent stalls selling sausages, wine, and fresh seafood, bread, and produce [e3.2.13.7] are housed in addition to artfully arranged pyramids of spices (Figure 3.95) and arrays of sweets (Figure 3.96). Restaurants offer fish and chips, donuts and snacks, and deli products [e3.2.13.8]. The market contains an art gallery and food sector training school. But it is that which lies below ground that adds special value to the experience of visiting the Markthal.

Figure 3.94.

Figure 3.95.

Figure 3.96.

Built on top of a tenth-century farm which evolved into the fourteenth-century polder village of Rotta and then gave rise to the city of Rotterdam, the Markthal advertises itself as being the place where the past and the present come together under one arch. The reason for this is that, unique among food markets, the Markthal contains an archaeological museum. The 'Time Staircase' (www.detijdtrap.nl/en) is a permanent exhibition in the escalator stairwell (Figure 3.97) [e3.2.13.9] that leads down to the parking garage where tourists leave their vehicles [e3.2.13.10] or gather for guided tours (Figure 3.98). Each of the four parking levels corresponds to a different time period, so through descending the escalators, which display a timeline of significant dates in the city's development [e3.2.13.11], is like participating in an archaeological dig. Hundreds of artifacts – jewellery, weapons, utensils, food remains – are exhibited, vividly portraying this location's rich palimpsest of medieval life (Figure 3.99) [e3.2.13.12]. Signage, in both Dutch and English, informs about history and the artifacts displayed [e3.2.13.13].

Figure 3.97.

Figure 3.98.

Figure 3.99.

Although a handful of celebratory online videos exist of the market from the members of the general public, professional architecture critics seem to be less enthusiastic. Bevan (2015), for example, thinks the Markthal is 'emblematic of Rotterdam's inability to conceive and sustain a sense of urban life'. He is critical of the overall theory that great market halls equate with great food culture, citing that for every Madrid, Melbourne, and Montreal, for which this is true, there are recent examples, such as Leeds and Newcastle, where fancy architecture is unrelated to fostering notable cuisine. In his opinion, all that Markthal contributes to Rotterdam's food culture is to provide 'a very visible statement of investment in the idea of it'. The problem is that the location is really a sterile space with none of the customary bustle and messiness that are a characteristic trait of traditional food markets. Despite the claim that some 20% of stall-holders are collectives or family businesses, the site doesn't feel like a real market at all. It is basically an impressive food court surrounded by shops exhibiting luxury items, such as chocolates, nuts, and fancy meats, with only a single greengrocer selling produce which has been polished. For Bevan, the Markthal is like London's Borough Market, more about destination shopping and dining than it is about acquiring foodstuffs. Because of this, he considers it to resemble Barcelona's 'tourist-infested' La Boqueria. With this criticism in mind, in order to experience the feel of a true food market, it becomes necessary to exit the Markthal and to return to the open plaza smack in front of its main entrance. There, next to a subsurface storage area that demonstrates the Dutch passion for urban cycling [e3.2.13.14], is the weekly farmers' market where one can jostle with crowds of locals as they shop for produce, seafood, cheeses, and flowers [e.3.2.13.15].

Every time following a period of absence when once again I visit the Low Countries, I'm struck with the impression that this must certainly represent the apogee of Western Civilisation. It is early on a Saturday morning in late summer when I leave my floating hotel and walk through the historic harbour. I admire the over-the-top modern architecture that resembles a child's construction project of assembled building blocks, and also at the way that the pavement between and on either side of a set of trolley rails has been replaced by permeable pavement and grass. The first sight of the famous Markthal stops me dead in my tracks. For rarely have I seen such an incredible building that so well showcases the edgy imagination of modern architecture without crossing over into the realm of ostentatious silliness. For a few minutes I circulate among the temporary stalls of the outside farmers' market but soon find the allure of the Markthal building to be irresistible. As is my customary practice when visiting the great cathedrals of Europe, I begin with a slow walk around the building. It is still quiet at this time in the morning and only a few patrons are spotted at the outside tables of the cafés or residents looking down from their outside balconies. Circumambulation completed, I enter the hall and am immediately awestruck by the expanse of the space and the enormity of the Horn of Plenty painting. With my head cranked backward in order to gaze upward, it is truly like remembered visits to such places as the Sistine Chapel or Hagia Sophia. I spend more than half an hour examining each illustration of plants and animals in turn, sometimes using, just as I do when visiting cathedrals, my small monocular lens for magnification.

It is when next I ascend to the mezzanine walkway and stare down onto the market that I begin to feel my initial infatuation beginning to wane. For it is the sounds and smells of a typical food market that are so strangely absent here. Back on the ground floor, an eerie quietness prevails. Compared to the scene in the outside farmers' market, fewer people wander about the food hall with grocery bags or pull-carts. Many of those present seem to be, like me, camera-toting tourists captivated more by the architecture and artistry than by the foodstuffs offered for sale. There is an inauthentic, Disneyfied or *The Truman Show* unreality to the place. And certainly, the presence of several supermarkets and chain stores seems egregiously counter to the spirit of market culture as heralded and sought after by so many 'foodies'.

My unease over the cultural sterility of market hall is quickly forgotten once I descend the escalators and the Time Staircase. Here, I linger in front of the exhibition windows while eavesdropping on the conversations of several parties of locals and French, Italian, and English-speaking tourists, who gesture towards and comment upon the displayed artifacts. But once the basement level is reached, I again feel uncomfortable. Never have I seen such an immaculate parking lot. The floor is so remarkably clean of any traces of vehicles that one can imagine actually eating off its surface. I have flashbacks to visits to Kyoto, where city workers respond to complaints from locals about autumn leaves falling on the ground, and to Singapore, where one cannot see a piece of litter in that antiseptic city. It is only when I return to the clamour and clutter of the outside farmers' market that I find a happy return to gritty reality. I stare out over the antiquated tent awnings of the simple stalls and back toward the superlative Markthal and frown in confusion at my mixed emotions of equal parts admiration and apprehension.

3.2.14 Case study lessons

- Saturday open-air markets enliven town squares (Plaza de Armas) and downtown city cores (Haymarket), as do permanent outdoor markets for the latter (Kensington Market).
- It is possible for markets to provide space for both temporary tables of regional farmers as well as for permanent retail booths (Feria Alcalde José Sandoval Gomez, St. Lawrence Market, Mercado do Bolhão, Markthal).
- Onsite food kiosks and restaurants play a crucial role contributing to the overall experience of visiting markets (Goods Shed, Halifax markets, St. Lawrence Market, Feria Alcalde José Sandoval Gomez, Mercado do Bolhão, El Mercado Central, Mercado de La Boqueria, Borough Market, Markthal).
- Attention to the architecture of market buildings is a critical feature of their reputation and success, including those imaginatively repurposed from previous usages (Goods Shed, Halifax markets), or those created especially for the purpose, be this either through their preservation as heritage structures radiating old charm (Eastern Market, St. Lawrence Market, Mercado do Bolhão, El Mercado Central, Mercado de La Boqueria, Borough Market) or as newly constructed, high-visibility edifices (Feria Alcalde José Sandoval Gomez, Markthal).

- Visitor experience is enhanced by the presence of art, such as murals depicting bygone days (Kensington Market, Halifax markets, St. Lawrence Market) or grandiose celebrations of food (Markthal), as well as sculptural installations (Kensington Market, Haymarket).
- Long-established markets are sites of historic importance in civic history, something which can be celebrated through onsite museum-like displays (St. Lawrence Market, Markthal).
- Food markets offer experiences beyond those simply related to the acquisition of foodstuffs, in that they serve as social venues of multicultural diversity of the type sought after by tourists, to such an extent that guided visits can be specifically established for non-locals (Eastern Market, St. Lawrence Market, Mercado do Bolhão, El Mercado Central, Mercado de La Boqueria).
- Achieving the delicate balance between the interests and shopping needs of locals with those of visitors and tourists is acknowledged as a challenge (Halifax markets, Eastern Market, El Mercado Central, Mercado de La Boqueria).
- Successful markets serve as vehicles for urban regeneration through fostering a flourishing of retail shops and hospitality service establishments that spills over into the immediate vicinity (Plaza de Armas, Kensington Market, Haymarket, Eastern Market, El Mercado Central, Mercado de La Boqueria, Borough Market), as well as setting up a charitable trust to support local community initiates (Borough Market).

References

Abbott, R.M. 2008. Uncommon cents: Thoreau and the nature of business. Green Frigate Books/Libri, Farringdon, UK.

Allaby, M., G.K. MacDonald and S. Turner. 2021. Growing pains: small-scale farmer responses to an urban rooftop farming and online marketplace enterprise in Montreal, Canada. Agriculture and Human Values 38: 677-692.

Anonymous. 2012a. Eastern Market development plans. Available at: www.detroiteasternmarket.com.

Anonymous. 2012b. Top ten reasons why Eastern Market is and will remain Detroit's most 'true green' neighborhood. Available at: www.detroiteasternmarket.com.

Barnett, L. 2011. The supermarket growing food on its roof. The Guardian 9 March.

Barton, B. 2013. Britain's marvelous markets. The Official Magazine, Britain. Travel, Culture, Heritage, Style. 81: 81-84.

Bevan, R. 2015. Markthal in Rotterdam by MVRV. The Architectural Review, 5 January.

Bhatt, V. and L. M. Farah. 2016. Cultivating Montreal: a brief history of citizens and institutions integrating urban agriculture in the city. Urban Agriculture & Regional Food Systems 1: 1-14.

Bibeau, E.A. 2020. The elephant in the market: unearthing provincial policy alignments and perceived challenges to grow the impact of the farmers' market sector in Nova Scotia, Canada. M.A. Thesis, Dalhousie University.

Bohn, K. and A. Viljoen. 2016. Urban agriculture: designing the productive city. In: CITIES (ed.) Farming the city: food as a tool for today's urbanisation. Trancity, Haarlem, the Netherlands.

Bond, D. and R. Feagan. 2013. Toronto farmers' markets: towards cultural sustainability? Journal of Agriculture, Food Systems, and Community Development 3: 45-60.

Brooks, K. 2017. 'Biggest artwork in the world' hits the Netherlands, dubbed 'Sistine Chapel of Rotterdam'. The Huffington Post, 12 June.

Brunet, M. 2012. Hope Blooms in the North End. Halifax Local Connection, 15 October.

Buck, N.T. 2018. The urban bioeconomy: Extracting value from the ecological and biophysical. Journal of Environmental Planning and Management 64: 182-201.

Bucklain, F.A. and N. Azab. 2020. Taif botanic conservation center. Journal of Critical Reviews 7: 311-315.

Carroll, M. 2012. Market performance: environmental stewardship as a key driver in the adaptive reuse of a massive building on the piers of Halifax's waterfront. Canadian Architect 14: 28-33.

Cheveley, M. 2021. Alternative vs. conventional food networks: a geospatial analysis in relation to neighbourhood sociodemographic characteristics in Montreal. M.A. Thesis, McGill University.

CITIES (ed.) 2016. Farming the city: food as a tool for today's urbanisation. Trancity, Haarlem, the Netherlands.

Cockrall-King, J. 2012. Food and the city: urban agriculture and the new food revolution. Prometheus Books, Buffalo, NY, USA.

Coles, B. and P. Crang. 2011. Placing alternative consumption: commodity fetishism in Borough Fine Food Market, London. In: Lewis, T. and E. Potter (eds.) Ethical consumption: a critical introduction. Routledge, London, UK.

Corum, V., M. Rosenzweig and E. Gibson. 2001. The new farmers' market: farm fresh ideas for producers, managers, and communities. New World Publishing, Auburn, CA, USA.

Crespi-Vallbona, M. and D. Dimitrovski. 2017. Food markets from a local dimension. – La Boqueria (Barcelona, Spain). Cities 70: 32-39.

Crespi-Vallbona, M., M.D. Perez and O.M. Miroi. 2017. Urban food markets and their sustainability and the compatibility of traditional and tourist uses. Current Issues in Tourism 22: 1723-1743.

Cretetla, A. and M.S. Bvenger. 2016. Food as creative city politics in the city of Rotterdam. Cities 51: 1-10.

Curtin, J. 2006. North Downs Way. Trailblazers Publications, Hindhead, UK.

Czerniak, J. (ed.) 2002. Case: Downsview Park Toronto. Prestel Publication Verlag, London, UK.

Deeb, E. 2012. History of Detroit's historic Eastern Market since 1891. Available at: www.detroiteasternmarket.com.

Delgado, H.F. 2012. De lo proyectado a lo construdio. El Mercado Central de Valencia. Revista de Expresión Grafica Arqvitectónicca 17: 114-125.

Dickau, J., J.M. Pilcher and S.K. Young. 2021. If you wanted garlic, you had to go to Kensington: culinary infrastructure and immigrant entrepreneurship in Toronto's food markets before official multiculturalism. Food, Culture and Society 24: 31-48.

Dimitrovski, D. and M. Crespi-Vallbona. 2017a. Urban food markets in the context of a tourist attraction – La Boqueria Market in Barcelona, Spain. Tourism Geographics 20: 397-417.

Dimitrovski, D. and M. Crespi-Vallbona. 2017b. Role of food neophilia in food market tourists' motivational construct: the case of La Boqueria in Barcelona, Spain. Journal of Travel and Tourism Marketing 32: 475-487.

Esteve, A.J. and R. Arnandis-Agramunt. 2020. Touristification in the Central Market of Valencia: Fact or fiction? In: de Almeida, C., A. Quintano, M. Simancas, R. Huete and Z. Breda (eds.) Handbook of research on the impacts, challenges, and policy responses to overtourism. IGI Global, Hershey, PA, USA.

Ferrando, A. and S. Tealdi. 2013. Food markets in the belly of the city – Barcelona la Boqueria. Stefilm International, Turin, Italy.

France, R.L. 2011. Veniceland Atlantis: the bleak future of the world's favorite city. Libri, Farringdon, UK.

Francis, M. and L. Griffith. 2011. The meaning and design of farmers' markets as public space. Landscape Journal 30: 2-11.

Froese-Stoddard, A. 2012. Exchanging culture?: Opportunities for intercultural interaction at the Halifax International Market. Journal for Undergraduate Ethnography 2: 1-13.

Fuertes, P. and E. Gomez-Escoda. 2020. Suppling Barcelona: the role of public market halls in the construction of the urban food system. Journal of Urban History 20: 1-19.

Goodstein, J. and K.C. Turino. 2015. Haymarket (images of America). Arcadia Press, Mount Pleasant, SC, USA.

Gorgolewski, M., J. Komisar and J. Nasr. 2011. Carrot city: creating places for urban agriculture. Monacelli Press, New York, NY, USA.

Haberman, D., A. Canter, V. Rinner, L. Pancrazi and F. Martellozzo. 2014. The potential of urban agriculture in Montreal: a quantitative assessment. International Journal of Geo-Information 3: 1101-1107.

Hall, C.M. and L. Sharples. 2008. Future issues and trends: Food events, festivals and farmers' markets. In: Hall, C.M. and L. Sharples (eds.) Food and wine festivals and events around the world. Elsevier, Amsterdam, the Netherlands.

Harries, M. and L. Heder. 2005. The touch of water: artists working with water and communities. In: France, R.L. (ed.). Facilitating watershed management: fostering awareness and stewardship. Rowman & Littlefield, Lanham, MD, USA.

Hawkins, D. 2011. FARM:Shop – East London's radical experiment in food growing and community building. The Ecologist, 14 January.

Henderson, J. 2014. Hope Blooms to build new greenhouse next to north-end garden. CBC News 30 June.

Hesterman, O.B. 2016. Local food, fresh and fair. In: CITIES (ed.) Farming the city: food as a tool for today's urbanisation. Trancity, Haarlem, the Netherlands.

Hoogstraten, K. 2019. Growing a rooftop revolution: repurposing urban roofs for increased food production in Montreal, QC. M.A. Thesis, University of Manitoba, Canada.

Jans-Singh, M., P. Fidler, R.M. Ward, R. Chaudhary and M.J. DeJong. 2019. Monitoring the performance of an underground hydroponic farm. International Conference on Smart Infrastructure and Construction.

Jans-Singh, M. K. Leeming, R. Chaudhary and M. Girolami. 2020. Digital twin of an urban-integrated hydroponic farm. Data-Central Engineering. Cambridge University Press, Cambridge, UK.

Johnson, J. 2010a. Designs on food. Azure, May issue: 65-75.

Johnson, M.L. 2010b. Market forces: commercial realtors hot for high rents see new Yorkville in funky Kensington. Now Magazine, 21 March.

Kilcup, F. 1998. Halifax Farmers' Market: chasing the dawn. Recipes and recollections from Canada's oldest farmers' market. City Market Halifax Cooperative.

Lambie, C. 2013. Fish seller vacates farmers' market. The Chronicle Herald, 12 February.

Langton, A. and K. Cockshull. 2013. All change in the glasshouse. The Horticulturalist 22: 17-20.

Lewis, J. 2011a. London's Lea Valley – a century of growing. The history of the Lea Valley Growers' Association from 1911 to 2011. Libri, Farringdon, UK.

Lewis, J. 2011b. Regeneration and innovation: invention and reinvention in the Lea Valley. Libri, Farringdon, UK.

Li, N. 2015. Kensington Market: collective memory, public history, and Toronto's urban landscape. University of Toronto Press, Toronto, Canada.

Mas, J. and O. Ubide. 2011. Boqueria Gourmand. Viena Edicions, Barcelona, Spain.

McChrystal, R. 2015. Urban agriculture is bringing food production back into our cities. Elite Business Magazine.

McLean, H. and B. Rahder. 2003. The exclusionary politics of creative communities. Canadian Journal of Urban Research 22: 90-110.

Nordhal, D. and C. Fisher. 2016. Public produce: food by the people, for the people. In: France, R.L. (ed.) Integrated urban agriculture: precedents, practices, prospects. Libri, Farringdon, UK.

North, A. 2012. Processing Downsview Park: transforming a theoretical diagram to master plan and construction reality. Journal of Landscape Architecture, Spring Issue, 1-19.

O'Reilly, J. 2017. Planting seeds for healthy youth: exploring parents' perceptions of a community-based program in Halifax, Nova Scotia. M.A. Thesis, Dalhousie University.

Petts, J. 2008. The economics of urban and peri-urban agriculture. In: Viljoen, A. (ed.). CPULs: Continuous productive urban landscapes: designing urban agriculture for sustainable cities. Elsevier Architectural Press, Amsterdam, the Netherlands.

Prytherch, D.L. 2003. Urban planning and a Europe transformed: the landscape politics of scale in Valencia. Cities 20: 421-428.

Queiros, S. 2019. A disappearing world: the ever-expanding frontier of gentrification through the eye of Porto's historic centre long-term residents. In: Albert, A. and N. Benach (eds.) Gentrification as a global strategy. Routledge, London, UK.

Radovich, T. 2011. Heirloom vegetable varieties: a promising future for past treasures. Hani'Ai/the Food Provider, Dec 2010/Jan-Feb 2011.

Raine, K.D. 2014. Improving nutritional health of the public through social change: finding our roles in collective action. Canadian Journal of Dietetic Practice 20: 1-25.

Rawsthorn, A. 2011. Making food seriously local. The New York Times, 18 September.

Richler, J. 2010. Food fights: jockeying for fresh fruit and vegetables reveals middle-class Torontonians at their competitive, grasping worst. Maclean's Magazine, 10 April 2010.

Romero, D., L. Bashford, K.K. Heng and R. Zelaya-Cavalcante. 2014. Assessing food availability and food deserts in the City of Toronto with respect to Fresh City Farm's current customers. York Space, York University, UK.

Selwood, J. 2003. The lure of food: food as an attraction in destination marketing in Manitoba, Canada. In: Hall, C.M., L. Sharples, R. Mitchell, N. Macionis and B. Cambourne. (eds.) Food tourism around the world: development, management and markets. BH Publishers, Nashville, TN, USA.

Seth, C., S. Pittoello, R. Gueli and A. Singh. 2019. Integrity economics: business supports for plant-based diets. In: Kevany, K.M. (ed.) Plant-based diets for succulence and sustainability. Routledge, London, UK.

Shaw, J.L. 2005. Resistance admist disorganization: understanding the nature of community organizing in Toronto's Kensington Market. M.A. Thesis, University of Toronto, Canada.

Smith, E. 2018. Tiny farms grows giant crop in old Dartmouth shipping container. CBC News Nova Scotia, 21 July.

Sommer, R. 1989. Farmers' markets as community events. In: Altman, I. and E.H. Zube (eds.) Places and spaces. Plenum Press, New York, NY, USA.

Sommer, R., J. Herrick and T.R. Sommer. 1981. The behavioral ecology of supermarkets and farmers' markets. Journal of Environmental Psychology 1: 13-19.

Spurr, B. 2013. Halifax kids' salad dressing does well in the Den. The Chronicle Herald, 4 November.

Stephenson, M. 2012. Lots of tangled roots behind Seaport Farmers' market issues. The Chronicle Herald, 24 May.

Terek, D. 2011. Shop owner offers little bit of France in Eastern Market. The Detroit News, 23 February.

Webster, G. and L. Dawson. 2012. Futuristic farm shop grows food in synthetic veg patch. CNN News, 13 April.

Chapter 4
Healing connections: hope and education

4.1 Community sustaining gardens

Anecdote

Personal experiences of food confraternity. My maternal grandmother arrived in Canada as a Polish immigrant in the 1920s and spent almost forty years working a farm in south-eastern Manitoba situated at the interface between the boreal forest and the fertile prairies. In her old age, while a resident in a dense neighbourhood in downtown Winnipeg, she nurtured a small kitchen garden in her backyard with loving attention, proudly harvesting and using its produce, from which she prepared wonderful meals to stuff into her grandchildren upon every visit. During one such occasion more than four decades ago, I watched through a window while she was picking produce. An obviously destitute and homeless person appeared along the back alley and stopped to have a conversation with her. After several minutes, my grandmother went back to her plot and pulled up a cluster of carrots and pulled off several handfuls of small tomatoes and handed them to the vagrant who bowed his head in gratitude before strolling off. Later, when she returned inside, I asked my grandmother about the interaction. She replied that good, fresh food was hard to come by in the area and that it was her duty to provide it to any in her neighbourhood who were in need.

Book

Street farm: growing food, jobs, and hope on the urban frontier, by Michael Ableman, is a case history of one of the most inspiring urban agriculture stories, that of Vancouver's 'Sole Food Street Farms.' This is an organisation with a goal of gardening society by growing wellbeing, weeding out despair, and sowing hope. Although Vancouver has an international reputation for being a green city of prosperity, the harsh reality is that its Downtown Eastside is the poorest postal code in Canada, and one that is plagued by the highest per capita rates of hepatitis and HIV and opioid deaths in all of North America. The strength of the book is in its impassioned, first-person storytelling about what the author refers to as a 'very simple idea: rehabilitation through meaningful work':

DOI 10.3920/978-90-8686-938-1_4, © Wageningen Academic Publishers 2022

> Growing food is high on the list of meaningful work. It brings us down to real basics; it allows us to see cause and effect play out daily, to see how our action or inaction creates interaction. We feed the soil that feeds the plants that feeds us. We begin to sense how working with plants and soil and sunlight and rain and wind can heal us.
>
> Good farmers know that to be successful we must be immersed in a consummate love affair – with the land, with the living soil, and with the broader community that eats our food. We know that this work alone will not heal the folks who come to work with us. We have no illusions that a farm or good food or real work will help everyone get healthy. But the raw biology, the simplicity and physicality of the work, the magic of seeds emerging, plants thriving, food being harvested and filling boxes and bellies, in places where previously there was nothing but hardscape and trash and rubble: This is recovery. Experiencing it will touch and change you (p. 63).

Eschewing scholarly referencing, Ableman instead spills considerable ink on providing glimpses into the lives of raw desperation of the many street people whom he and his farms have employed over the years; their ongoing struggles with substance abuse, mental anguish, and emotional turbulence, all under an umbrella of how, at least for a while, the growing of food on vacant lots has been a way of nurturing a sense of fulfilment in lives that can be vacant of purpose other than day-to-day survival. And if the words were not powerful enough, the book contains many photographs that show the destitute and homeless. Throughout, the emphasis is on how Street Farms and its produce markets provide jobs and create a 'farmily' of shared purpose. That said, Ablemen is a clear-eyed realist rather than a rose-tinted romantic; and above all, a farmer. He is no fan of charity, admitting to be much more impressed when he is complimented on the taste of his carrots rather than the good works his projects do in the community. Frequent referral is made to 'beautiful food', and the fact that the organisation was named top food producer in the region by a prominent magazine is stated with justifiable pride. Ableman is critical that most of what is professed to be 'urban agriculture' is actually just urban horticulture. Agriculture, he insists – the ability to produce substantial volumes of food – is a much more complex enterprise than anything occurring in garden-sized projects. This seems to have trickled down to his workers who bridle when people refer to them as gardeners rather than farmers. He also makes the point that many Canadian restaurants that brag about their menus containing locally sourced food in the middle of winter are guilty of fraud.

The process of transforming abandoned and debris-laden urban land into vibrant street farms to grow artisan-quality fruits and vegetables is recounted in detail. The team soon realises that some land is simply beyond their ability and resources to regenerate due to limitations caused by topography, access, and contamination. With regard to the latter, they soon reach the conclusion that paved land is the best to work with as it isolates soil contamination. In consequence, their projects make use of raised-box planting beds filled with imported soil. These have the added bonus in being non-permanent and therefore able to be moved on short notice. The latter is important as Sole Food Street Farms purposely selects high-visibility sites with the trade-off being that such locations are often in the eye of real estate developers. One lesson is that whereas land is of course a necessity for the growing of food, ownership of that land is not. It becomes necessary to be reconciled to the ongoing challenge of contested land use. This is particularly true when farms are built atop hectares of pavement, which, given our automobile-obsessed society, means that such sites are attractive to those who believe that cities can never have enough space in which to park cars. To counter this, Ableman provides the useful calculation that a standard 9-foot × 18-foot parking space of 162 square feet can produce 450-500 pounds of food in a four- to six-month growing season, which, with an average retail value of $3 per pound will generate $1,500 revenue per annum.

Other tribulations are shared with aplomb, including issues of security and theft, experiments with different construction materials for planting beds, and the omnipresent challenges of dealing with municipal bureaucracy, as for example, the fact that tunnel hoop-houses do not exist as a specified typology in the city's building permit documents. One interesting anecdote recounts the struggle they faced in needing to prove that runoff from their urban farm was not the source of elevated faecal coliform bacteria counts detected in a nearby waterbody.

One of the great strengths in the book is its reliance upon numerous photographs to tell the story, several of which occur as double-page spreads. The contrast between the city hardscape of high-rise buildings and elevated roads with the arrays of green plants in plastic growing tubs which occupy half the surface area of a vast parking lot are memorable. In a chapter on urban agriculture's role in rewilding cities, Ableman dreams of a future in which agricultural parks focused on orchards will abound, thereby converting the urban heat sink into edible calories: 'Imagine walkways winding through fruiting trees, vegetables, small livestock, and grains providing education, fresh food, and jobs for neighbourhood residents. Consider the ecological, social, economic, and educational value of a farm within a park' (p. 153).

Sites

4.1.1 Cultivate London (London, UK)

As L. Johnson (2010: 135) reminds, 'community gardens are, essentially, participatory landscapes.' One such is run by 'Cultivate London,' an independent social enterprise and charity that is dedicated to creating employment opportunities in horticulture at the same time as engaging local communities in projects that transform derelict sites into vibrant greenspace for improving health and wellbeing. What is particularly interesting is that the organisation began as a charity with the aim of addressing youth unemployment *then* selected horticulture as the business model for training 'because of a growing interest in local food, and the therapeutic benefits associated with working with plants in a hands-on outdoor setting' (www.cultivatelondon.org; Nardi 2014). The organisation works with businesses, city councils, and community groups toward a goal of fostering urban wildlife, reducing climate change, and promoting other social and environmental outcomes. Cultivated outcomes include helping individuals to build confidence while learning skills, and helping communities to build social cohesion through shared visions and hands-on work related to United Nations Sustainable Development Goals of sustainability, biodiversity, and organic horticulture.

Cultivate London was founded in 2010, and at time of my visit four years later, they were actively tending four sites. One of their first projects in turning vacant land into verdant growing space was their flagship community garden in the West London borough of Hounslow. Here, on a vacant property that had once been a garden but which had become 'impenetrably overgrown', they worked to remove brambles and accumulated rubbish. The garden was restored with help of specialists who had worked on restoring the Chiswick House Kitchen Garden (see Chapter 6.1), and several plastic hoop-houses were erected to grow salad plants for sale to farmers' markets and local outlets.

A large plastic banner strung up on a wire fence facing the sidewalk denotes the urban garden and explains Cultivate London's goals of 'growing together' through addressing employment, land regeneration, and climate change concerns [e4.1.1.1]. A series of photographs show stages involved in the clean-up of the site (Figure 4.1) [e4.1.1.2]. Peering through the bars of the fence reveals a narrow sliver of property occupied by an abandoned hoop-house, the uncovered frame of another, a pile of wood mulch, rolls of screening, partially opened bags of potting soil, several wheelbarrows, and a jumble of wooden pallets scattered about (Figure 4.2) [e4.1.1.3].

The year following my visit, the site was transformed into a training garden and a space for social events and for those seeking a tranquil refuge from urban life. Photos on the website depict an attractive space with ornamental plants, a garden shed, and outdoor wash basins and cooking facilities. Today, Cultivate London's landscaping and gardening services, working under the mandate of 'we unlock underused spaces in London for the benefit of community cohesion and environmental benefit,' have created greenspace in eighty locations (including a housing development with 'regen' in its name). Some of these projects include designing edible gardens for apartment tower courtyards and rooftop allotments. Ongoing activities focus on building community engagement through consultations and in running workshops about hands-on outdoor learning, food literacy, healthy diets, and horticultural therapy.

Figure 4.1.

Figure 4.2.

As I leave the Tube station, I reflect that it is a pleasant change to be searching out an urban agriculture project that is located in the west, instead of the east, part of London. When I arrive at the site, however, I find myself to be remarkably underwhelmed by what I see. Despite it being the peak of the spring growing season with other urban agriculture projects in full swing, the present site is forlorn. I find it difficult to balance out the statements made on the website read earlier and on the large sign just now with what I see in front of me. A wheelbarrow that has been loaded with mulch and rolled from the storage pile only to be abandoned just in front of the hoop-house has the look of sudden departure as if in a real life a crime scene or the set of an alien abduction movie.

I am somewhat disheartened by my visit, wondering if this is another well-intentioned project that has folded. Could it be, I wonder, due to the fact that the organisation is (was?) one pertaining to cultivating employment rather than plants? [I am happy to learn, years later, upon researching updates for writing this book, that not only has the organisation not folded, but that it has thrived and expanded in terms of building meritorious projects throughout London. Obviously, my visit came at an inopportune time in the evolution of Cultivate London. This underscores that many of these organisations, even those of demonstrable success, undergo periods of uncertainty, challenge, and rethinking with respect to how their sites are managed.]

4.1.2 The Stop's Earlscourt Community Garden and Food Centre (Toronto, ON, Canada)

'The Stop' is one of the most impressive social enterprises located anywhere with a primary focus on creating a just food system for urban dwellers. Emblazoned on the cover of their 2011 annual report, famous chef Jamie Oliver is quoted as saying that 'I've travelled all around the world and I've never seen anything like The Stop. Every city should have one'. The Stop is guided by three ideologies: (1) to confront hunger it is necessary to go beyond simply handing out food; (2) food access and food security constitute basic human rights (Levkoe 2006); and (3) empowering people to grow food and help one another helps communities build resilience (Cockrall-King 2012). They achieve this through providing programs for the urban underprivileged that help 'people find respite from their problems, and a welcoming environment where they can get food when they're hungry, learn new skills, meet friends, help their neighbours and find their voice' (Saul and Hewitt Smith 2011). The Stop's mission is to use urban agriculture and community building as means to confront hunger and poverty (www.thestop.org). The organisation began in the mid-1970s and moved to their central food centre in Davenport West in 1997. This is a district of Toronto that is notable for having more than half of its single adult residents living below the poverty line. The Stop's first community garden started the following year, with that in Earlscourt Park breaking ground two years thereafter. A decade on, and after receiving several prestigious awards, The Stop opened its high visibility 'Green Barn' (see Chapter 7.1).

The Stop's 'Food Centre' is housed within the ground floor of an apartment block on a busy street (Figure 4.3). This is where the immediate needs of people are met 'with dignified and respectful community programs and services – all centred around fresh, nutritious, and nourishing food.' Here, in addition to being able to access one of Canada's first food banks, people can participate in drop-in sessions, take classes on social wellbeing, and engage in a community kitchen that provides instruction about on how to maintain a healthy lifestyle.

Earlscourt Park is located several kilometres away from The Stop's Food Centre and contains all the elements expected in an urban greenspace: children's playground, wading pool, off-leash dog-park, sports fields, and band-shell. Here too, are 743 square metres of community gardens operated by The Stop, in which they remark with humour that they have to tend to the problem of 'devilish squirrels.' Lettuce, kale, cabbage, and root vegetables such as radishes, carrots, and turnips are grown along with eggplants, squash, leeks, various types of red peppers, and herbs such as parsley and garlic. The hundreds of kilograms of annual harvest are split between the approximately twenty volunteers and The Stop's food bank and other food programmes. The gardens are suggested to provide a comfortable sense of security in lives that are often characterised by uncertainty. As one volunteer gardener expressed, the gardens represent 'a community [that] has longevity because it is a place for stability and long-lasting relationships as well as a safe place' (Loiselle-Gascon 2013).

Figure 4.3.

After a few moments of uncertainty, I spot a throughway that provides access from the residential street to the community garden. A colourful hand-painted welcome sign greets visitors (Figure 4.4) while another informs that the project is sponsored by The Stop and the Toronto Department of Parks and Recreation [e4.1.2.1]. The gardens are sandwiched between a back lane of private garages and a series of playing fields [e4.1.2.2]. The site amenities present are few and simple, comprised of an open-fronted work shed, a seating bench, and a cinderblock planter (Figure 4.5). Several water drums and a wooden tower of cryptic purpose are situated in the vegetable plots [e4.12.3]. A notice board [e4.1.2.4] contains postings about drop-in sessions, free English classes, and The Stop's 'Healthy Food Philosophy' [e4.1.2.5]. The latter states:

> The Stop Community Centre's philosophy recognises food as an essential part of life, necessary to maintain the good health of body, mind and spirit.
>
> We believe that access to food is a basic human right. Accordingly, we aim to distribute as much food as possible to our community. However, we do believe that some foods should not be distributed, and therefore we decline some of the most unhealthy donated food we receive.
>
> As much as possible, we strive to provide fresh, locally produced foods that are seasonal, minimally processed, affordable and accessible to all. We respect the health of the people who work the land and water to produce food, and acknowledge that we are part of an interdependent food system. Whenever possible we use our purchasing power to support growers, producers and retailers who share this philosophy.
>
> We recognise that food plays an important part in the cultural diversity of our community and work to meet those needs through our programming. In this process, we nurture community through a shared food experience.
>
> We welcome opportunities to learn about, grow, cook and eat food together. We promote equally the importance of healthy nutrition and of the pleasure that comes from choosing and eating food that we love. In the food that we serve and promote, we are constantly looking for the terrain where 'nutritious' and 'delicious' intersect.
>
> Through innovative programming and public education we work to promote food security at all levels. Our philosophy of food security is based on the belief that change must happen at the individual, community, and policy level. Adequate income to buy food, a food system that promotes healthy eating and sustainable agriculture are all necessary components of food security.

The Stop and Earlscourt Community Gardens have been the subject of two university theses. Levkoe (2006) investigated how participation in gardening enabled the development of strong civic virtues and critical perspectives for transformative adult learning by allowing individuals 'to influence policy makers; to increase their level of political efficacy, knowledge, and skill; [and]

Figure 4.4.

Figure 4.5.

to directly challenge anti-democratic forces of control'. He found that the development of a food security curriculum in local high schools helped attract parents to join various land-to-mouth courses in which they learned about their dependence on the corporate food system. But it was the collective experience of directly participating in the act of gardening at Earlscourt and all the affiliated celebratory activities therein, such as the seasonal festivals, cultural days, and volunteer parties, which were most important to fostering individual learning. Interviewees expressed the importance of social interactions and the value of feeling connected to something that was larger than themselves.

A follow-up study by Loiselle-Gascon (2013) addressed the role that the Earlscourt Community Garden played in place-based environmental education through the act of bringing people closer to nature and community through collective food production. Of particular interest was how gardening can be used as a means for circumventing the prevalence of 'nature-deficit disorder' that plagues society (see Chapter 4.2). Nevertheless, she believes improvements can occur, noting that the absence of group work limited community building on a day-to-day basis and that the communal plots meant that there was a lack of consistency and stewardship for individual plants. These in turn were thought to hamper the development of feelings for being part of Nature's processes.

It is a June day when I visit the Earlscourt Community Gardens. Given the extremely positive experience of the hustle and bustle of The Stop's Green Barn at the Artspace Wychwood location (see Chapter 7.1), I find the present site to be underwhelming. There is nary a gardener present; nor any birds or other wildlife. What is there, is silence. Just me, the ghosts, and the forlorn traces of their past activities. Even the vegetable plots seem to have been given only a smattering of horticultural attention. The contrast between these community gardens and the commercial ones at Fresh City Farms seen the day before in Downsview Park (see Chapter 3.1) is striking, calling to mind as it does the criticism of the director of Detroit's Eastern Market (see Chapter 3.2) about the capabilities of such socially-based growing enterprises. Indeed, as I look around, I find it difficult to believe that enough produce is being grown here to supply the wonderful work of the organisation in terms of its food bank and food programmes. As I sit down and record my notes, I wonder if The Stop has of late refocused the majority of its gardening activities to those at its high visibility Green Barn location. Looking out again at the site, I remind myself, however, that the purpose of these community gardens is certainly as much to do with producing community through social interactions as it is in producing quantities of harvested vegetables. As is the case for many environmental restoration projects which I have studied and written about previously, it is the *process* of participation that is often more important than the products that ensue. Feeling more generous in this reappraisal, I sit back and imagine the place brimming with the conjured ghosts of milling gardeners during one of its celebratory events.

4.1.3 D-Town Farm (Detroit, MI, USA)

The Detroit Black Community Food Security Network (DBCFSN) is a coalition of organisations and individuals working together to build food justice and food security in Detroit's black community by: (1) influencing public policy; (2) promoting urban agriculture; (3) encouraging co-operative buying; (4) promoting healthy eating habits; (5) facilitating mutual support and collective action among members; and (6) encouraging young people to pursue careers in agriculture, aquaculture, animal husbandry, bee-keeping and other food related fields (www.d-townfarm.com). As described on its own Wikipedia entry, the organisation was founded in 2006 by an anti-capitalist activist motivated by the goal of achieving 'healthy food, healthy planet, healthy people'. The purpose of the DBCFSN and its signature 'D-Town Farm' is described on a sign at the entrance to the site [e4.1.3.1]:

> D-Town is a community self-determination project of the DBCFSN ... and sits on 2.01 acres in the city of Detroit's Rouge Park. We grow a variety of fruits and vegetables using organic, sustainable methods. Additionally, our bee-keepers maintain bee hives for pollination of our crops and to produce honey. We also raise various edible mushrooms.
>
> In 2007, the last major supermarket chain left Detroit, leaving a void that has caused many to characterise much of the city as a 'food desert.' However, the vast amounts of vacant land in the city of Detroit provide the opportunity for its residents to grow significant amounts of the food that we consume and to contribute to revitalising the local economy. D-Town Farm is a model of how unused and underutilised land in the city of Detroit can be used to address the lack of access to high quality produce that many Detroiters experience.
>
> D-Town Farm is also a teaching and learning site. As you walk through the farm please stop at each of our 11 learning stations to learn more about the environment, and the work we do at the farm. We hope you are inspired by your visit.

Two previous sites had been briefly farmed before the present location of a former city-owned tree nursery in Rouge Park was leased for a dollar in 2008, allowing the development of what became the largest urban farm in Detroit. Thirty different types of fruits, vegetables, and herbs are grown and sold at an onsite produce stand on weekends and through farmers' markets, wholesalers, and black-owned restaurants. DBCFSN operates a youth development program and conducts adult workshops about how to achieve food security and healthy lifestyles through gardening. Farm tours are offered and an annual harvest festival occurs with concerts, lectures, and demonstrations. The farm operates with a full-time manager and an army of volunteers, most of whom are women. A major goal has been to help to reconnect people to the land and its bounty. This was not an easy task due to the stigma held in the African American community around agriculture through its connection to plantation slavery and the deep historical trauma of being dispossessed of land in the southern states by the Ku Klux Klan.

D-Town Farm is situated in an urban 'ecotone,' one of those blurry boundaries that exist between different ecosystems in the landscape. The dividing line is the street, notable for its wide central medium strip of mowed grass. The west side is occupied by a row of small, well-maintained homes with no signs of the property vacancies that characterise the blight affecting other parts of the city. Across from the residential neighbourhood is Rouge Park, an area of woods and fields in the riparian zone of a stream. It is here, marked by the presence of parked cars and behind a locked gate, where the urban farm is located [e4.1.3.2]. A welcome sign [e4.1.3.1] identifies the site as belonging to the DBCFSN and describes the farm's purpose. Situated near the street entrance to the site are several hoop-houses which are used to sprout plants before their subsequent replanting (Figure 4.6) [e4.1.3.3]. Massive compost piles demonstrate the practice of sustainable farming techniques. The farm sprawls into the distance as a wide field of fenced-off plots bordered on either side by the dense woods of the park (Figure 4.7) [e4.1.3.4].

Figure 4.6.

Figure 4.7.

The D-Town Farm was founded specifically in response to the gentrification of Detroit and is based on the need for African Americans 'to have control over all aspects of their lives, including food' (Anonymous 2020). The farm is therefore about more than growing crops; it is regarded as growing community-based resistance as a strategy for initiating political change. Linked in spirit to the Black Panther Party, the farm empowers 'people of colour to resist the system that disadvantages them' (Anonymous 2020). As White (2011) states, the D-Town Farm allows 'activists [to] participate in the revival of a city mired in racism and poverty, and all but abandoned by politicians, the automobile industry, and the merchants and supermarkets who once served Detroit's residents'. The overriding purpose of the farm is to educate members to assume larger roles in urban agriculture discussions in Detroit. This is a response to the reality that most of those who presently do so are whites who exert a control disproportionate to their population base. Given that 83% of Detroit's population is black, the DBCFSN believes it is their right to be in leadership rather than be dictated by outsiders. This belief in self-reliance is key: whites need to offer support, but only from the position of occupying the back seat. As the D-Town Farm director states: 'In most cases, community garden work is being done in black and Latino communities mostly by young white people with a missionary mentality', something he refers to as being part of 'white supremacy' (Ladner 2011: 187). Many of those interviewed by White (2011) express that their efforts in growing food is a way for them to 'demonstrate self-determination toward political and economic liberation'. D-Town Farm is therefore a counterpoint to what the DBCFSN considers to be a disturbing trend across the country wherein whites are moving into black neighbourhoods and building urban farms and gardens. The organisation's underlying philosophy is based on a 'right to farm act' which declares that no authority should create laws to regulate agriculture in Detroit.

Long before my arrival to the site and before anyone had heard the phrase 'Black Lives Matter,' I was well aware that, as White (2011) succinctly states it, Detroit is 'ground zero for numerous negative social conditions and social ills'. From Vogel (2008; see Chapter 7.2), I knew of the 'white flight' to the suburbs and the consequent racial polarisation in the urban core of Detroit, resulting in it being the poorest city in America, and one that perennially teeters on the cusp of declaring bankruptcy.

Our arrival to the D-Town Farm is coincident with the imminent departure of one of the founders of DBCFSN. And so, after informing him of my research, I have the opportunity to ask questions about the project. It is during this brief conversation that the word 'capitalist' is mentioned several times as well as the Marxist belief expressed that owned property is a colonial idea in contrast to the right of everyone on the planet to take land and grow food on it in order to survive. Land, he lectures, is a partner and cannot be owned; rather it should be used as a source of resistance against a system of oppression and white supremacy in which the memory of African American agrarian culture is steeped. 'We've been here for generations and it is not appropriate for whites to come in and tell us what and how to do things', he intones. As I leave the site, I reflect that in some instances food can be regarded a means not just an end to itself, its growing constituting a wilful political act.

4.1.4 Regent Park Health Centre Community Garden (Toronto, ON, Canada)

Community health centres have operated in North America for more than half a century. Unlike primary care centres, the mandate of community health centres is to address the root causes or 'upstream' determinants of health. Their focus is on providing non-clinical, preventative care to marginalised individuals in lower-income neighbourhoods, such as that of Regent Park in the downtown eastside Toronto (Collins *et al.* 2014). Regent's Park contains Canada's largest social housing project, and is located several kilometres distant from the Riverdale Farm (see Chapter 6.1) and the St. Lawrence Market (see Chapter 3.2). A member of Ontario's Alliance for Healthier Communities, the 'Regent Park Health Centre' was established in 1973 and is a non-profit, community-based organisation focused on improving the health of local residents. With a slogan of 'Removing barriers. Improving Lives', it works to ensure equitable health outcomes and social justice through community engagement (www.regentparkchc.org). A wide suite of programmes are offered for immigrants and youth, including one on nutrition that is augmented by dietitian counselling.

Along with two other social service charitable organisations operating in the area – 'CRC' and 'Fred Victor' – the Regent Park Health Centre manages community gardens (Anonymous. 2006). Some of these gardens are fenced and found adjacent to buildings, with most becoming active in the evenings through a cadre of dedicated female gardeners (Wakefield *et al.* 2007). The results of in-depth interviews of participating gardeners revealed that their motivations were based on reducing their monthly food bills, particularly in terms of cultivating cultural foods that are rarely found in grocery stores at reasonable prices. Other reasons included the importance of the gardens as places for socialising and establishing neighbourhood identity. Gardening was regarded as a win-win activity in terms of providing a source of physical exercise and mental relaxation (including a break from television watching) while producing healthy food. Wakefield *et al.* (2007) also found that gardeners derived a sense of pride from an activity that they perceived as a benefiting the community as a whole, through 'serving as an impetus for broader community improvement and mobilization'.

The community garden I elect to visit is located on Dundas St. East, a busy urban thoroughfare along which I cycled hundreds of times during my residency in the city while a doctoral student at the University of Toronto. A sign affixed to a chain-link fence in the shadow of nearby towering apartment blocks identifies the community garden, as does another sign stating that the produce grown herein contributes to a food bank [e4.1.4.1]. The small lot squeezed beside a church is chockablock full of individual growing plots, each about the size of a dining room table and demarcated by wooden stakes, strung cordage, and erected netting (Figure 4.8) [e4.1.4.2].

Figure 4.8.

It is early on a Saturday morning in June when I visit the garden. As I stroll around the site, I am struck by how well Wakefield *et al.* (2007) hit the mark in their appraisal that 'overall, the gardens conveyed a sense of lushness and abundance'. Indeed, the garden is so brimming with produce wrested from every square metre of ground that it is difficult to imagine how its tenders can even negotiate their way around the individual cheek-by-jowl plots. Rarely have I seen a more demonstrable example of how sunlight, water, and soil can conspire to generate bounty in a dense urban setting.

4.1.5 Aghios Dimitrios Municipal Garden (Athens, Greece)

As the first decade of the new millennium drew to a close, Greece faced a perfect storm of societal problems variably described as a 'crisis' or a 'nightmare' (Kolokouris 2015; Moran and Fernandez de Casadevante 2016). The financial collapse, bank bailouts, public indebtedness, and harsh austerity measures led to simultaneous reductions in both employment and public services. Massive demonstrations ensued, calling for social transformation. One such transformation that arose was the rapid development of urban agriculture through the reclamation of public spaces as community gardens. This was a somewhat surprising epiphenomenon given that, as urbanisation really only took place in Greece after the Second World War, there is no real tradition of urban farming. As well as providing a source of immediate sustenance, the municipal garden movement was envisioned as a way to cultivate social responsibility and solidarity as growers had to donate at least 10% of their harvest to their local food bank or soup kitchen. The sites are therefore regarded as 'social gardens' with food grown for both self-consumption and for sustaining particularly vulnerable groups. Other goals, according to the Urban Nature Atlas (www.una.city), include improved health and wellbeing, beneficial microclimatic alterations, re-established connections with nature, educational opportunities for youth, and land regeneration. The gardens are also looked upon as playing social roles in terms of providing a place to meet and improve self-esteem brought about by shared work. As such, the gardens 'impact not only food access, but also social cohesion, education and employment' (Moran and Fernandez de Casadevante 2016). By being more than just a means to cope with economic hardships through the growing of one's own food, the gardens are therefore 'also and above all a way to rebuild social bonds and reshape a space/place of belonging and self-esteem in a rapidly changing world' (Anthopoulou 2016). Through serving as places of memory and recreation of the rural past, the gardens represent sites of lived experience (Partalidou and Anthopoulou 2016).

One of the first municipal gardens to be established in Athens with European Union and city budget funding was that of Aghios Dimitrios, located in a dense part of the city and comprised of 45 plots on a hectare of land beside a graveyard. Every two years those in the vulnerable categories of being unemployed, low income, retired, and single-parent or large families compete to be awarded a plot. Residents, of whom almost 80% are either unemployed or low income pensioners, manage their own gardens on a day-by-day basis, augmented by weekly advice and assistance provided from municipal workers or collaborating organisations.

In addition to commenting upon the obvious economic relief ensuing from growing their own food, participants interviewed by Anthopoulou (2016) likened the municipal garden to being a valued member of their family and in some cases a reminder of the village life of their youth. Some individuals enjoyed the physicality of the act of being in contact with the earth, thereby allowing them to 're-sprout along with [their] vegetables'. Whereas others championed the socialising and the feeling of belonging, as for example, 'The garden is my life; my support; my daily routine; a part of my identity; a school in creation. Even if the crisis ends I will still be there. I have discovered a new way of life'. It is therefore simplistic to regard the community gardens of Aghios Dimitrios as just an emergency action to thwart poverty and food insecurity. Rather, these 'urban gardens above all trigger processes that redefine the identity of individuals/growers themselves' by instilling a sense of being part of a collective that stewards the land in a sustainable manner (Anthopoulou 2016).

Signs mark the entrance to the community garden [e4.1.5.1]. Here, adjacent to a phalanx of three- and four-storey apartment dwellings, the growing space is partitioned with screen and mesh dividers, the dense plantings of vegetables suspended by a cat's cradle of strings and wires strung between hundreds of stakes (Figure 4.9) [e4.1.5.2]. The lush verdancy of the site is truly remarkable [e4.1.5.3], a testament to the benefits of both the Mediterranean climate and the skilled attention of the neighbourhood gardeners [e6.1.5.4]. Vegetables, some easily recognisable [e4.1.5.5], offer an appealing contrast to the view of the nearby apartment blocks [e4.1.5.6]. Several sections admix the edible landscape have been given over to beautiful flowers (Figure 4.10) [e4.1.5.7]. Requisite compost bins [e4.1.5.8] appear to be the only shared utilities offered. The site is popular among Athenian social workers and cultural geographers interested in studying the response of locals to the financial crisis [e4.1.5.9].

Figure 4.9.

Figure 4.10.

Self-reliance born of necessity, I reflect, while watching the gardener tend her plot with careful attention. However, there is also something else that is immediately apparent. I know that the municipal gardens around me were conceived of as a form of social resilience during the Greek crisis. But there is no indication of any of impression of 'crisis' discerned and displayed in front of me. Instead, the gardener is enthusiastically chatting away with the professionals with whom I am visiting the site. Her pride of horticultural accomplishment exudes from every animated gesture as she points out different plants while presumably described their growing. I can only guess, however, since, to paraphrase Shakespeare's idiom, 'it is all Greek to me'. Noting my silence, the gardener asks me a question, to which I gesture incomprehension until my colleague intercedes. After more words spoken between them – which I later learn was when she was informed that I was a visiting professor from Canada with an interest in community gardens – the gardener bends down to pick and then hand me several plum-shaped cherry Roma tomatoes. I thank her and bite into the tiny red gems, nodding appreciatively and telling her through my colleague's translation that they are the sweetest such that I have ever tasted. The huge smile that spreads across her face is worth the price of the Istanbul to Athens plane ticket that has gotten me here.

4.1.6 Brightmoor Alliance Youth Garden (Detroit, MI, USA)

Brightmoor is a neighbourhood in northwest Detroit that was established in 1922 to provide cheap housing for workers at the nearby, then rapidly expanding Ford motor factory. Today 'Blight-More', as it has come to be pejoratively known, is a microcosm for Detroit in decline. Its present population of 13,000 residents is only a third of what it once was at its peak, with a quarter of that loss taking place in the first decade of the present century. This has resulted in high levels of land abandonment and vacancy following the demolition of derelict homes. Present-day conditions are bleak, with vacancy and poverty rates approaching 50%, and the neighbourhood being declared a hot-spot for drug dealing and murders. Pride (2016) refers to Brightmoor as a 'hyper-abandoned and disinvested urban neighborhood', noting its unique hybridisation of rural ambiance and urban concrete jungle, and citing a writer for *The New Yorker* who stated that 'much of Brightmoor matches what Detroit looks like in the popular imagination – an alarming amalgam of city dump, crime scene, and wild prairie' (p. 48).

The 'Brightmoor Alliance' was founded in 2000 as a coalition of fifty neighbourhood organisations to address the poor housing, high rates of crime, and vacant land. Its vision was to work toward creating a neighbourhood community 'where people of modest means can live, learn, commune, and work in a safe, healthy, culturally diverse, and sustainable environment' (www.brightmooralliance.org). The organisation believes that, like Thoreau (see Chapter 3.1), prosperity involves 'more than just money, but the whole of people's lives'. Prosperity can be achieved through personal power, collective action, and public policy based on neighbourhood involvement. This takes the form of beautification projects organised by block clubs, childhood education (including tutoring and support for school attendance), as well as providing resources for residents in terms of information, financial support, and utility assistance. Early on in its existence, the Alliance recognised the neighbourhood to be a food desert, and so listed more than a dozen food distribution locations under title of 'food pantry assistance'.

Coincident to my visit, Brightmoor Alliance had partnered with the 'Community Development Advocate of Detroit' to create a land use and quality of life plan, the *Brightmoor Restore the 'Moor Report* (Anonymous 2014). Based on a goal of transforming the neighbourhood into a healthy, safe, and supportive environment for families, the plan includes green infrastructure and plant-based initiatives such as community gardens. Future land-use maps envision making use of vacant parcels to generate land-based jobs and living opportunities related to farming. These include: (1) developing the concept of urban homesteads that are suitable for country living in the city with land that can be used for gardening and farming; (2) developing green ventures for farming, horticulture, forestry, and aquaculture; (3) creating 'naturscapes' in the form of landscaped parks and urban wilds; and (4) reinforcing the importance of reclaiming outdoor space with gardening as a crime deterrent. Featured in the document are photographs of bounty baskets filled with vegetables, a goat, and the successful 'Brightmoor Youth Garden,' the latter promoted as the seed upon which to grow.

The Brightmoor Youth Garden began in 2006 in a single city lot and expanded to six lots within three years. Its purpose it to use gardening to teach life skills, encourage academic achievement, employ youth, all as part of a strategy to combat rising crime in the neighbourhood through growing a sense of responsibility (Pride 2016). By integrating with an after-school art programme, the gardens serve to beautify the neighbourhood with floral plantings and landscape design, and to foster a sense of pride among residents while at the same time producing healthy food.

The Brightmoor neighbourhood displays the same symptoms of urban decline that characterise much of the troubled city of Detroit (see Chapter 7.2). Here, however, it is reassuring to see the transformation of vacant lots into garden plots through rows of growing stakes, drip-irrigation tubing, and raised wooden growing beds [e4.1.6.1]. What is special about the Youth Garden is the pride of the community that shines through in the form of attention paid to the ancillary details of aesthetics. Most notably in this regard is a long wooden structure decorated with brightly illustrated panels that runs along the sidewalk to demarcate the border (Figure 4.11). At one terminus, a large, two-metre tall sheet of painted plywood has been erected that is emblazoned with a painting of a pumpkin and the following words: 'Brightmoor Youth Garden. This garden is by kids and for kids. Please respect our profit. Look, don't pick!' [e4.1.6.2]. Some of the illustrated panels depict raindrops [e4.1.6.3], which at first is puzzling until one enters the garden and sees the backside of the wooden structure. Doing so, one realises that the painted wooden panels are there to protect a series of elevated cisterns that harvest rainwater from the structure's roof and use it for irrigating the proximal plants [e4.1.6.4]. Other details include wooden sculptures of butterflies (Figure 4.12) as well as scarecrows and an attractive wooden entrance gateway [e4.1.6.5].

Since the time of my visit, a combination of both formal (community-based development) and informal (do-it-yourself) urban gardens now exist in Brightmoor, totalling an area of more than ten thousand square metres (Pride 2016). This includes the 'Farmway,' a series of urban farms that have replaced vacant and blighted lots. At the same time, public art from youths has also expanded to include decorating numerous structures in the neighbourhood. The project is regarded as a success for building social capital in terms of allowing residents to participate in healthy activities that foster solidarity and appreciation for their neighbourhood. As well as providing a fresh supply of healthy food, the gardens serve as gathering places where residents build rapport and create bonds. The net result is that some residents have been convinced to stay and others have been encouraged to move into Brightmoor.

Figure 4.11.

Figure 4.12.

For his doctoral thesis, Pride (2016) interviewed many residents, some of whom clearly acknowledge the important role played by agriculture in the regeneration of their blighted neighbourhood:

> [People] can go right up here, down this street here, bring their family and pick tomatoes or hot peppers or whatever's there. See, that's adding to the neighbourhood, not taking away. We come together on these things, we become closer as a community. And we can feel good about Brightmoor (p. 74).
>
> Through gardening you can learn so much. And this is the main reason why we make our garden available to the entire community. We want them to come here to see how to work together, how to grow good food that they can put into their bodies and feel good. And it works. Oh, yes, gardening is a great way to teach responsibility, how hard work and teamwork leads to beautiful things. There are so many different aspects to gardening that has to be right in order for things to turn out the way you want them to. You have to take the soil, take care of the soil in just the right way. You have to follow through with all the small details. You can take these lessons and apply to them to any aspect of your life and find that they help with problem solving, or getting difficult tasks done. I believe with hard work and teamwork we can rebuild this community (p. 76).
>
> The change we're trying to achieve here can really only be accomplished with the cooperation of residents living here who want to do something, something that shakes this community to its core. So we have to learn to work together, right? The only way we can do this is by working together. The community gardens, and we have a few, are places where we can build those skills. When we learn how to cooperate the right way, because we can do it ourselves, we will make this community better and safer and a place where people want to live. I have no doubt (pp. 76-77).

Pride (2016), however, raises a concern arising from the fact that the people who are moving into Brightmoor to take part in urban farming are mostly white, middle-class professionals; those whom he refers to as 'green-hipsters'. Community change therefore comes with the risk of 'eco-gentrification' which can push up property values and thereby displace that portion of the population who are the most economically vulnerable.

Youth gardens are often ancillary addendums tucked into the corners of larger urban agriculture projects. Not so here. The fact that the present garden is, in its entirety, managed for and by youths is what makes the site exemplary. Moreover, the inclusion of youth-drawn art in their gardening plots of plants is memorable. But it is something else entirely that produces my strongest impression of the site. My guide and I are the only souls present in the garden on the day of our springtime visit. But no means do we have the site to ourselves. For there is a constant twittering of multitudinous birdsong that permeates the place. We stop talking and walking and for several moments simply enjoy the avian chorus.

We wonder as to whether they are greeting us or perhaps expressing their displeasure at our invasion of their territory. Later, toward the end of our visit, we watch as several birds make their way from their nests in the surrounding glade of trees to land upon open patches of soil in the garden and there to grub about for worms for their voracious young. The lesson couldn't be clearer: urban agriculture = urban biodiversity.

4.1.7 Southside City Farm (Providence, RI, USA)

The following brief description of the 'Southside City Farm Somerset Community Garden' is from the Chapter 'Designing urban agricultural forms: history, education, proposals and projects' in the book *Integrated urban agriculture: precedents, practices, prospects* (France 2016: 173):

> Providence, Rhode Island is bisected by the elevated interstate highway and contains areas that are challenged by high unemployment, abundant vacant land and boarded up homes. The South Side Community Land Trust was started in 1981 by a group of community activists with an interest in rehabilitating different parts of the city. The purpose was to work with neighbours, many of whom were immigrants from South-east Asia and West Africa and whom did not have access to land. The main goals were to teach people how to grow food, to locate and secure land, and to give resources to recent immigrants. And the values that were hoped to be instilled by urban farming included social justice, sustainable small-scale agriculture, education, self-governance, and collaboration and community building (www.southsideclt.org).

Community land trusts are based on the belief that land should be available in perpetuity to the people who use it; in other words, they offer a model for retaining ownership of land that is tended by and for the community. Whereas most community land trusts are focused on affordable housing, the 'Southside Community Land Trust' is the first and only one in the United States that is specifically dedicated to urban agriculture (Brown and Brush 2018). Although urban agriculture has grown exponentially in recent years, a frequent problem is insecurity of tenure and lack of strategies for the long-term protection of growing spaces in the face of speculative market forces. Yuen (2014) investigated community land trusts, including Southside's, concluding that such entities provide 'locally a controlled approach to land use that fosters community activism and engagement while responding to evolving market conditions and neighbourhood needs'.

The Southside Community Land Trust provides access to land to enable social justice, economic security, and community building. They provide education to deepen understanding of the world and to strengthen capacity for problem solving. And they facilitate people growing food with sustainable methods 'essential to our responsible participation in the restoration and preservation of our ecosystem'. The organisation has an annual budget of half a million dollars, employs seven full-time staff, has thousands of participant members, and manages a dozen community gardens throughout Providence and other cities in Rhode Island. Workshops, community events, and city-wide urban agriculture initiatives are also offered. The organisation belongs to a like-minded consortium of NGOs dedicated to strengthening local agriculture through providing assistance to farmers. In this light, they support a network of CSAs, restaurants, and farmers' markets by encouraging the population to always buy local if food can't be grown by one's self. With four decades of building and creating a sustainable and fair local food system, the land trust is deeply involved in shifting from a food supply that is faceless and far away to one where people assume control:

> By reimagining our relationship to what makes us function at a core level – the nourishment we give our bodies – we can reconnect with the environment and one another. Growing food isn't just about working in the garden: it's about being outdoors, playing in the soil, chatting with neighbors, and enjoying super-local produce. Joining a community garden, or organising your neighbors to start one, is a great way to save money, make new friends, and learn more about the natural world.

Located on a hectare in the heart of downtown Providence, the inaugural property functions as a commercial farm, and outdoor classroom, and an example of establishing a thriving centre of urban biodiversity. When it began in the 1980s, the surrounding neighbourhood was 'considered so blighted that the city was selling lots for only $1' (Cobb 2012). This led to boarded up houses, abandoned businesses, and vacant lots. One of the latter, which had once been a car dissembling shop, was purchased and transformed it into a shared garden. Recent immigrants worked to clear the land and were then allowed to farm their plots. The result was that the community garden 'not only improved residents' diets and economic self-reliance, it stabilised a neighbourhood weakened by arson, neglect and crime. Today, the site is the oldest urban farm in Rhode Island and demonstrates successful horticulture practices. More than two hundred families participate in gardening the plots, three-quarters of whom are people of colour, many being immigrants from the Developing World (Cobb 2011). Produce is sold at farmers' markets and local restaurants, and is donated to soup kitchens and food pantries. The site also serves to strengthen community ties. Given that there are few opportunities for people of different races and nationalities to work together, the gardens are a living anthropology of multiculturalism (Kirk 2006). A summer school is run to instruct about plant growing, soil health, pest management, and salad preparation, and culminates in a big, end-of-the-summer block party.

The organisation's signature urban farm is located in a tough neighbourhood in which the project's toilets have to be locked to prevent their use as drug injection sites. The boarded-up homes which surround the gardens signify the life challenges and consequent financial hardships faced by many in the neighbourhood (Figure 4.13) [e4.1.7.1]. In addition to a charming hand-painted sign that identifies the gardens and another that informs of the sponsoring organisation [e4.1.6.2], a third notice board explains growing methods, including the use of weather-protective protective cloth coverings, seafood-based nutrient fertilisers, spray soap to reduce bugs, and cover crops to improve soil health [e4.1.7.3]. A vacant lot is brimming with vegetables, some growing under protective cover [e4.1.7.4]. The bulk of the community gardens, however, occur in a portion of a city block that is separated from the street and sidewalk by a perimeter fence [e4.1.7.5], with plantings flush to abutting occupied and abandoned homes [e4.1.7.6]. Despite the very low project funding, site managers decided to use local art students and sculptors to elaborate upon the traditional chain-link fence motif (France 2016). The resulting whimsical fence, though small in size, is designed as a cheery, welcoming gesture to the community (Figure 4.14) [e4.1.7.7]. Inside the central area, individual plots are demarcated with wood stakes and screens (Figure 4.15) [e4.1.7.8], some of which have been set up with careful attention as private enclaves [e4.1.7.9]. The verdant bounty being produced is impressive [e4.1.7.10]. There are familiar New England vegetables of green beans, tomatoes, and lettuce, as well as plants native to other regions or countries, such as okra, peanuts, and fancy squashes (Kirk 2006). Site infrastructure for the gardeners include a mulch pile, picnic working table, rain barrel, and equipment shed [e4.1.7.10].

Figure 4.13.

Figure 4.14.

Figure 4.15.

When a one-time mayor of Providence visited the site, he expressed surprise at how much food could be grown in such a small space and how it seemed to be a win-win model that could be adopted elsewhere. This good first impression led to the generation of a series of reports on urban agriculture planning for the city (including for establishing community gardens in public parks), and on developing a sustainable and local food system, as well as to changes in zoning ordinances and bylaws in order to facilitate the growing of food for sale on private properties, the so-called 'Lots of Hope' program (Brown and Brush 2018).

It has been a long drive to reach downtown Providence from where I am staying with friends in the neighbouring state of Massachusetts. After waiting for more than half an hour after my pre-arranged rendezvous time with one of Southside's directors, I realise that I've been stood up. And so, nonplussed, I undertake my tour alone. As I stroll around, taking photos, it is impossible not become impressed by the site's overt productivity. Despite the welcoming bit of artistry displayed on the entrance fence, two interactions indicate that my presence is unwanted. The first is with a gardener who arrives and begins work while I'm walking about. When shouting over the locked fence as to whether she might be the Southside director with whom I am to meet, my inquiry is met with a harsh 'No' followed by a terse 'Don't know her. I'm busy'. Later, when I am taking several photographs of the boarded-up homes that surround the garden, a window in an adjacent dwelling opens and an angry occupant demands to know what I am doing. Normally one who is quite careful when it comes to photographing sensitive subjects, I certainly understand the upset that can be caused by 'ruin porn'. Realising that a long academic explanation would only serve to exacerbate the situation, I apologise and remove myself in chagrin, vowing to be more careful in the future to absolutely make certain that no one is about to take umbrage when photographing the sites of neglect that so often give birth to urban agriculture projects.

4.1.8 Kipos3 Community Garden and Urban Vineyard (Thessaloniki, Greece)

The northern Greek city of Thessaloniki did not escape from the same social-economic crisis that plagued the entire country (Kontothanasis 2017). Modelled after the 'hidden vineyards' in Paris, Thessaloniki established its own urban grape orchard in 2013 on the one-time site of a former taxi stand that had been converted into a neighbourhood park. Expenses were covered by the city government, with management provided by a nearby university school of viticulture. Four indigenous grape varieties are grown and harvested by citizens, with production and bottling taking place on an estate outside the city, and all wine being donated to charity. The site has become popular with visiting school groups and tourists.

The decision to create the Kipos3 Community Garden adjacent to the vineyard was a deliberate act. Locals, who had not been consulted in the top-down process involved with establishing the vineyard, had become upset at the loss of their neighbourhood park. Following the wish of the residents whose homes surround the central square, an aesthetic and edible landscape was created by a university landscape architecture studio with an end for encouraging the re-use of the civic space. An important idea was for the site to become a co-owned urban 'commons' and incubator of socio-ecological knowledge while simultaneously establishing a shared cultural identity (Gavrilidou and Ritou 2016; Oureilidou 2018). It was hoped that, with its focus on regeneration, Kipos3 would play a role in redefining open space planning and the possibilities of everyday life in Thessaloniki. Food in this regard is used as the 'carrot' to effect social and physical change by building resilience in times of crisis (Gavrilidou *et al.* 2015).

The gardens are modest, with 15 raised beds, a common toolshed, and a shared plot for herbs and aromatics. Water is provided by the municipality. The area is actively tended by more than a dozen gardeners and yields four kilograms per square metre (Kontothanasis 2017). Kipos3 is regarded as a 'social space, a park, the evening walk, a 'foodscape', a hobby, a meeting point, the everyday coffee hour, a landmark for the city, [and] even an 'encyclopaedia' of different nature in the urban environment' (Gavrilidou and Ritou 2016). In this respect, the designers believe that Kipos3 is as much a symbolic initiative as it is a simple garden, offering as it does, a turning point in the way that urban landscapes are regarded in Greece. In this regard, Oureilidou (2018) considers such projects to demonstrate how the social capital of urban agriculture can contribute to urban regeneration.

The community gardens and vineyard are located in a densely inhabited section of the city and occur in a public greenspace that is encircled by three- and four-storey apartment blocks [e4.1.8.1]. Vegetables, many being staked tomato plants [e4.1.8.2], occur in square box planters. These are nicely laid out on a sea of grass (Figure 4.16) which, together with the adjacent large grassed field and palm trees [e4.1.8.3], harkens the site to having been a park for recreation. Amenities for gardeners include a cute toolshed [e4.1.8.4] and a seating area situated around a table made of wood slats balanced upon a tractor tire base (Figure 4.17). It is the adjacent urban vineyard, bordered with a rustic wooden fence (Figure 4.18) [e4.1.8.5] and perimeter plantings of ornamentals [e4.1.8.6], that makes the site so memorable. Rows of young vines are suspended by stakes and wires and sustained in the hot climate through drip irrigation [e4.1.8.6]. The visual contrast between a specialised form of agriculture, viticulture, that is so emblematic in the minds of many to the bucolic Mediterranean countryside, and the surrounding residential streetscape of apartment blocks, is as sharp as it is spectacular (Figure 4.19) [e4.1.8.7].

Figure 4.16.

Figure 4.17.

Figure 4.18.

Figure 4.19.

Staked tomatoes and lettuce rows growing in urban settings are so commonplace as to be almost banal. But the conversion of a portion of a city park in a central square into a mini-vineyard is something that is truly special. Upon arrival, I smile at the surreal scene in front of me. The very idea of an 'urban vineyard' defies logic; it is an oxymoron. I hop over the wooden fence and walk admix the rows. Several times, I bend down to lightly cup a hand around a cluster of green grapes and to ever so gently lift them in order to feel their weight. At such times, when the surrounding buildings are obscured by the dense foliage of the vines enveloping me, it seems as if I could be living some sort of Arcadian dream à la *Under the Tuscan Sun*. Once more I smile, this time while contemplating that if only Diane Lane – the actress in the filmed version of that bestseller – were now present with a glass of offered wine, then all would be perfect indeed.

4.1.9 Common Roots Urban Farm (Halifax, NS, Canada)

Thirty years ago, the province of Nova Scotia produced more than 60% of its food, an amount that has now fallen to only 8-12%, which has instigated a rise in urban agriculture in the region (Melrose 2013). 'Common Roots' is a non-profit organisation that was founded to 'promote a holistic approach to health and build capacity for urban framing and gardening by empowering our community to grow and eat healthy food and connecting people with restorative landscapes' (http://www.partnersforcare.ca). Their pilot project farm was established in 2012 as interim land-use on hospital land located right in the heart of downtown Halifax. On this one hectare site, which had once been a high school and was slated for future redevelopment related to the planned expansion of the hospital, the regional health care board wished to have a project that would incorporate elements of health combined with those of education and urban agriculture. The urban farm contains 195 community plots, communal areas that are open to the public, and a market garden that produces vegetables and flowers for donation and sale, in addition to a vibrant urban greenspace. Activities on the farm reflect the organisation's commitment to social, environmental, and cultural sustainability, with the pilot project being looked upon as an incubator for establishing similar projects elsewhere in the city. The idea of an incubator farm is that it will function as an educational wellspring to enable the community to learn that successful farming is part science and part art, with an incredible amount of hands-on management (Melrose 2013).

The goal is to create a multifunctional location that is not only agriculturally, but also socially and ecologically, productive. Toward this end, the site is worked by a dedicated group of staff, summer students, and volunteers from numerous groups. Resources and work donated by charities, including the Kinsmen Club, created many of the amenities on the site, so that 'visitors to the farm can relax and enjoy the beauty of the garden and perhaps be inspirited watching the plotters at work.' Self-tended plots are available for a nominal fee to interested groups and individuals with the understanding that community gardeners will provide at least eight hours of work to the general upkeep of the farm. Comprehensive instructions are provided to community gardeners. Common Roots also runs a CSA tied into a local food bank with members' fees supporting farm operations in return for receipt of vegetables (Gillis 2013). CSA membership also supports the farm's donation of produce to the food bank, for which individuals obtain a tax receipt for charitable donations. Also, once a week at its on-site farm stand, the public can purchase kale, peas, beans, tomatoes, cucumbers, squash, and bouquets of flowers supplied from the 'social enterprise garden.' Compost tea is used to increase soil microbial activity, as is rock powder and seaweed products to provide soil nutrients. Common Roots produced a manual, *Urban farms everywhere. Implementation guide* (http://www.cch.novascotia.ca), which contains information on land-use legalities and logistics, design patterns and principles, tips about practicing social inclusion, and an entire suite of programming concepts, budgeting details, and fundraising ideas (including operating a market stand, planting donors' trees, scavenger hunts, seed/bulb drives, art shows, photo contests, concert hosting, and seasonal festivals). One popular event is the pumpkin smash festival held every November 1st, wherein people are encouraged to bring their own Halloween Jack-o'-lanterns from home and sledgehammer them into compost for growing the next year's crop.

The three main areas of the urban farm are: (1) the market garden, which is maintained by farm staff; (2) community plots, which are rented to individuals, and (3) edible landscaping, which includes 'nibble plots' for visitors to graze. And because the farm is located on one of the city's busiest corners, these visitors are many, with more than a dozen people an hour strolling through at certain times of the day. It is hoped that after-hour food theft – which the organisation views as a sort of success – will be directed to these nibble beds. Given its proximity to hospital, attention was devoted to making the farm accessible to the most fragile populations. In this regard, a series of wheelchair-accessible garden beds propped up on stilts were built. Based on the concept of horticultural therapy (see Chapter 7.2), the farm is looked upon as a 'healing oasis for hospital patients'. Families of patients sit quietly among the plants while contemplating the condition of their loved ones, patients stroll through seeking solace in the profusion of flowers, and hospital workers simply enjoy a peaceful respite from their hectic work lives. And if that were not enough, Common Roots has partnered with regional immigration services to establish the 'Deep Roots' programme to enable recent arrivals to adapt what they know of farming from their homelands to their new Canadian climate while at the same time allowing them to develop a sense of belonging to networks of support (Drudge 2019). Immigrants from Syria, Tanzania, Bhutan, Rwanda, and Burundi have participated in the programme. A woman from the latter grew lenga-lenga, a type of amaranth not available locally: 'The food, it takes care of my body. It is like medicine and it connects me with the Creator. It is also social. I get to help other people. We meet, we talk, we are happy. It is amazing. Without food, we are nothing' (d'Entremont 2019). The original website contained numerous testimonials of those who became inspired by the process of planting, growing, and harvesting food to improve community health. For example, one out-of-town visiting university student commented:

> Common Roots Farm was super-inspiring. The story of how land was reclaimed and purposed for helpful, healthy, active outdoor education to help families and community members get their hands dirty and experience a new way of gardening is so unreal. Especially when mixed with immigrants and restaurant owners who are able to share their knowledge with those around them including the farm coordinators themselves. I don't think I have ever left a field trip so light and determined to become and stay involved in community farming.

What makes the Common Roots Urban Farm [e4.1.9.1] so exceptional is its location, accessibility, and auxiliary structures. The farm is located in the downtown core of the province's capital city and is situated immediately beside the region's largest hospital [e4.1.9.2]. As shown on the large map displayed onsite [e4.1.9.3], people can enter the gardens from numerous access points. Also highlighted on the map are the diversity of site's elements, including the therapeutic walk from the hospital, entry garden, bike parking spots, community garden plots, greenspace, outdoor classroom, children's garden, greenhouse and nursery beds, compost area, main plaza and performance area, market gardens, therapeutic gardens, and edible walking paths. Produce is grown in large plots and small beds bordered by wood or brick [e4.1.9.4], both aided through a network of garden stakes and strings for climbing plants [e4.1.9.5]. Beautiful flowers (Figure 4.20) are intermixed with edibles, with seating benches provided for repose [e4.1.9.6] to enjoy the smells while watching pollinators at work [e4.1.9.7]. A shelter provides a weather-protected work area [e4.1.9.8], where sign boards announce daily chores and harvest details [e4.1.9.9]. Nearby are a hoop-house, picnic worktable, and watering station [e4.1.9.10], as well as a compost pile [e4.1.9.11] and wooden shelter near prepping beds used for mixing mulch and soil [e4.1.9.12]. A signpost sports the names of sponsors stencilled into wooden planks [e4.1.9.13].

Considerable attention has been paid to developing an impressive suite of amenities for improving the aesthetics and multi-functionality of the Common Roots site. An interesting art installation is designed as a sort of pergola (Figure 4.21) [e4.1.9.14]. A gathering area with triangular planters is covered in an artsy shade canopy (Figure 4.22) [e4.1.9.15]. A children's play area contains a sandbox with seating for supervising parents (Figure 4.23), where also several teepee frames are constructed from tree limbs that have been wrapped in ribbons of colourful material [e4.1.9.16]. Here too is a large sign [e4.1.9.17] that describes the design and creation of the children's garden:

> The Impetus: As Common Roots Urban Farm grew, so did the number of children coming to the farm. We wanted a space where kids could fearlessly and actively play with soil, tools, and plants.
>
> The Design: The first phase of this project was funded through Clean Foundation's 'Celebrating 25' fund. Rhea Dawn Natural Landscapes guided a series of design workshops attended by kids, youth, parents, and farm volunteers.
>
> The Build: Phase II was funded by the Walmart-Evergreen Green Grants program. This allowed us to hire four youth leaders to guide the building process. Children have been involved at every step of the garden's design and creation.
>
> The EarthLoom: We asked the Art Bikers to visit to help us put art in the Garden. They suggested building an EarthLoom (http://www.earthloom.org). Feel free to weave in some weeds, flowers, or yarn and remember how plants produce fibre and colour that humans have woven for centuries. Thanks to the 4C's Foundation for supporting the Art Bikers.
>
> The Result: The Children's Garden Playscape invites children to actively play and to learn about plants, gardening, and nature. Suggestions are very welcome! Feel free to take a self-guided tour through the various sections of the garden, including both edible and native pollinator plants, and earthloom for weaving natural materials, a mud kitchen for play, a willow tunnel and sitting nooks for both children and adults to relax in.
>
> Thank you: Developing interest in the natural world and the skills for growing food are gifts that will last. We are happy to be able to provide this space for young gardeners and their families. Thank you to everyone who was involved in this project, from design to creation!

A note on the children's garden image: The image on this sign represents play and curiosity with the mask adding a theatrical element. This is to honour a theatre teacher, Nancy Marshall, whose classroom was once where the children's garden is now.

Elsewhere on the site, a large teddy bear constructed of circular hay bales is another structure of artful whimsy (Figure 4.24).

Figure 4.20.

Figure 4.21.

Figure 4.22.

Figure 4.23.

Figure 4.24.

Everything written above about this project should really be in past-tense. In 2019, the Common Roots Urban Farm was removed in order to make way for the long-intended hospital expansion. Although this was always to have been the fate of the original 'incubator' farm, there is still no denying the feeling of loss of what was one of the most comprehensive, best designed, and ideally situated of all community sustaining gardens, anywhere. A new farm has been set up on a much smaller parcel of vacant land located at some distance away from the central core of the city. Community plots, communal beds, and market gardens are provided at this new location, along with a series of community programmes, skills development workshops, with flowers and vegetables being grown for sale and donation (http://www.commonrootsurbanfarm.ca). A second and more interesting site has recently been established across the harbour on the grounds of another hospital that provides mental health and addiction treatment counselling services for patients. According to information on the current website, fertiliser, tote bags, relishes and picked vegetables, bouquets, and salad mixes can be purchased.

My initial visits to the Common Roots Urban Farm generated remarkably different experiences. The first occurred late one evening following attendance at a theatre show and several days after listening to a local radio broadcast that had mentioned that some of those participating in the guided walks being offered at the site had elected to do so blindfolded. The reason for the latter was, according to the guide, in order 'to explore our senses, to come back into the present moment and to experience the world around us' by paying attention to other senses such as focusing on the smells of the garden. And so, all alone in the garden at this late hour, I sought out various straight sections of the pathways where I could safely stroll with eyes closed and pieces of tissue jammed into my ears, all the while deeply breathing in the fragrances wafting up on this hot summer night. It was an enjoyable experiment in that for several moments I could actually imagine myself in a backyard garden somewhere out in the remote countryside. Perhaps, I reflected as I settled into my car for the long drive back home, there is something to be said about visiting urban gardens at night-time. After all, Thoreau loved middle-of-the-night peregrinations about his Concord, Massachusetts environs, and Dickens was famous for his jaunts as a *flâneur* through the streets of London in the wee hours.

Another visit to the urban farm, this time with eyes wide open, occurred during the middle of the day. At this time, I wheeled a relative from her room in the adjacent hospital where she was recovering from an aneurism. This was her first time outside in more than a week and her gratitude at having survived was obvious in the way she revelled in the effusive display of life – plants, insects, birds – in which she found herself immersed. Watching this, it was easy to believe the claims made by enthusiasts that urban gardens function as experiential places of emotion in addition to simply being physical spaces where food is grown.

4.1.10 Case study lessons

- Gardening provides employment opportunities which contribute to individual and societal wellbeing (Cultivate London; Regent Park Health Community Garden; Aghios Dimitrios Municipal Garden).
- In terms of creating places to foster emotional wellbeing, association of gardens with healing centres makes perfect sense (Common Roots Urban Farm).
- Regenerative urban agriculture flourishes on the reuse of brownfields such as abandoned spaces (Cultivate London; Aghios Dimitrios Municipal Garden) and post-industrial landscapes (Southside City Farm) as well as on established open spaces such as parks (D-Town Farm; The Stop's Earlscourt Gardens and Food Centre).
- 'Social gardens' combine focus on both the growing and distribing food (The Stop's Earlscourt Gardens and Food Centre; D-Town Farm; Aghios Dimitrios Municipal Garden; Southside City Farm) as well as offering instruction about healthy lifestyle choices (Cultivate London; The Stop's Earlscourt Gardens and Food Centre; D-Town Farm).
- Community sustaining gardens address social justice based on a concept that food access is a basic human right (The Stop's Earlscourt Gardens and Food Centre; D-Town Farm), particularly in terms of building resilience in times of socio-economic hardship (Aghios Dimitrios Municipal Garden; Kipos3 Community Garden and Urban Vineyard).
- In addition to fostering civic virtues, food can be employed as a political tool (The Stop's Earlscourt Gardens and Food Centre; D-Town Farm).
- Youth and children's gardens can be integrated into larger community sustaining gardens (Brightmoor Alliance Youth Garden; Common Roots Urban Farm).
- Urban agriculture sustains urban biodiversity (Brightmoor Alliance Youth Garden; Southside City Farm).
- High visibility gardens serve to bolster community identity and pride while at the same time thwarting gentrification and displacement of impoverished residents (Brightmoor Alliance Youth Garden).

- Art installations contribute to neighbourhood beautification (Brightmoor Alliance Youth Garden; Southside City Farm; Common Roots Urban Farm).
- Establishment of a land-trust can ensure the long-term use of the property for urban agriculture (Southside City Farm).
- Imagination and innovation can expand the type of farming that can take place within cities (Kipos3 Community Garden and Urban Vineyard).
- Community sustaining gardens are social places for multi-cultural mixing as much as they are horticultural spaces (Southside City Farm; Common Roots Urban Farm).
- Consideration of site aesthetics in terms of architecture (e.g. shade structures) and landscape architecture (e.g. flowers) substantially raises the profile of gardens, as does delivery of a suite of fun programmes (Common Roots Urban Farm).

4.2 Educational farms

Anecdote

Personal experiences of a childhood at the rural-urban interface. Charleswood, a suburb on the outskirts of Winnipeg, was a wonderful place in which to grow up. Tucked into the farmland at the southwest corner and separated from the rest of the city by the Assiniboine Forest, the largest urban forest in Canada, the suburb was ideally positioned to take advantage of both urban and rural benefits. A half-hour bus ride would bring one to the downtown core, whereas a short walk down our residential street would bring one to the edge of suburbia and the start of the limitless horizon of prairie grain fields. School friends lived in both splashy new houses on cul-de-sacs and in rustic multi-generational farm homes. And of course, every suburban property was characterised by one-third to one-half of its lot area being dedicated to a vegetable garden. Such gardens, I soon realised, offered benefits far beyond supplying foodstuffs for our table. Of course, there was the simple 'miracle' of planting seeds and watching the shoots creep skyward, a rate that was sometimes quantified as part of a horticultural competition amongst siblings. Much to my enjoyment and my parents' consternation, backyard gardens also offered sustenance to animals venturing out from the surrounding woods to nibble plants (France 2004). And there was no better place in which to sequester oneself during neighbourhood games of hide-and-seek than between the rows of corn that were present in every garden. Each spring and autumn a fellow would drive by in an old-fashioned red farm tractor selling his services for rotor-tilling the kitchen gardens. Ten doors down the street was an operating mink farm where one could view animals in cages and learn that they would be killed for the owner's livelihood. Several blocks and a short bicycle ride away was a property right at the edge of residential homes and the large prairie farms. Here, one could feed chickens and, for a modest price, even ride around the property atop a pony kept just for that purpose.

Today, sadly, much of this easy access to bucolic landscapes and lifestyles is gone. The woods behind the family home that once contained numerous tree-forts and which was inhabited by all manner of imaginary creatures culled from the novels of Edgar Rice Burroughs and the biographies of David Livingstone and other explorers, are no more. They were torn down decades ago for a new subdivision whose miniature lots are exclusively covered with manicured grass lawns and ornamental flower beds. The large backyards of the older homes which were once family gardens have been subdivided and sold to developers who have managed to squeeze in additional homes. The golf course across the street, once filled with a menagerie of night-time creatures after golfers had relinquished the links and retired to the clubhouse, is now bisected by a busy road. The mink farm down the street is long since gone, for who wants to drive (few seem to walk anymore) past a place where animals are being killed while on their way to pick-up plastic-wrapped sandwich meat at the supermarket. Neither horses nor other livestock inhabit the onetime suburban farm which is now itself been subdivided and built upon. In short, opportunities for children growing up in this Winnipeg suburb to ever see a farm animal or to grow their own vegetables, much less to play in the woods, even if their fear-filled parents would permit them to do such today, have vanished. No red tractor drives along the street to mark the change of seasons. All connectivity to the area's agricultural roots has been erased.

Book

Last child in the woods: saving our children from nature-deficit disorder, by Richard Louv, became a national bestseller in the United States, was the recipient of the 2008 Audubon Medal, and has been translated and published in a handful of languages. It is also one of the most depressing and possibly even terrifying books to be published in recent years.

Louv cites a plethora of studies demonstrating the essentialness of natural settings for healthy child development vis-à-vis stimulating the complete spectrum of senses when informal play and formal learning are integrated. He believes that children should experience the imperfect world of nature 'filled with loose parts and possibilities, with mud and dust, nettles and sky, transcendent hand-on moments and skinned knees' (p. 97). Education about natural history needs to instil feelings of wonder and joy, excitement and mystery. Furthermore, research has shown that the 'restorative environment' of what Louv calls 'nature's ritalin' may be the best way to stem the rising epidemic of ADHD amongst today's youth.

But sadly, all is far from well in the relationship that children have with nature. Although familiar with threats to the global environment, they remain functionally illiterate about their immediate natural surroundings. In short, they suffer from what Louv terms 'nature-deficit disorder.' The problem is that nature has become an abstract concept, not a physical reality. One reason is the irrational fear that 'helicopter parents' (i.e. those who always hover about their offspring) have of the dangers they imagine lying in wait to harm their children. Remarkably, in some jurisdictions, free-range playing has even become illegal. Even when outside play occurs, it is much different from the days when parents chased kids out the door with nothing more than 'make sure you're home for supper'. Today, childhood imagination is drowning under a regiment of structured activities. In their overly programmed young lives, children no longer play on their own; instead they go on parent-supervised 'play-dates'. The net result of all this is, as Louv states, that the physical boundaries of children's lives have become smaller and smaller.

As well, the experience of nature has, over the last century, Louv believes, progressed from 'utilitarianism to romantic attachment to electronic detachment' (p. 16). An environment of litigious fear (at least in the United States) has meant that in many grade school curricula, computer or laboratory research activities about nature have replaced field trips and experiential education. Remarkably, up to 40% of American elementary schools have even eliminated or are now considering to eliminate outdoor recess breaks. Screens have replaced streams. In other words, 'that which cannot be Googled does not count' (p. 68). Children have electronic lifestyles in which they can identity Japanese cartoon characters but not a single one of the plants and animals outside their homes. Many do not consider anything in the physical world worth watching compared to what can be found on television, or worse, in a video game. As one fourth-grader intoned: 'I like to play indoors better 'cause that's where all the electrical outlets are' (p. 10).

Throughout the book, Louv uses agricultural examples, championing eco-village planning, urban farms, therapeutic gardening, edible landscaping, community gardens, and farmers' markets. He states that those readers over forty are the last generation of North Americans to personally know relatives, usually their grandparents, who farmed. Most of today's youth have lost familial and cultural links to the recent agrarian past. We have become severed, both physically and also increasingly, conceptually, from our food's origins, as Louv playfully states in one section heading: 'For the young, food is from Venus; farming is from Mars' (p. 20). Within the lifetime of those now in their middle age, suburban vegetable gardens have transitioned from necessity to recreation, as we buy more and more shrink-wrapped, lab-produced food. Most of us better understand the workings of machines than we do animals. At one time our knowledge and relationship with farm animals was based on hands-on husbandry; today it is it is intellectually-based. Urban dwellers, particularly children, fuelled by watching nature documentaries, either feel overly protective of animals or fear them. The idea of animals being well cared for in order to enable the production of goods and services, as occurs on family farms, is totally foreign.

New education programmes need to be developed to reintroduce and reengage children with the wildness about them for 'healing the broken bond between our young and nature' (p. 3). Louv visited a school in California where fourth to sixth graders were part of a class that planted, tended, and harvested tiny garden plots. There, the children confessed to being surprised that the food they grew in the school gardens tasted better than that their parents bought in supermarkets. The teacher described the experience: 'For us, the garden has been much more than simply planting vegetables and taking care of them. It's been a bonding experience. When we go to the garden as a class at the end of the day, there is a strong feeling of shared joy and peace no matter how hard the day has been' (p. 262).

Children must develop a respect for where food comes from. Louv calls for reviving the 'farm school' idea of the founder of the Montessori education system, wherein adolescent students spend part of a year operating a working farm. He also would like to adopt an education programme from Norway in which farmers work with educators to create curricula based in both classroom and farm: 'If, at times, as a form of subsidy, government can pay farmers not to plant crops, surely it can pay them to plant seeds of nature in the next generation' (p. 221). One good news story is the city of Austin's purchase of a farm, which they renamed 'Pioneer Farm,' turning it into a living history museum for agriculture and natural history education. Louv believes that more opportunities where children can interact with farming animals and lifestyles are sorely needed. Some examples of these are introduced below.

Sites

4.2.1 Kentish Town City Farm (London, UK)

What is most remarkable about the 'Kentish Town City Farm' is its setting in the London Borough of Camden. The immediate surrounding neighbourhood is densely populated, with development consisting of modern row houses, sterile apartment blocks, and older brick terrace houses [e4.2.1.1]. Notable is the dearth of trees or other vegetation. Kentish Town City Farm was founded in 1972 on a sliver of derelict land adjacent to a rail line bisecting the area [e4.2.1.2]. Being England's first city farm, the site has played an important role in the development of the city farm movement across the country. As 'vestiges of village life in the United Kingdom' (Cockrall-King 2012:110), these farms bridge between the countryside and city and are based on the belief that establishing human-animal relations is formative to fostering feelings of connectivity to nature within urban settings (Smith 2019; Tarajko-Kowalska 2017).

Late eighteenth-century illustrations of the area portray a bucolic scene with a river flowing through fields occupied by livestock. All this changed with the Industrial Revolution and the breaking up of the large estates, and the arrival of the railways and manufacturing. Like so many areas of greater London, Camden suffered greatly during the Blitz, resulting in thousands of damaged buildings which when demolished, leaving patches of vacant land that persisted into the early 1970s. It is at this time that a group of local theatre artists and youth community workers discovered this piece of land tucked behind terrace houses. They soon transformed the site's derelict buildings into a horse-riding school and its weedy open space into allotment gardens.

Today, the Kentish Town City Farm advertises itself as 'offering lifelong learning, outdoor therapy and education to children, adults with special needs and anyone needing respite from the strains of urban life' (www.ktcityfarm.org.uk); all this by way of getting one's hands dirty while learning and caring for animals in a tranquil place that is also a haven for wildlife. A horse and pony club still exists that offers low-cost riding lessons to local children as well as providing therapeutic rides for visitors with special needs. In addition to school visits, the farm runs a work experience programme for youths and provides opportunities for adults to volunteer as a way to contribute positive, community-supporting work. Like all such institutions in the United Kingdom, the farm is chronically underfunded and relies upon support from the National Heritage Lottery Fund as well as from ongoing charitable and legacy donations. Interestingly, another line of revenue apart from admission fees is the sale of manure compost to gardeners in the area. Much of the site's history is celebrated on the website, including archival photographs and film footage – including a visit by Prince Charles in 1979 – and audio interviews from the time of the farm's establishment.

One enters through a walled gate squeezed between homes on a residential street (Figure 4.25). A large sign on an abutting house proudly announces the location to be all about animals [e4.2.1.3]. Inside, several brownstone buildings serve as the farm offices, a small gift shop, and a toolshed with a mural [e4.2.1.4]. Operating in the same vein as what North Americans would call a 'petting zoo,' children interact closely with wandering waterfowl [e4.2.1.5], the latter drinking from small rills that drain the site [e4.2.1.6]. Horses, cows, and enormous but friendly pigs are housed in stables and barn stalls [e4.2.1.7]. Murals, obviously drawn by children, decorate the high brick walls surrounding the farm [e4.2.1.8], and a repurposed mailbox serves to receive donations beneath a sign reading: 'PLEASE HELP to KEEP the FARM GOING. It Costs over £1,000 per mth. To Feed the Animals. YOUR DONATION WILL HELP – *Thank you*!' [e4.2.1.9]. Further away, goats and sheep graze and lounge underneath a railway arch (Figure 4.26). Nearby are located privately tended allotment gardens, the plants arranged in wooden planting boxes [e4.2.1.10]. These might be important for establishing neighbourhood goodwill in terms of placating concerns raised about the smell and noise given that the rear windows of abutting homes look right down onto the farm.

Figure 4.25.

Figure 4.26.

Although I have been a regular visitor to London for decades, the present research project of visiting the city's city farms takes me into regions of the metropolis far removed from those frequented by tourists. And so here I am, starting my survey by walking through long stretches of nondescript residential streets located near Hampstead Heath. Expecting a broad open space for the farm, I am shocked when I find myself actually walking right past its tiny entrance, which is nothing more than a gap in the wall sandwiched between two brownstones. Once inside, I am greeted by the pleasant scene of laughing toddlers interacting with squawking waterfowl. When the latter begin to drink the runoff of the morning's rain shower from the drainage rill, I have an urge to intervene and save them from the contamination I know them to be ingesting. However, I am soon cheered up again by an inquisitive hog who leans over a railing to give me a good sniff, no doubt in hope that I might have an apple or some other treat to share. I smile at the inventive whimsy of the mailbox repurposed to receive donations. But it is the overt compactness of the farm that is its most interesting feature: how tightly it is sandwiched between abutting homes along one side and the adjacent railway along the other. The use of a railway arch as a shady stable giving the animals a reprieve from the urban heat island effect is a particularly imaginative use of space.

4.2.2 Hackney City Farm (London, UK)

Inspired by the success of the Kentish Town City Farm, the neighbouring borough of Hackney created its own city farm in 1984 with the aim 'to give youngsters and adults the opportunity to experience farming' by allowing them the 'opportunity to get up close to a range of farmyard animals, see, smell and plant vegetables and other food plants; and learn new skills to live a healthier, happier life with a lower environmental impact' (www.hackneycityfarm.oc.uk). Hackney is noted for being one of London's most ethnically diverse and impoverished boroughs, and the city farm exhibits a rich history of land use: a market garden in the early nineteenth century, then a brewery and manufacturing hub during the first part of the twentieth-century, before use by a road haulage company in the 1940s. Transforming what had become a derelict and littered lorry park, the 'Hackney City Farm' is situated within the larger Haggerston Park, and proudly announces its presence through prominent welcoming signage [e4.2.2.1]. It has become a 'thriving popular community resource' with older buildings being renovated to house animals. Here are found all the usual barnyard suspects of such city farms, whose presence affords urban children the opportunity to enjoy and learn about goats, sheep, pigs, geese, donkeys, rabbits, in addition to emboldened chickens who give the impression that they run the place [e4.2.2.2]. Children also learn by forming intergenerational relationships with elderly volunteers who share stories and knowledge about countryside living and their own farm upbringing.

Hackney City Farm is situated in a densely populated neighbourhood where dwellers in apartment buildings and row housing can gaze down on the site [e4.2.2.3]. Once past a narrow alleyway (Figure 4.27) and paved forecourt [e4.2.2.4], the farm pervades a much more open space feeling with none of the claustrophobia of the afore-visited Kentish Town City Farm. Vegetable gardens are prominently located in attractive raised beds, some having mulch or stone walkways and wicker decorations [e4.2.2.5]. A flower-shaped sign states that 'The vegetables are grown by our hardworking garden volunteers, all the produce is either sold in our farm shop or in the café ... yum yum!' [e4.2.2.6]. Benches enable gardeners to enjoy a respite from their labours. Manure compost is stored onsite [e4.2.2.7], and a small roof-top apiary produces honey. The gift shop contains branded merchandise (mugs, tote bags, t-shirts) and all manner of children's toys (Figure 4.28). Here too visitors can buy fresh produce and honey from the gardens and beehives, the idea being that upon returning home children can consume food about which they have learned at the farm. Financial support is provided by a core grant from the municipality in addition to the Heritage Lottery Fund.

The farm provides successful mix of community and commercial enterprises. A programme of hands-on skill courses are offered and school groups are given guided tours each week. Licensed office workspace is offered to charities or is available to rent by other organisations. An award-winning 'Frizzante Café,' which advertises with its own prominently placed sign at the entrance to the farm [e4.2.2.8], promises rustic food, good wine, and candlelit shared seating while listening to live jazz and other music in the evenings (Figure 4.29). Hurst (2012), who studied the spatial use of the farm for her thesis, is critical that the café's prices are too high for the borough's underprivileged and that its operations are not tied to the farm's environmental and food system aims. Other enterprises include: a hub to rent bicycles and a location to host bike-related events, straw-bale building workshops, a pottery studio, fine arts courses, private animal handling and grooming sessions (at the steep price of about £50 per hour), workshops in creating energy conservation sculptures, and the Hackney Yoga Project, which is promoted as a safe and respectful space for refugee and asylum-seeking women to practice yoga. The farm's therapeutic and supportive programming, focusing on providing opportunities for self-betterment and empowerment of the challenged and marginalised, is particularly meritorious (Hurst 2012). Future plans call for increasing the growth and harvesting of onsite produce. An orchard has been planted in the abutting park which is intended for harvesting and educating about the potentials of growing food in cities. A recent addition (subsequent to my visit) has been the creation of a series of thematic

Figure 4.27.

Figure 4.28.

Figure 4.29.

nature trails in the woodland area. These are centred around a small pavilion that is designed to resemble a grain silo and constructed from recycled material. The site can be rented for private outdoor events and corporate workshops. Through these and other multifunctional endeavours, Hackney City Farm hopes to achieve greater financial independence while at the same time acting 'to improve the local environment, enhance community cohesion, [and] act as a hub of social enterprise,' all the while maintaining its primary mission of education.

Several times I've had to consult my map while walking from the Tube station through the uninspiring residential streets of Hackney. But it is not until a street sign is seen pointing the way to the 'Hackney City Farm' that I'm certain I'm in the right place. Simultaneously, I'm impressed that the latest use of the site for urban agriculture is considered to be permanent enough to warrant formal announcement on a municipal lamppost. Once inside the farm, I admire the effort employed to build the raised planting beds in shapes other than the typically boring rectangles or squares. I gaze up at the abutting apartment blocks and imagine how nice the view must be of the attractive garden from their windows, especially when accompanied by a morning cup of tea or an evening glass of wine. This makes me wonder too if a garden-view premium exists in terms of real estate prices for such homes. As it is the middle of the day, there is no jazz music is being played in the café. The location is nevertheless of sufficient interest for me to buy a cup of chai and to linger. Tables are occupied by hectic mothers trying to feed packed lunches to their children who seem to be more interested in dashing about, obviously still wired-up after a morning spent visiting and touching the various farm animals. Later, I'm puzzled by my visit to the farm shop. For whereas most of merchandise for sale is agricultural or natural history in theme, some, such as the plastic fire and dump trucks, racing cars, and tugboats, seem oddly non sequitur to the site's educational intent. Outside again, I bid adieu to the precocious chicken who has been following (or guiding?) me about, and make my way back to the Tube station from which to travel to the next city farm on my tour of such through London's boroughs.

4.2.3 Vauxhall City Farm (London, UK)

Located across the river from Westminster in the historic borough of Lambeth, the 'Vauxhall Pleasure Gardens,' as described in an historic sign [e4.2.3.1], were created in 1661 and reached their apogee of popularity during the Regency Period (Southworth 1941). There and then, well-turned-out Londoners and tourists would gather to listen to live music, examine art installations, and imbibe refreshments, all while promenading about pavilions, fountains, and formal gardens [e4.2.3.2]. The gardens, which figured prominently in famous novels of the time, such as *Vanity Fair*, were closed in 1859 and replaced by housing. By the middle of the twentieth century, the area had fallen into a state of dilapidation, having never fully recovered from the devastation of the Blitz. Since 1976, the site has been returned to park with broad lawns, recreation fields, and, in one corner, nestled beside a church (Figure 4.30) [e4.2.3.3], the 'Vauxhall City Farm'. The farm was established by local activists in 1977 on a parcel of land left vacant following demolition of buildings and now receives up to fifty thousand visitors a year (www.vauxhallcityfarm.org). The farm heralds its impressive visitor and programme statistics in a series of onsite displays [e4.2.3.4]. The popularity of the farm has led to numerous online videos posted by enthusiastic visitors, creation of its own Wikipedia page, and published articles by architects and social workers (Doron 2005; Pretty *et al.* 2013). Testimonials of volunteers are displayed in several onsite locations [e4.2.3.5] as well as being posted on the website, including, for example, the following: 'I was first interested in volunteering at the farm because I love animals! However, when you're at the farm, you quickly see how much value it adds to the community. It's an inclusive environment, filled with positivity, so not only is it a great place to be, but it's great to be able to contribute and support it.'

Vauxall City Farm's vision is to contribute toward building a city where wellbeing is emphasised through a wide range of educational, recreational, and therapeutic programmes that:

- Support children and disadvantaged people to boost their confidence and aspirations via the development of personal, social, life and practical skills.
- Promote emotional, mental, social and physical health and well-being.
- Develop environmental awareness and action.
- Strengthen community cohesion.
- Create enjoyment and recreational opportunities for those from our local and wider communities.

Animals are housed in paddocks beside the church [e4.2.3.6]. In addition to chickens, goats, ducks, rabbits, guinea pigs, sheep, ponies, and turkeys [e4.2.3.7], the farm also contains more unusual species such as chinchillas, rats, aviary birds, and alpacas [e4.2.3.8], the latter being extremely popular with visitors (Figure 4.31) [e4.2.3.9]. The farm purveys a message of sustainability by informing about composting and with buildings that are constructed of recycled materials, including rammed earth-filled tires [e4.2.3.10]. The latter are also imaginatively used to create planters [e4.2.3.11]. Of particular note are the wonderful views provided of the farm's verdant gardens [e4.2.3.12] and grazing livestock (Figure 4.32) contrasted against the flashy high-rise office buildings in this busy section of downtown London.

Figure 4.30.

Figure 4.31.

The farm has established partnerships with leading education organisations, and offers diverse programmes, which include meeting different animals, learning about wool, the 'micro world' of invertebrates, milking lessons, 'land to lunchbox' classes in growing and nutrition, summer and school break immersions, 'farmer for a day' experiences, a young farmers' club that meets on Saturday mornings to learn about caring for animals and exhibiting them at agricultural fairs, and various employment classes for teenagers [e4.2.3.13]. Other programmes are offered for adults and special needs individuals. A horticultural therapy group has as its mandate 'extending the joys of growing as a way for self-healing and mental wellbeing.' The riding centre features therapy sessions with ponies and horses. And a spinners group uses traditional techniques to spin wool from the sheep and alpacas wherein the material is dyed through use of plant extracts that are grown onsite.

A sign unabashedly announces the operating costs of the farm [e4.2.3.14]. Fundraising activities range from corporate donations to individual adopt-an-animal and legacy giving opportunities. The Vauxhall City Farm was successful in obtaining a national community fund prioritised toward children, ethnic minority groups, refugee and asylum seekers, older people, and individuals disadvantaged by social and economic change. The range of revenue-generating endeavours offered is as imaginative as it is broad. Taking advantage of the presence of trained veterinarians, people can board pets at the farm when they go on holiday. Farm animals can be hired for off-site use in professional films, photoshoots, and theatre productions, as well as for schools, birthday parties, and fairs. Egg-to-chick classroom incubators can be rented by schools [e4.2.3.15]. In addition to the riding school, 'pony parties' can be arranged for private groups. The 'Old Dairy Café' offers snacks, and the shop sells seed pollinator packs, birding kits, and gift vouchers for animal interaction activity packages. Two rooms are available for rent for meetings, functions, and events. Private tours and birthday parties can be hosted, with opportunities for guided interactions with different animals. The website does an admirable job toward fostering a sense of community: news updates are provided on animals that have been given names, short articles are presented on such topics as guerrilla gardening, the benefits of green spaces for dementia, and of cross-generational gardening, facts about much maligned rats, as well as personal stories from long-time workers.

In addition to a series of small raised planting beds [e4.2.3.16], the farm harbours allotment gardens (Figure 4.33) [e4.2.3.17]. Despite the bucolic tranquillity of the site, all is not well in this little sliver of urban Arcadia. Allotment gardens have been tended by locals since the farm's inception. However, the farm, as part of its push for education outreach, now wishes to use the gardens as a living classroom to teach children and families about the role of different plants in healthy nutrition. This change in land use has caused great umbrage in the neighbourhood, with long-time locals bemoaning the shift in the farm from a modest and community-oriented operation to a larger and more business-like venture (Brown 2020). They argue that growing in the new garden will now be done by the staff and visitors rather than by residents of the neighbourhood. As a result, the farm's ties to the local community will disappear. It is certainly a thorny issue. With the increased interest in locally produced produce and urban agriculture, there is a waiting period of well over a year to obtain an allotment in London. But at what point do the interests of the few – presently the allotments are tended by less than a dozen individuals – trump the pedagogic aims of the farm to educate the many?

Figure 4.32.

Figure 4.33.

Unlike similar sites I've visited on this tour of London, the Vauxhall City Farm is situated in a familiar area of the city, though I am somewhat embarrassed at having previously walked within a few blocks of its location without knowing of its existence. Once at the farm, I find the juxtaposition of the animals and plants with the surrounding buildings to be particularly appealing. The contrast between nature and human artifice is made memorable by hearing the sonorous sound of Big Ben wafting across The Thames. As this is the last city farm I am to visit on this trip, I stand transfixed, revelling in the capstone experience of being able to see farm animals prick up their ears when they hear the sound of that famous bell that is so quintessential of England's capital city. Later, when wandering about the farm, it is impossible not to break out in a big smile when watching children interact with the cute-as-a-button alpacas. Not so, however, when I take a photo of an allotment gardener who is piqued at the intrusion.

Fleeing from said fellow's withering glare, I return to the central courtyard area and encounter the piles of manure. As at other city farms, I wonder about the opinions held by the neighbours concerning this product of animal husbandry. It is not so much the smell that might be an issue, since it seems, at least on the day of my visit, to be restricted to within a few metres of the accumulations. But what about the hordes of associated flies that fill the air? The thing about flies is... well, they fly. I swat away some of the pesky insects which are hovering about my head and look up at the nearby apartment blocks while wondering if any residents therein are ever bothered by the flies when they sit on their balconies overlooking the farm.

4.2.4 Surrey Docks Farm (London, UK)

The 'Surrey Docks Farm' is located on the south bank of the River Thames in the greater London borough of Rotherhithe. It was established more than forty years ago on a derelict dockyard that was described as 'looking as mournful and miserable as a battlefield'. The initial goal was 'to try and raise livestock and produce food from what was regarded as wasteland' (www.surreydocksfarm.org.uk). The project began when an individual used the abandoned site to pasture her goat and turned a dilapidated building into a milking barn and henhouse (Peters 1979; Anstey 1992). Soon, local children were coming by to help and unemployed dock workers began to plant allotments, all of which led to the location to be referred to as 'The Farm'. When the original site at Greenland Dock was redeveloped for housing in 1986, the farm was relocated to a nearby hectare of former shipyards. Its new mandate was to provide opportunities for people to learn about farming and food production by becoming actively involved in the ongoing operations of the site. In addition to its own website, filled with useful information about history, programmes, and site management, the farm also has a Wikipedia listing. With more than fifty thousand visitors a year, it is perhaps no surprise that one can also view online videos filmed by parents of their children playfully interacting with the animals.

Today, the farm is a well-known and waymarked (Figure 4.34) [e4.2.4.1] destination beside the popular 'Thames Path' [e4.2.4.2], with stunning views offered of the cityscape of condos and office towers across the river at Canary Wharf and the Docklands (Figure 4.35). As with many of London's city farms, apartment blocks abut the agricultural buildings, food gardens, and animals (Figure 4.36). The latter include a herd of milking goats as well as sheep, cattle, pigs, donkeys, bees, and chickens [e4.2.4.3]. Waterfowl gather in an attractively designed pond located beside the riverside walk [e4.2.4.4]. This is very much a working farm, wherein visitors circulate around machinery, trucks, and bustling employees. Horticultural areas include an orchard, herb and dye gardens, vegetable plots from which fresh produce is sold to the public, a permaculture garden, and raised planting beds for both the community and children (Figure 4.35) [e4.2.4.5]. Facilities include a dairy milking barn, a working blacksmith's forge, an apiary, a shop, and adult education rooms. The juxtaposition of all this agricultural activity – especially the animals stabled in the paddocks or milling about in the central yard – in relation to the surrounding apartment blocks is striking (Figure 4.36). The farm also houses a Frizzante Café [e4.2.4.6] and studio space for pottery and yoga classes.

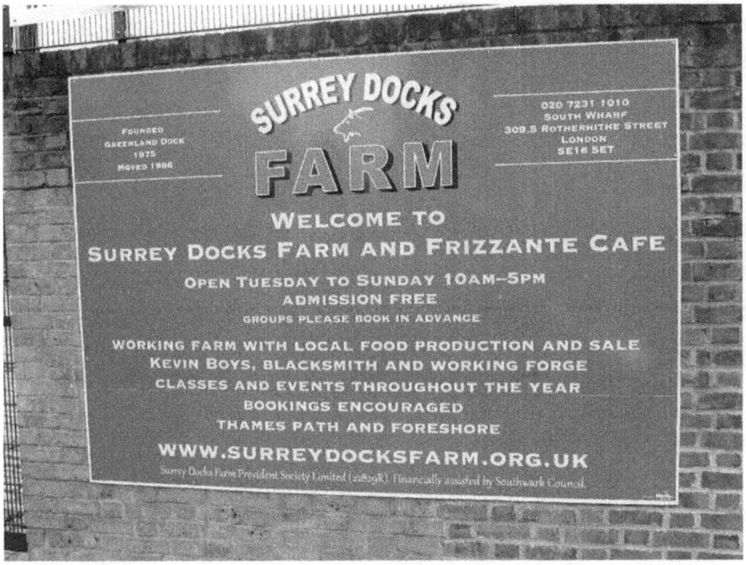

Figure 4.34.

Figure 4.35.

Figure 4.36.

Considerable attention has been paid to environmental education. The site operates a compost bio-digester, and buildings have solar panels and green eco-roofs, from which rainwater is harvested for irrigation [e4.2.4.7]. A small area has been set aside as an unmanaged, urban wild. Nettles and logs are kept for invertebrates, wildflowers are grown for pollinators, hedges are provided for birds, boxes have been built for bats, and the pond was created for amphibians. The farm has partnered with a local biodiversity organisation, and the website keeps tabs on unusual visiting species. One of the farm's most memorable design details is the inclusion of artistic latticing on the metal fences (Figure 4.37), as well as whimsical installations in several of the gardens [e4.2.4.8].

Figure 4.37.

Much has changed since my initial visit to the Surrey Docks Farm a decade ago. A trail has been established to highlight, through use of six information panels, the site's eventful history in terms of its progression from a shipyard, to a timber storage wharf, to a river ambulance service centre, to a quarantine facility, to a river fire station during the Blitz. Bronze sculptures of a family of pigs and a donkey commemorate the transference through the city streets of the original farm's livestock to the present location in 1986. The biggest addition has been the award-winning architectural renovation of several buildings (Beaumont 2020; Mairs 2020) in line with the general redevelopment in the neighbourhood, all part of the municipality's advertising of the area as being attractive for young families. A final improvement has been to allow access to the farm from a new entrance on the recreational Thames Path.

An extensive education programme exists based on the wish 'to improve our local environment, build community, and benefit people's health and wellbeing.' Schools from across London make use of the fully equipped classroom for classes on topics such as apple tasting and pressing, and animals and their babies. School tours are tailored to the National Curriculum, and a series of clubs have been formed for different age-groups to learn more about the workings of the site [e4.2.4.9]. Private 'farm animal experiences' can also be arranged for children or for special needs adults to allow intimate, hour-long, one-on-one encounters with goats, sheep, donkeys, ponies, rabbits, and guinea pigs. Adult workshops are provided on such topics as horticulture techniques, cooking, seasonal eating, herbal medicine, beekeeping, natural dyeing, and container and vertical growing. Large public gatherings occur at the time of spring and Christmas fairs in addition to an autumn harvest festival.

Surrey Docks Farm is managed by a registered charitable organisation, with funding provided from the Docklands Development Corporation and the National Lottery. Like some other city farms, such as Hackney, imaginative means are employed to boost the revenue needed to maintain operations: (1) manure and compost are sold to local gardeners; (2) the spiffy, new River Room, which offers attractive views of The Thames, can be rented for private functions; (c) a mobile farm [e4.2.4.10] can be hired for fairs, picnics, fundraising events, and even live nativity scenes; and (d) classroom incubators for hatching chicks can be rented by schools.

Having conducted considerable research on the health of Canadian lakes, I am tickled to see the sign 'Canada Water' that greets me as I emerge from its eponymous Tube station. The walk along the Thames Path is so enjoyable I almost forget that I am in Rotherhithe for a research purpose. Eventually I spot a city sign pointing the way to 'Downtown. City Farm. Business Centre.' I find the combination of wording so appropriate to my interests in urban agriculture, that I will later use a photo taken of the sign on the cover of my *Integrated urban agriculture* book (France 2016). I am enamoured from the moment I enter the Surrey Docks Farm. The charming details of the metal-sculpted leaves and animals worked into the fences is a reminder that incorporation of art need not be restricted to only flashy projects created by landscape architects, i.e. small details can, and often do, make large impressions. The presence of the green eco-roof and rain harvesting system is also a lesson that sustainable design should be part-and-parcel in all urban farms.

My visit provokes two strong, but contrasting, impressions. First, the positive one. Perhaps it is the large central courtyard filled as it is with animals happily milling about that causes me to appreciate, to a degree exceeding that experienced at other sites, the sharp contrast between the scene of farmyard rusticity and that of modern urbanity in the form of the surrounding apartment towers. Second, several negative ones. Once again when visiting such sites, I am struck by the unfortunate missed opportunity resulting from no information being presented about former land uses. Thus, in the absence of site interpretation about greater London's only city farm situated beside the river, most visitors, including those with whom I talk, have no idea about the rich maritime heritage of the location. Then, as I look out past the lovely pond to the proximal Thames Path, I am puzzled by the absence of an attempt to link the two together. I go as far to communicate these comments to several members of the staff. It is with much interest I later learn that both my issues have been admirably addressed in the latest iteration of the farm.

4.2.5 Mudchute Park and Farm (London, UK)

'Mudchute Park and Farm' is located in the London borough of Tower Hamlets' Isle of Dogs. The site had been derelict land arising from the dumping of dredging spoil when the nearby docks were constructed more than a century ago. Sadly, no interpretation of this fascinating site history is offered at the site (although information is provided on the official website, www.mudshute.org), an embarrassing absence for a location with an educational mandate. To obtain such information one must visit the nearby basin of the 'Millwall Outer Dock. 'Today, bordered by new housing development, this excavated basin is quiet and empty [e4.2.5.1]. Not so in the past, according to its interpretive sign [e4.2.5.2]:

> The early 1860s were prosperous times for heavy industry on the Isle of Dogs. The marshy pastureland, previously grazed by cattle, was being swallowed up by iron foundries and shipyards. In response to this economic boom, Millwall Docks were planned with visions of quaysides lined with shipyards and warehouses ... Only one dry dock was built were six had been planned.
>
> By the time the new dock opened in 1868, the Isle of Dogs first industrial phase was over. Engineering, manufacturing and food processing took over from heavy industry ... Neighbouring factories filled the air with aromas of spices, pickles, paint, chemicals and jam. More recently the empty quaysides have proved attractive to the printing industry with a number of London's major printers moving here form cramped buildings in Fleet Street.
>
> Inventors, with ideas to improve efficiency, were also attracted to the docks ... [One such] invented a new dredging system which rapidly removed silt from the docks onto a settling bank behind you called Mudchute. Once the smelly silt had settled and stopped 'wobbling like a jelly', as some locals recall, the Mudchute became a fertile allotment site, noted for superb rhubarb! The land is now home to Mudchute Park and Farm.

And so it is to Mudchute Park and Farm that I venture. Throughout much of the 1900s, the site existed as an unmaintained urban wild that was popular among birdwatchers, amateur naturalists, exploring children, and the homeless. When slated to be developed for high rise apartment blocks in the 1970s, the community protested and insisted that a 'people's park' be created instead [e4.2.5.3]. Farm animals and horses were introduced, trees and plants installed by volunteers and teams of corporate volunteers, and local schools were encouraged to use the site for natural history lessons. Today, the site includes a local nature reserve, the farm, an education centre, wildlife ponds, allotment gardens, and a park with recreation fields, a fitness area, a bowling green, and formal gardens [e4.2.5.4]. The site is governed by a board of trustees of stakeholders representing varied local interests, and has 'steadily built a reputation for providing a variety of educational and leisure activities on a London-wide basis' based on the following objectives (www.mudchute.org):

- to preserve the natural environment of the Mudchute area;
- to promote the opportunity of recreation and leisure on the Mudchute for local people, especially the young;
- to advance the education of local people about countryside activities;
- to promote animal welfare through hands-on engagement with the Farm;
- to provide education and training for young people;
- to promote conservation both of the natural and the built environment.

Situated on 13 hectares in the heart of east London, Mudchute Park and Farm has the advantage of space absent from most other city farms in the capital [e4.2.5.5]. Indeed, it is the largest city farm in Europe. Children can directly interact with animals in the farm courtyard, including horses [e4.2.5.6] which are available for riding lessons through the equestrian centre. A series of dollhouse-sized dwellings along a 'residential' alleyway provide habitation for smaller animals, identified with signs such as 'Ferret Square,' 'Guinea Mews,' and 'Ducks Avenue' [e4.2.5.7]. But it is the opportunity to observe various species of ungulates in the large grazing fields bordered by wooden fences that is what is most special about the farm [e4.2.5.8]. This is particularly the case with respect to the wonderful views provided of the nearby office towers and expensive condos contrasted with the animals (Figure 4.38). Elsewhere onsite, a series of allotment gardens are diligently tended by locals [e4.2.5.8].

Figure 4.38.

The Mudchute Farm is an extremely popular destination for student education. The Muddy Boots Nursery [e4.2.5.9] functions as an after-school club. During the day, visiting school groups and preschoolers, when not taking lessons inside a classroom crammed full of teaching aids [e4.2.5.10], enjoy the outdoors either under supervision or as 'free range' individuals (Figure 4.39). Given the expansiveness of the site [e4.2.5.3], use is made of directional signposts with simple illustrations of animals as navigation aids [e4.2.5.11]. The Mudchute Kitchen, once run by the omnipresent Frizzante organisation, but now by a private family, serves soups, homemade cakes, gelato, coffee, and tea for breakfast and brunch (Figure 4.40) [e4.2.5.12]. A small shop sells candy snacks and generic, animal-themed toys and uninspiring souvenirs (Figure 4.41).

Figure 4.39.

Figure 4.40.

Figure 4.41.

Operating as a registered charity that solicits sponsorship donations [e4.2.5.13], and due to being well connected to public transport, Mudchute Park and Farm generates additional income through an extensive events programme that enables it to operate as a free resource to the community. Conference rooms can be rented for corporate meetings or training workshops, as can fields for specific recreational pursuits or group picnics. The kitchen and courtyard can be booked for private and corporate dining or summer parties and BBQs. Other special events which can be hosted include birthdays, christenings, Christmas parties, exhibitions, murder mystery evenings, award dinners, product lunches, team building days, and fêtes associated with Easter, Christmas, and Halloween. Adult courses are offered on topics such as keeping chickens, building drystone walls, and working with stained glass.

The most impressive aspect of visiting Mudchute Park and Farm comes from appreciating the vastness of the site and knowing that everything under one's feet is land that was artificially created more than a century ago with dumped spoil from the nearby dock excavations. The views past the rustic wooden fences and across the rolling fields of green toward the shiny glass towers of the Docklands redevelopment project are stunning. Rarely have I ever seen such a marked contrast between country and city juxtaposed in such close proximity. Despite the drizzle I find myself pausing to appreciate the view again and again and again. I'm reminded of the famous landscape art project of 1982 by Anges Denes, '*Wheatfield – A confrontation*,' in which she planted and harvested a field of grain in lower Manhattan within the shadow of the towering skyscrapers (see France 2016). There is also a memorable auditory component to my experience. My own walkabout over, I tag along in the wake of a gaggle of young children and their supervisors, enjoying the former's surprised exclamations and loud laughter when they spot the llamas or when a precocious goat leaps upon a mound right in front of them. Later, in the kitchen, I eavesdrop on exchanges between mothers while they serve lunch to their rambunctious children who seem frustrated at having to eat when they could be outside seeing more animals.

4.2.6 Spitalfields City Farm (London, UK)

Located in London's East End in the borough of Tower Hamlets, 'Spitalfields City Farm' is close to the historic square mile of The City, and thus is in one of the most urbanised farm sites in the greater metropolis. Notable landmarks in the neighbourhood include the Old Spitalfield Market, in which food products were sold as early as the eighteenth century, a Hawksmoor church of architectural significance, and Brick Lane and The Ten Bells pub, known for being frequented by victims during Whitechapel's 1888 'Autumn of Terror' [e4.2.6.1]. The farm is situated beside both a church and a playground, and is overlooked by apartment buildings and office blocks [e4.2.6.2]. It was established in 1978 on a derelict site that had been a former railway goods depot by a group of locals who had lost their allotment gardens to development and were eager to transform the 'wasteland'.

Spitalfields City Farm has become a popular destination for those seeking a reprieve from city life, with one journalist, for example, waxing on about the site being an enclave of peace and tranquility in the midst of the 'encroaching chaos' of 'gentrification and behemoth buildings' that have 'shot up like weeds' in the vicinity. He goes on to state that it is the proximity of the farm to the financial skyscrapers of The City that makes for such an astonishing surreal experience: 'Imagine that you're in the centre of London surrounded by concrete, steel and smoke. All of a sudden you look to your left and what's that? Yes, a drove of donkeys frolicking through the pasture'.

Scholars have also been interested in the farm, using it for their research. Loukola *et al.* (2020) studied nest-site selection of solitary bees; McCartney and Maliphant (2018) used the farm as a case study in how not-for-profit organisations can contribute to urban regeneration; and Liu *et al.* (2017) investigated the use of the farm's allotment gardens for community building and as therapy spaces. One interviewee as part of the latter research project commenting that 'I do feel a strong bond to the animals ... They are like my family ... if I am stressed out I go and cuddle a goat.' And a handful of papers and theses have been produced focused on the farm's managerial uses of digital technology.

The farm's goals (www.spitalfieldscityfarm.org) are:

- To give people the opportunity to become members of their community and encourage participation in the development of the farm.
- To explore and promote the benefits of healthy and sustainable lifestyles.
- To sustainably manage natural ecosystems for the benefit of local wildlife and as a resource for education and recreation.
- To provide children and young people with a safe, stimulating and inclusive space in which to play and to connect with nature.

Bold signage announces one's arrival at the farm (Figure 4.42) [e4.2.6.3]. Visitors make use of a map and signposts to navigate about the complex site [e4.2.6.4] (Figure 4.43), which, although only a half a hectare in size, contains many elements of interest. Signs immediately inform about the need to wash one's hands when touching animals and that food cannot be taken into the farm, and also display children's paintings [e4.2.6.5].

Figure 4.42.

Small paddocks house the animals characteristic to such city farms: sheep, rescue hens, ferrets, donkeys, cattle, and pigs [e4.2.6.6], in addition to ducks milling around a fenced-in pond [e4.2.6.7]. Cute signage for children helps them to distinguish between sheep and goats, while posted fact sheets for adults inform about rare breeds [e4.2.6.8]. An 'activity pack' can be picked up which contains a small brochure for children that contains an identification key of rare breeds at the farm that can be coloured, as well as a series of quizzes and games. Other nearby signs mention the costs involved in caring for the animals and solicit for donations [e4.2.6.9]. A handful of animals are mentioned by their individual names on the farm's website, which is set up like an online dating page wherein descriptions are given of each's 'hobbies,' such as searching for treats or encouraging back scratching. Two exotic New Zealand pigs are humourously named 'Holmes' and 'Watson' due to their inherent intelligence.

Considerable greenspace exists on the farm. Verdant plots team with bountiful produce [e4.2.6.10], and in one case (Figure 4.44) [e4.2.6.11], a sign informs about the plot's creation and purpose [e4.2.6.12]: 'This juvenile [i.e. recently planted] garden was still covered in concrete in March 2013. With excellent design, minimal cost & minimal environmental impact it has been recovered with 4 tonnes of compost diverted from landfill and nitrogen fixing trees. In a few years the carbon negative and flood resistant drought resistant forest area will be even more productive than the veg plot'. The emphasis placed on creating a miniature woodland area for urban wildlife is commendable. Raised beds of edibles are provided for instruction, including a special area set aside for a garden club [e4.2.6.13] and a community garden contained within planting boxes [e4.2.6.14].

A tea hut provides snacks (Figure 4.45) [e4.2.6.15]. Wood is used to create sheltered walkways and work areas (Figure 4.46) [e4.2.6.16]. The view contrasting the rusticity of the farm's treehouse behind which a neighbouring office tower is being constructed is dramatic [e4.2.6.17]. No big, fancy buildings exist on this city farm. Instead, its maintenance area contains a small wooden office surrounded by potted plants and a set of hoop-houses filled with nursery plants, some of which are replanted onsite while others are offered for sale (Figure 4.47) [e4.2.6.18].

Figure 4.43.

Figure 4.44.

Figure 4.45.

Figure 4.46.

What makes Spitalfield City Farm so impressive is its stellar job in educating about non-food-related issues relating to urban biodiversity and sustainability. A bike lane leads to the site [e4.2.6.19] from where access is given by a green lane lined with eco-pavers to enable stormwater infiltration and bordered with an allée of crumbling posts and logs repurposed as habitat for invertebrates [e4.2.6.20]. A wildlife pond that is home to newts is nestled into unkempt shrubbery [e4.2.6.21]. Near a beehive, a 'bug hotel' is composed of hundreds of pieces of wood to provide habitation for solitary insects [e4.2.6.22]. One perimeter fence has been planted with an environmental hedgerow (Figure 4.48) and is accompanied by a sign [e4.2.6.23] explaining its purpose:

> This juvenile [i.e. recently planted] biofriendly / pollinator friendly / nitrogen fixing / carbon fixing / super food drought resistant / flood resistant hedge will protect this garden from prevalent South Westerlies, increase veg production, prolong the growing season, provide birds to control slugs & snails, increase local food production, increase food sovereignty, promote self-reliant active lifestyle choices, diminish lifestyle disease ...

Figure 4.47.

Figure 4.48.

A 'butterfly bank' is sandwiched between another perimeter fence and an exclusion railing, with a sign stating that these plants are for wildlife and not farm animals [e4.2.6.24].

Spitalfields Farms' attention to site aesthetics is also meritorious. An artistically designed wetland tub is surrounded by eco-pavers and potted emergent plants (Figure 4.49) [e4.2.6.25]. Cheerful, brightly-coloured signs occur in several locations [e4.2.6.26]. The use of a red phone booth as a weather shelter for displaying fair ribbons is a cute addition (Figure 4.50). A lovely totem pole features carvings of farm animals and children [e4.2.6.27]. Of special note is an art project, 'The Talking Quilt,' which was created in 2011 as part of the Connected Communities research project to showcase the role of technology in building local interactions through active art. During a series of workshops at the farm more than eighty participants integrated hand-printed and painted panels of fabric into which were embedded audio files that could be activated by touch. The final product, which has gone on to be displayed throughout the East End of London, including in a museum, is therefore an interactive, 'talking' piece of art that is a container of experiences and a reservoir of memories (Weir 2011):

> Each audio file is edited from unstructured interviews with participants talking about their experiences, memories, and ideas about food, food growing, and community in East London. To hear the quilt, the viewer/listener must get close to and touch it to trigger and overlay the audio narratives. It is possible to switch on and overlay up to 150 voices in random sequence, enabling the listener to open up an endless number of discussions and conversations about food, community and the role of the City Farm. The quilt doesn't work as a typical archive, its narrative is ono-linear, produced collaboratively, and is shaped by the actions of the viewer. Moreover, a new connection is created between the makers and the viewers' of the quilt, and a new link between Spitalfields City Farm and the environment in which the quilt is being exhibited. As the quilt is 'played' it becomes a testament to the diversity of opinion about what community is and can be, our shared humanity through our need to be nourished, and our ability to find common ground through the act of growing and eating food.

Figure 4.49.

Figure 4.50.

Attention is given to providing visitors places for relaxation in the form of tables, benches, and solitary seats around the farm [e4.2.6.28]. 'It's a nice place to almost not interact with each other,' one interviewee commented, continuing 'You don't have to; you just sit and interact with the place and with nature ... people don't feel the pressure to chat, but they probably interact in a quieter way' (Liu *et al.* 2017).

When not taking lessons inside in a comfy classroom (Figure 4.51), children can participate in nature-themed activities designed to make them become inspired and creative about the natural world [e4.2.6.29]. 'Families Go Wild' programmes have been created for entire families. Intensive youth work experiences provide opportunities to look after animals, tend gardens, and contribute to site maintenance. Interactive guided tours exist to 'meet, stroke and feed a range of furry friends whilst learning interesting facts along the way,' such as about best techniques in animal welfare and historical changes in husbandry practices. Adult workshops are offered on making paper pots and the benefits of creating gardens based on the concept of 'companion planting' wherein flowers and herbs are intermixed with edibles as a natural way to attract pollinators and dissuade unwanted insects (see Chapter 5). A mobile farm transports animals to offsite functions [e4.2.6.30], including the staging of fairy tales come to life, as for example, 'Little Red Hen' and 'Three Billy Goats Bluff'.

At one time a shop used to sell items such as animal magnets and badges as well as soaps made from goat milk or lavender and other herbs. Capitalising on the farm's proximity to the financial hub of The City, corporate team days are hosted, which, in 2017, contributed six thousand hours to the upkeep and general maintenance of the site. Weddings and private parties are hosted as well as 'company away days' through the hired use of a safari tent or yurt [e4.2.6.31]. It is even possible to rent the entire farm with catering, bar, music, and entertainment provided. Like all city farms, revenue generation is a constant worry. Lack of funding almost forced the farm to close in 1987,

Figure 4.51.

and for much of its existence it has faced the threat of development. Recently, volunteers have accused the Board of Trustees for not being transparent about cutting programmes for youth in the face of ongoing financial struggles (Koneh 2020).

I find myself completely bowled over by the attention given by the farm to 'ancillary' environmental concerns. Both head and camera are twisted every which way so to ensure that nothing is missed. I find a small plastic container and fill it up with water from the ornamental wetland. I then walk over to the green pathway and, as a group of families arrives, demonstrate the infiltrative capabilities of the eco-pavers. I smile when several parents thank me, thinking that I am a staff member. I crawl over some logs and, on hands and knees, part the cover of plants to examine the wildlife pond. There, I spend half an hour watching several frogs and a newt in the water. Later, I cautiously examine several openings in the bug hotel but am disappointed to not find any of the cavities occupied. I wonder if it is perhaps too early in the spring for such. Elsewhere, several minutes are spent watching butterflies visiting flowering plants in the perimeter hedgerow.

Observational tour complete, I do one final perambulation, this time spending moments in contemplation while sitting upon each of a half a dozen seats and benches situated throughout the farm. The site is indeed a peaceful refuge from the hectic whirl that exists outside the green enclave. It is therefore with some trepidation that I take my leave of Spitalfield City Farm and re-enter the 'madding crowd' of downtown London. The bustle of Old Spitalfields Market is overpowering, and so I jostle past the hordes to have a pint at The Ten Bells which is much less crowded than I remember from a previous visit of decades ago when it had controversially marketed itself to tourists by capitalising on its notorious infamy of being associated with violence against women. There, I read the fact sheet provided me by a member of the farm's management group about The Talking Quilt project. By doing so, I become even more impressed by the ingenuity of the Spitalfields City Farm as well as feeling unfortunate at not being able to participate in the haptic experience of touching and listening to the quilt as it was presently out on tour. What a remarkable place, I think as I drain my pint and move on to visit the next city farm.

4.2.6 Case study lessons

- Educational city farms can exist alongside or in city parks (Hackney City Farm, Vauxall City Farm, Mudchute Park and Farm), be squeezed into the densest of urban neighbourhoods (Kentish Town City Farm); they can also be situated on derelict land that has been regenerated from previous usage (Kentish Town City Farm, Hackney City Farm, Surrey Docks City Farm, Mudchute Park and Farm, Spitalfields City Farm).

- Additional revenue can be obtained through the sale of manure and compost (Kentish Town City Farm, Surrey Docks City Farm), the rental of facilities or the entire site as venues for corporate or private functions and events (Hackney City Farm, Vauxall City Farm, Surrey Docks City Farm, Mudchute Park and Farm, Spitalfields City Farm), and from the boarding of pets (Vauxall City Farm).
- Visitor experience can be enhanced by presence of a café and shop (Hackney City Farm, Vauxall City Farm, Surrey Docks City Farm, Mudchute Park and Farm, Spitalfields City Farm), in addition to attention paid to site aesthetics in the form of professional or amateur art installations (Kentish Town City Farm, Surrey Docks City Farm, Spitalfields City Farm) and inclusion of sequestered locations for relaxation (Spitalfields City Farm), as well as from the presence of rare or exotic animals (Vauxall City Farm, Mudchute Park and Farm) and interpretive information about the history of the site (Surrey Docks City Farm).
- A city farm's outreach and educational message can extend beyond the physical bounds of its physical space through a mobile farm which can be hired to 'loan' animals for offsite events (Vauxall City Farm, Surrey Docks City Farm, Spitalfields City Farm).
- In addition to providing opportunities to interact with animals in a sort of 'petting zoo' and offering a suite of instructional courses, city farms enable urbanites to see and in some cases to learn about vegetable growing through gardens created for pedagogy or therapy, in addition to simple allotments (Hackney City Farm, Vauxall City Farm, Surrey Docks City Farm, Mudchute Park and Farm, Spitalfields City Farm).
- City farms can educate about wider topics related to sustainable design through innovative architecture and landscape architecture (Surrey Docks City Farm, Spitalfields City Farm), and through the adaptive reuse of materials in green construction (Vauxall City Farm).
- Areas set aside as natural habitat to educate about and foster urban biodiversity are an important pedagogical role played by city farms (Surrey Docks City Farm, Spitalfields City Farm).

References

Ableman, M. 2016. Street farm: growing food, jobs, and hope on the urban frontier. Chelsea Green Publishing, White River Junction, VT, USA.

Anonymous. 2006. Seeds, soil, and stories: a pilot study of community gardening in southeast Toronto. Urban Gardening Research Opportunities Working Group (UGROW). Centre for Urban Health Initiative. University of Toronto, Canada.

Anonymous. 2014. Brightmoor restore the moor report. Community Development Advocates of Detroit, Detroit, MI, USA.

Anonymous. 2020. D-Town farm: a modern-day manifestation of an ongoing historical battle for equality. Available at: www.historyengine.richmond.edu.

Anstey, J. 1992. The Surrey Docks City Farm in London. Environmental Conservation 19: 73-74.

Anthropoulou, T. 2016. Urban gardening in Greece. Not just food but a sense of belonging and self-esteem. In: Keshavarz, N. and R. Fox-Kämper (eds.) Urban allotment gardens in European cities: future, challenges and lessons learned. European Science Foundation, Strasbourg, France.

Beaumont, E. 2020. Holding on to Arcadia at Surrey Docks Farm. Architect's Journal, 26 May.

Brown, K. and S.D. Brush. 2018. Edible Providence: Integrating local food into urban planning. In: Cabannes, Y. and C. Marocchino (eds.) Integrating food into urban planning. University California Press/FAO, Rome, Italy.

Brown, J. 2020. The battle at Vauxhall City Farm. Available at: www.wicketleeks.riverford.co.uk. 28 May.

Cobb, T.D. 2012. Reclaiming our food: how the grassroots food movement is changing the way we eat. Storey Publishing, North Adams, MA, USA.

Cockrall-King, J. 2012. Food and the city: urban agriculture and the new food revolution. Prometheus Books, Buffalo, NY, USA.

Collins, P.A., S.J. Resendes and J.R. Dunn. 2014. The untold story: examining Ontario's community health centres' initiatives to address upstream determinants of health. Healthcare Policy 10: 14-29.

D'Entremont, Y. 2019. How a new Halifax urban farm is planting deep common roots. The Halifax Star, 13 October.

Doron, G. 2005. Urban agriculture: small, medium, large. Architectural Design 75: 52-59.

Drudge, E. 2019. Cultivating community. Ecology Action Centre, Halifax, Canada.

France, R. 2004. Blurring the border between nature and culture. In: Vince, S. W. (ed.) Forests at the wildlife-urban interface: conservation and management. CRC Press, Boca Raton, FL, USA.

France, R.L. (ed.) 2016. Integrated urban agriculture: precedents, practices, prospects. Libri, Farringdon, UK.

Gavriilidou, E. and M. Ritou. 2016. KIPOS3 project: a garden for Thessaloniki. In: Keshavarz, N. and R. Fox-Kämper (eds.) Urban allotment gardens in European cities: future, challenges and lessons learned. European Science Foundation, Strasbourg, France.

Gavriilidou, E., H. Kleinmann, E. Ourelidou and S. G.Z. Zafeiropoulos. 2015. Urban agriculture in Thessaloniki: an academic project meets reality. Ri-Vista 02: 60-70.

Gillis, J. 2013. Common Roots harvest benefits Parker Street Food Bank. Available at: www.cdha.nshealth.ca.

Hurst, K. 2012. Designing for food, community and multi-use space: lessons learned from grass-roots urban agriculture. M.L.A. Thesis. University of Guelph, Ontario, Canada.

Johnson, L. 2010. City farmer: adventures in urban food growing. Greystone Books, Vancouver, BC, Canada.

Kirk, L.M. 2006. Growing community. The Providence Journal, 22 October.

Kolokouris, O. 2015. Urban gardening in Greece – A new form of protest. Green European Journal 11, 3 July.

Koneh, N. 2020. Volunteers protest over fears of changes at city farm. East London Lines, 18 November.

Konthothanasis, G. 2017. Social practices of urban agriculture in the metropolitan region of Thessaloniki. Environmental Sciences 38: 666-673.

Ladner, P. 2011. The urban food revolution: changing the way we feed cities. New Society Publishers, Gabriola Island, Canada.

Levkoe, C.Z. 2006. Learning democracy through food justice movements. Agriculture and Human Values 23: 89-98.

Liu, P., P. Gilchrist, B. Taylor and N. Ravenscroft. 2017. The spaces and times of community farming. Agriculture and Human Values 34: 363-375.

Loiselle-Gascon, M. 2013. Nature and community experience in community gardening. M.A. Thesis, York University, Ontario, Canada.

Loukola, O.J., E. Gatto, A. Cttijar-Islas and L. Chittka. 2020. Selective interspecific information use in the nest choice of solitary bees. Animal Biology 70: 215-225.

Louv, R. 2008. Last child in the woods; saving our children from nature-deficit disorder. Algonquin Books, Chapel Hill, NC, USA.

Mairs, J. 2020. Surrey Docks Farm: renovation of formerly derelict city farm buildings breathes new life into riverside Rotherhithe. The Evening Standard, 16 December.

McCartney, M. and A. Maliphant. 2108. Localism, community involvement and the social economy. What are the opportunities and challenges facing the not-for-profit sector in term of regeneration? Journal of Urban Regeneration and Renewal 11: 343-354.

Melrose, J. 2013. Common grounds. Available at: www.thecoast.ca.

Moran, N. and J.L. Fernandez de Casadevante. 2016. Urban agriculture springs up in Greece. City Farmer Greece. Urban Agriculture Magazine 28: 31-34.

Nardi, M. 2014. Cultivate London, where agriculture and business merge. Available at: www.massimonardi.it.

Oureilidou, E. 2018. Planning the cultural and social reactivation of urban open space in Greek metropoles of crisis. In: Egoz, S., K. Jorgensen and D. Ruggeri (eds.) Defining landscape democracy: a path to social justice. Edward Elgar Publications, Cheltenham, UK.

Partalidou, M. and T. Anthopoulou. 2016. Urban allotment gardens during precarious times: from motives to lived experiences. Sociologia Ruralis 20: 1-18.

Peters, H. 1979. Docklandscape. Private Publication.

Pretty, J., C. Wood, R. Bragg and J. Barton. 2013. Nature for rehabilitating offenders and facilitating therapeutic outcomes for youth at risk. In: South, N. and A. Brisman (eds.) Handbook of green criminology. Routledge, London, UK.

Pride, T. 2016. Resident-led urban agriculture and the hegemony of neoliberal community development: eco-gentrification in a Detroit neighbourhood. Ph.D. Thesis. Wayne State University.

Saul, N. and G. Hewitt Smith. 2011. Why The Stop matters. 2011 annual report. The Stop Community Food Centre, Toronto, Canada.

Smith, H. 2019. How are we with animals: understanding connection with nature in urban settings through multispecies ethnography. Ph.D. Thesis. University of London, UK.

Southworth, J.G. 1941. Vauxhall Gardens: a chapter in the social history of England. Columbia University Press, New York, NY, USA.

Tarajko-Kowalska, J. 2017. Functions and basic principles for spatial organization of educational and recreational farms with animals in European cities. Technical Transactions 11: 53-68.

Vogel, S. 2008. Detroit (re)turns to nature. In: France, R.L. (ed.) Handbook of regenerative landscape design. CRC Press, Boca Raton, FL, USA.

Wakefield, S., F. Yeudall, C. Taron, J. Reynolds and A. Skinner. 2007. Growing urban health: community gardening in south-east Toronto. Health Promotion International 22: 92-101.

White, M.M. 2011. D-Town farm: African American resistance to food insecurity and the transformation of Detroit. Environmental Practice 13: 406-417.

Weir, M. 2011. Connected communities through food: the development of community supported agriculture in the UK/The talking quilt: 'A moment in community'. Available at: www.spitalfieldscityfarm.org.

Yuen, J. 2014. City farms on CLTS: how community land trusts are supporting urban agriculture. Lincoln Institute of Land Policy Land Lines, April.

Chapter 5
Healing art: beautifully grown food

5.1 Art-filled gardens

Anecdote

Personal experiences of visualising the seasonal round of medieval farming. As a field ecologist I have always felt fortunate that my work is dictated by the progression of the seasons. How sad it must be to reside in the north-temperate zone and have an office or other type of indoor job whose day-to-day activities are completely dissevered from the rotation of the seasons, being exactly the same in the middle of August as they are in the middle of February.

It is a few months before the start of the pandemic and I am in the Museé de Cluny in Paris for the first time since my initial visit there nearly four decades before. Still, I find myself just as awestruck upon viewing the famous unicorn tapestries a second time. In the gift-shop I am happy to find what I have been in search of for years: a collection of postcards comprising paintings from an illuminated manuscript – in this case the fifteenth century 'Les Très Riches Heures de Duc de Berry' – that illustrate the monthly cycle of farming activities. Once back home in Canada, I take the collection of cards, which had cost less than eight euros, and have them scanned and printed with the enlarged images mounted on gold-framed plaques, a process that costs more than three hundred dollars. But I have that which I've been desirous of: a changing wall calendar, each month a beautiful painting illustrating another aspect of the seasonal round of medieval farming in the continent of my ancestors.

Each of the twelve 'Labours of the Month' depicts details of the activities of peasants and the delights of nobility taking place in a foreground behind which looms an imposing chateau. Surmounting each illustration is a hemisphere showing a solar chariot and the degrees of the zodiac. The activities shown include livestock rearing and slaughtering, wood gathering, field tillage and crop planting, picnicking, hunting and falconry, forays *en plein air*, cutting and piling of hay, harvesting grain with scythes, picking grapes, winter seeding, and feasting. The colours are radiant and the artistry is superb. And so it is with great pleasure that I now look forward to switching one plaque with the next in the sequence as the months progress, celebrating the seasonal round through the beauty of medieval art.

DOI 10.3920/978-90-8686-938-1_5, © Wageningen Academic Publishers 2022

The book

The cultivated landscape: an exploration of art and agriculture is an interesting collaboration by Craig Pearson, a professor in agriculture and environmental policy, and Judith Nasby, a curator and art gallery director. In the same spirit as Emilio Sereni's *History of the Italian agricultural landscape* (which, however, escapes detailed mention by the authors), the present book is designed to show how agricultural practice has changed over time. In this regard, the book would have perhaps been better subtitled something along the lines of 'An Illustrated History of Western Agriculture'. Whereas, as discussed previously (France 2020), Sereni's book reads like an extended power-point presentation, with small blurbs of text closely coupled to a series of individual paintings, *The cultivated landscape* contains long sections of rambling text unsupported by any images; and when the latter do occur, the associated parings can occasionally come across as being forced. There is an inescapable impression that the text was created first, into which augmenting images were then found and inserted by the art historian, rather than the other way around as was the methodology used by Sereni in his classic. Nevertheless, the scope of Pearson and Nasby's book is admirable for being truly catholic in that it spans three continents and more than a millennium, and deals not only with the pastoral landscapes signified by its title, but also with the wider topic of food systems in general. And it is for this reason that the book is noteworthy since it is a rare attempt to highlight the compelling visual component of agriculture.

Chapter One provides an overview of agriculture in Europe from the Middle Ages to the Renaissance. Bucolic scenes of peasants toiling away in fields of grain, rustic farmhouses nestled into tranquil valleys, and livestock wading in lovely streams are shown from paintings, engravings, and illuminated psalters. Although the book's emphasis is primarily on presenting an historical overview of agricultural topics, with the illustrations playing a supportive role, the much more interesting and novel aspect is to turn things around. Doing so, one comes to quickly realise just how influential Europe's agrarian roots were in terms of fostering its medieval and early-modern artistic creativity. Additionally, lithographs from later times that depict villa farms and architectural drawings of estates highlight the concepts of landscape designers.

Chapter Two takes the story to North America and Australia with presentation of plans for the early colonial settlements, quaint folk-art renderings, modern paintings, and bird's eye views of agricultural campuses, juxtaposed with paintings depicting Aboriginal peoples as hunters and harvesters. However, there is a dearth of images about agriculture's dark history in terms of widespread forest clearance and slave plantations. Chapter Three focuses on productivity and it is here where a series of interesting paintings is shown of the struggles of the landless poor and of farmers affected by war and poverty. Happier images are also presented of agricultural fairs with their livestock judging contests and of displays of bountiful harvest. The type of artistic creations shown are as diverse as is the subject matter, including magazine advertisements for farm machinery, Cold War propaganda posters, and illustrations of foreign development and the Green Revolution. This trend in showcasing artistic diversity continues in the next chapter which covers the environmental and social consequences wrought by intensive agriculture. Here are shown land art installations of devastated fields, digital prints of unhealthy and fantastical livestock, sculptures of disease-ravaged herds, and documentary photographs of social unrest. Again, the manner in which visual artists have responded to agriculture's deleterious effects is impressive. Chapter Five's text on sustainability is largely unsupported by accompanying illustrations and is overly preachy. The following chapter which deals with emerging concerns of the new millennium is much more interesting in demonstrating how artists have represented GMO 'Franken-foods' and industrialised food-processing plants. The final chapter's text is an overly romantic homage to earlier forms of agriculture, the saving grace being inclusion of art by Indigenous Peoples. The take-away message from the book is that agriculture and art, whatever their respective forms, need not be regarded as strange bedfellows.

Sites

5.1.1 Dalston Eastern Curve Garden (London, UK)

Although the role of artists in raising environmental consciousness is well established in the fields of innovative stormwater management (France 2005) and landscape restoration design (France 2008), their role in helping to bridge gaps within communities and between stakeholders in urban agriculture is much more limited (Wilde and Meier 2016). In point of fact, gardens are often ideal sites for artistic installations (Crouch 2003).

Dalston, a neighbourhood in the one-time gritty borough of Hackney in the East End of London, is being transformed into an artistic cultural centre for the greater metropolitan region. In addition to an influx of young artisans from across the United Kingdom, and consequent suggestions of incipient gentrification, the community still maintains its original racially diverse, working-class roots. Post-industrial land is as abundant in the area just as formalised, public greenspace is scarce. A pocket of land along the major road in the neighbourhood which has long hidden behind wooden fencing of a section of the abandoned Eastern Curve railway line provides a location for an interesting urban agriculture project. In 2010, following discussion among local residents and community cultural organisations, the site was opened as the 'Dalston Eastern Curve Garden' (www.dalstongarden.com). Capitalising on the linear nature of the strategic location, the small garden was envisioned as part of a greenway linking the Dalston Lane area to the Kingland Shopping Centre, itself built on the storage-yard of the former railway. In this respect, the garden and urban forest, designed by landscape architects, is a perfect example of Viljoen's (2005) concept of 'continuous productive urban landscapes'. Today, the space operates as an urban wild, a food garden, and as a park for relaxation and community-building activities. A *Lonely Planet* online posting refers to the garden as being 'typical of the kind of grassroots regeneration' occurring in Dalston. The site has received extremely brief mention in a few scholarly publications (Breitbart 2013; Polinna 2013; Vega *et al.* 2021).

A much-loved mural [e.5.1.1.1] near the entrance (Figure 5.1) bears witness to the legacy of community artists in the neighbourhood. Like Toronto's Artspace Wychwood Barns (see Chapter 7.1), an abutting building houses artists, studios, and a gallery. A boardwalk, constructed from old railcars and thus alluding to the site's history, runs along the back wall of the studio. This border of the tiny 0.2-hectare site is covered with artwork [e5.1.1.2] and planted with a native hedgerow [e5.1.1.3]. Mini-thickets of trees, selected to represent naturally colonising and wildlife habitat species, are planted throughout the site [e5.1.1.4] (which, based on online photos, have grown substantially over the decade since my visit). Honeysuckle is planted along the street-side fence to attract pollinators, and a series of bird-boxes and feeders, built and decorated by locals, help to encourage urban biodiversity. Beyond a lawn that is popular with loafing visitors (Figure 5.2), is an urban wild has been left for wildlife and for picking blackberries. A handful of raised planting beds, one guarded by scarecrows, are tended by locals [e5.1.1.5]. In homage to the borough's history of market gardens, a small greenhouse has been built in a corner to encourage the early seasonal growth of edible plants. Garden furniture scattered about the garden (Figure 5.3) [e5.1.1.6] is constructed from pallets recycled from a nearby market. Since the time of my visit, an outdoor stage has been erected to encourage productions by local artists.

The most prominent architectural feature is a four-arched, timber pavilion (Figure 5.4) [e5.1.1.7] which has become a popular gathering spot for members of the community who gather around tables for work or read quietly in chairs and couches while enjoying the lovely view (Figure 5.5). A small café offers beverages and free WiFi. A resident cat is on hand for 'keeping warm in colder months.' A small chapbook can be browsed through that describes the garden's development [e.5.1.1.8]. Roof runoff is harvested in reclaimed whiskey barrels for plant irrigation [e5.1.1.9]. A brick, wood-fired oven is used in community workshops to bake pizzas garnished with herbs and vegetables cultivated in the garden. The site is a multi-layered garden whose purpose seems to be more social than it is agricultural. In order to operate as community meeting place, careful attention has been paid to the design of the tightly constrained space and to the inclusion of artistic elements. The latter includes a large greeting mural that celebrates the history of the site [e5.1.1.1] and a beautifully crafted oak table (Figure 5.6). The result is a highly successful amalgam, creating an aesthetic garden for blending cultivation with conversation.

Figure 5.1.

Figure 5.2.

Figure 5.3.

Figure 5.4.

Figure 5.5.

Figure 5.6.

Online comments refer to the garden as an 'urban oasis' or 'wild haven'. The site manager mentions that in some cases it is possible to see an immediate physical manifestation of tension release the moment visitors enter the 'different world' of the garden. Given that the site's purpose is primarily tailored toward fostering relaxation in greenspace, it is intended to be accessible to everyone. As such, and unlike many city farms in London (see Chapter 4.2), it is not possible to rent any part of the garden for private functions. The garden is opened in the mornings for community support groups and in the afternoons for the general public. Educational workshops are provided on Saturdays. One such is the 'Inspired by Nature' programme for grade-school children, which is designed 'to nurture the next generation of custodians of the environment.'

Satisfied by my fine lunch at the FARM:Shop (see Chapter 3.1), I cross the street to visit the Dalston Eastern Curve Garden. From the start, even before reaching the entrance, I know I am about to visit a site where art matters. This is signified by the impressive mural on the side of the abutting four-story building. Admittance through the small doorway in the surrounding fence gives an impression that one is venturing into sort of 'secret garden' as in the eponymous children's story. Somehow the allusion to Frances Hodgson Burnett's famous novel is appropriate since its theme is about regeneration and rejuvenation. It takes but a moment inside the Dalston Eastern Curve Garden for me to realise the aptness of linking the two gardens this way. Before even undertaking my customary walkabout, I plunk myself down on a chair in the beautiful pavilion and spend half an hour gazing out over the scene of trees, planting beds, and portraits painted on the walls. Only a few souls are present and the contrast between the hushed tranquillity of garden and the din of busy streetscape outside is palpable. I look up as I write that last word in my field notebook and smile at its similarity to the word 'palatable'. Thinking once again about food – never far from my mind during these research tours – I get up and wander over to a planting bed and 'pinch' (as the Brits phrase it) a green onion upon which to munch as I contentedly begin to explore the site.

5.1.2 Fenway Victory Gardens (Boston, MA, USA)

Boston's Fenway Victory Gardens are America's oldest continually operational victory gardens remaining from the Second World War (www.fenwayvictorygardens.com), and in many respects are the gold standard among urban agriculture scholars interested in edible landscapes as community space (Lawson 2005; Nordahl 2009). Created in 1942 with fill from subway construction, the 6-hectare area of vegetable gardens were first managed by the Parks and Recreation Department until a separate entity, the Fenway Garden Society [e5.1.2.1], took over (although the land still remains part of the park system). The survival of the gardens in the face of development pressures over the last eight decades has been due to the efforts of a highly motivated and vocal constituency of both users and admirers. Today, the demand for the 150-square-metre allotments is huge, with some gardeners actually residing many blocks distant from the site itself.

The Victory Gardens are located in the Fenway, a linear park designed in the nineteenth century by Frederick Law Olmsted along the Muddy River [e5.1.2.2], an urban waterway that once drained the fens and is now chocked with the invasive common reed *Phragmites* [e5.1.2.3]. Nearby are some of city's most iconic buildings [e5.1.2.4]. The Victory Gardens are among the best loved greenspace in metropolitan Boston and are recognised as one of the world's most famous and successful fusions of cultivation and recreation.

The various entrances to the site are marked by floral planting, community news-boards, and various 'V' for victory sculptures [e5.1.2.4]. A large work area exists which includes a nursery, wheel barrows, compost piles, and an apiary [e5.1.2.5]. Many of the individual plots are humble and intended for growing food. Some are the typical allotment cheek-by-jowl jumble of ground plantings, whereas others utilise raised wooden beds [e5.1.2.6]. Other allotments are laid out in well-ordered rows [e5.1.2.7] or are overgrown in a chaotic verdancy that provides a wonderful contrast with the cityscape [e5.1.2.8]. But it is the park-like presence of trees [e5.1.2.9] mixed in with the jungle-like planting of flowers and flowering shrubs [e5.1.2.10], all so discordant with the nearby office towers (Figure 5.7) [e5.1.2.11], that makes the Victory Gardens so remarkable. In particular, through the artful design of peaceful refugia and ostentatious flower gardens, the spot is hugely popular with visitors and tourists alike. Clusters of flowers surround patios or border carefully laid out paths that lead to secluded chairs (Figure 5.8); several even sport their own water features (Figure 5.9). As Lawson (2005) writes: 'walking through its row of gardens is like visiting a garden exhibition ... [due to the presence of] perennial plantings, small fruit trees, furniture, and statues that express the individuality of the gardeners and give the garden a feeling of permanence.' The site therefore functions as a productive park wherein food cultivators, flower horticulturalists, and non-gardening *flâneurs* rub shoulders with each another. An annual social highlight that brings all together is the picnic run by the Fenway Garden Society.

Figure 5.7.

Figure 5.8.

Figure 5.9.

Other organisations periodically use the gardens to educate about urban agriculture and sustainability. During 2012, for example, a design competition was run by the architectural interest group 'Common Boston' (www.cbcb.commonboston.org). Six teams had four days in which 'to design and construct real projects for real communities' based on an urban agricultural theme that was researched through conversations with individual gardeners or the 'Fenway Garden Society' itself [e5.1.2.12]. A crowd of onlookers was ushered around while artists explained their installations [e5.1.2.13]. One installation educated about how during the war people were advised to grow food at home anywhere, including through us of 'WINDOW BOXES' (Figure 5.10), with an accompanying sign [e5.1.2.14] that read:

> The Fenway Victory Gardens were created in 1942 by the Roosevelt Administration as part of the war effort to supplement the nation's food supply during the Second World War. American families were encouraged to grow food at home anywhere they could (including WINDOW boxes) to feed themselves, thereby reducing the impact on the overall food supply so more could be sent to the Allied soldiers in Europe. Cities and towns took up the challenge as well, as this very land in the Fens was one of several plots throughout Boston – including the Common and Public Garden – that was used to plant fruit and vegetables. With mandatory food rationing in place, these gardens were essential in ensuring that there was enough food for all. Needed, by 1943, 20 million households produced more than 40% of the vegetables that were eaten stateside. While today only about 25% of the gardens grow vegetables, the legacy of the victory gardens is still keenly felt.

Another art installation was of a door [e5.1.2.15] and contained the following text [e5.1.2.16]:

> From its early beginnings as part of the Allied war effort, through the struggles of opposition and threats of development, to its varied educational programs and outreach to its neighbours, it is clear that these gardens have cultivated much more than just the flowers and vegetables seen at first glance. A strong sense of community and stewardship continue to flourish in our members year after year. Our DOOR is always open to new visitors, so please feel free to wander the paths, learn more about our history, converse with our gardeners, and experience this one-of-a-kind natural oasis in the heart of Boston.

Among the other design installations was a plant-filled toilet bowl (Figure 5.11) whose interpretive sign [e5.1.2.17] reviewed the precarious history of development threats to the location and warned that the Victory Gardens will never be 'FLUSHED' out of the Fens. A jumble of various items was shown [e5.1.2.18], representing the type of material which used to fill in the original fens, and which were discovered when victory gardeners started to dig in the area [e5.1.2.19].

One interactive installation was a pin-up board set up to encourage gardeners to 'take a seed, cultivate an idea' as part of a way to educate about the community-building aspect of urban agriculture (Figure 5.12) [e5.1.2.20]. Another interactive installation, called 'SOCIAL ROOTS,' contained burlap bags filled with plantings that were free for the taking (Figure 5.13), the idea being to remind that today's urban agriculture roots lie in the victory garden movement, which was based on the socially responsible sharing of resources.

Figure 5.10.

Figure 5.11.

Figure 5.12.

Figure 5.13.

It is always a nostalgic experience to revisit a city where one had once lived. As I stroll about the Victory Gardens on the present urban agriculture tour, I recollect my initial visit when I jogged the complete length of the Fenway on the very first weekend after my move to Massachusetts. I remember, too, the many professional meetings that I subsequently attended concerning 'daylighting' and other proposed interventions that might be employed for restoring the adjacent Muddy River.

My present visit has been timed to the art competition. Nothing could be better, I think, than circulating about one of the world's most historically significant and beautiful agriculturally-themed urban gardens than doing so with it being filled with art installations. It is a lovely sunny day and the crowd of onlookers are enthusiastic supporters, clapping after each artist's presentation about their project. Having published accounts by environmental artists, I am most appreciative of the interlinking of art and social-ecological concerns. At the same time, however, I find myself annoyed that the only permanent nod to the site's important role in the history of urban agriculture is the brief mention in the several large 'V' sculptures that mark the entrances. The most detailed information about the historical importance of the site is in the signage that accompanies the art installations; and these are but there only temporarily. To placate my criticisms over the embarrassing omission, I spend another hour admix the trees and flowers, enjoying the contrasts between the lazy sunbathers and the energetic joggers, as well as between the vegetative greenery and the shiny office towers.

5.1.3 Case study lessons

- Regardless of their size, regenerative art-filled gardens can be envisioned as being part of corridors of urban greenspace (Dalston Eastern Curve Garden; Fenway Victory Gardens).

- Swatches of grass provide lounging spaces for relaxing humans, with trees imbuing a park-like ambiance while at the same time providing habitat spaces for urban wildlife (Dalston Eastern Curve Garden; Fenway Victory Gardens).
- Presence of permanent murals (Dalston Eastern Curve Garden) and temporary installations (Fenway Victory Gardens) showcase how art can be inserted into gardens to enrich the experience of visitors.
- Planting plots can range from being humble boxes for growing edibles (Dalston Eastern Curve Garden) to elaborate landscaped refugia filled with flowers (Fenway Victory Gardens).
- Beautifully designed pavilions can be used for casual meeting spaces (Dalston Eastern Curve Garden) as can lawns for hosting annual picnics (Fenway Victory Gardens).

5.2 Artful gardens

Anecdote

Personal experiences of spying on cottage gardens. Can anything be more provocative of instilling a sense of place than the classic English country garden? The image of the thatched-roof cottage surrounded by its collar of blooming flowers while nestled into a quaint village of fieldstone buildings is instantly recognisable around the world from dozens of BBC-Masterpiece costume dramas and 'cozy' crime series. What a joy it is when on a narrowboat tour of the canals through the English countryside with my British-born father that we can observe the efforts of cottage gardeners. Being afforded the opportunity to view how people tend their private backyards gardens, as opposed to their showy front-yard exhibitions, is normally a rare occasion indeed. But from the viewpoint of the deck of a narrowboat slowly chugging its way along a canal adjacent to the edge of properties, there are many opportunities in which to indulge in such horticultural espionage. Each garden is a unique reflection of the owners' whims in terms of balancing beauty and bounty. The emphasis is on informality, the irony being that although the gardens are carefully designed, the impression conveyed is one of casualness in contrast to contrivance. Dense plantings of ornamental and edible plants are interspersed, flowers being used to keep weeds in check. Lovely gravel paths wind around clusters of fruit and nut trees, and wicker benches are arranged beside beds of vegetables. Herbs are juxtaposed beside hedges, and traditional roses and climbers scale trellises. The scene is often so perfect that, if the additional benefit of lavender or lilac fragrances are wafting in the spring air, my father and I make it our habit to temporarily moor beside such gardens while eating our lunch.

Book

The foodscape revolution: finding a better way to make space for food and beauty in your garden, by Brie Arthur, is based on the aim of bringing vegetable gardens out from their backyard isolation and making them integrated components of a property's high-visibility, aesthetically-pleasing landscape. The key, states the author, a graduate in landscape design and horticulture, is really no more complicated than in allowing veggies to share common space with ornamentals, sometimes even replacing the latter with attractive and nutritious consumables. In other words, 'foodscaping,' mixing bounty and beauty, is simply the process of enabling food to be grown alongside flowers interspersed with shrubs that provide an architectural framework for the edibles, all within the traditional landscape of turf grass front lawns. The process, she believes, is just as applicable to office buildings and school campuses as it is to private homes.

Arthur makes the point that in consequence of sprawl there are millions of hectares of American suburbs that have the potential to grow massive amounts of food. The book is richly illustrated with many colour photographs of edible plants growing *in situ* as well as brimming harvest baskets that are staged in front of lovely gardens from which they presumably came. Much is made of the ensuing benefits of foodscaping:

- Empowering individuals to learn and appreciate how food is grown.
- Increase local food sustainability while reducing food miles.
- Cost savings and reliance on healthy, ethically grown food.
- Absence of any empty space for colonising weeds should something prevent a food garden from being planted one year since trees, shrubs, and perennials will always be there.
- Dispensing with the absurd act of blowing away leaves, and instead collecting them to be added to garden mulch.
- Landscape multi-functionality in terms of wildlife habitat and absorbing runoff.
- Saved need to have to pay landscapers for maintenance.

That said, one notable challenge facing the aspiring foodscaper is that homeowner associations in high-end communities often have bylaws specifically forbidding food being grown in front yards.

The book is comprised of three sections. Part One builds upon a schematic plan of the author's own North Carolina property and highlights fruit and nut trees, edible meadow plants, and even some dabbling in hydroponics. Practical tips are provided on how to create various elements in what is referred to as 'the anatomy of a foodscape'. Plant species lists are provided, although no Latin names are given. Most of the discussion about design is based on creating what resembles a classic cottage garden such as one finds throughout the English countryside. Part Two is based on encouraging the human use of foodscapes through focusing on the 'edible ambiance' of garden 'rooms' by installing amenities such as patio pots and a fire pit. Part Three is an eclectic mix of advice on the harvesting, preserving, and processing of food, including providing selected recipes. In the end, it is a rare recreational gardener who will be uninspired from reading the book to reconsider the often-forced separation of edible and ornamental landscapes on her or his property.

Sites

5.2.1 Edible Estates (London, UK)

Brookwood Estate, a cluster of featureless high-rise apartment blocks located in the Bankside area of Southwark, London, is home to an artistic blending of urban agricultural form and function. Commissioned by the nearby Tate Modern Gallery, the 'Edible Estates' garden was designed specifically for its location, based on the idea of transforming front yards into creative vegetable gardens as part of a process of arts-based food education (Bieler 2017) and the Capital Growth greening programme (Viljoen and Bohn 2015). This is no marginal site, tucked away out of sight at the edge of town. Rather it is a visibly exposed triangular patch of land located in a dense city, surrounded by towering public housing blocks, where all gardening activities are a spectacle to be viewed from the balconies. Furthermore, the site is situated across the street from a school, and as such, is passed by hundreds of children each weekday. Significantly, the garden is designed to provide apartment dwellers with the opportunity to grow plants, while at the same time creating 'a beautiful urban amenity' in a neighbourhood comprised of disenfranchised individuals who have little sense of community spirit (www.fritzhaeg.com). Also, the garden dispenses with the English tradition of private allotments, replacing this with a 'holistic design that also integrates spaces where people may gather [in] a pleasure garden made up entirely of edibles.' The garden was actually part of the Tate Modern's 'Global Cities' exhibition, and is representative of the designer's ongoing rebellion against manicured lawns by replacing them with aesthetically pleasing edible landscapes (Haeg 2010). As Gorgolewski *et al.* (2011) state: 'The gardens are purposely located in places calculated to serve as vivid contrasts to their surroundings – surprise and shock are anticipated reactions and are intended to generate interest and curiosity about the transformations.' Initial fears of potential vandalism were thwarted by engaging school children at the early planning stages and by instilling a sense of ownership amongst adults in the tenements. The high visibility and openness of the site to passers-by opened a dialogue of shared advice about various growing techniques, which in turn led to a sense of protectionism for the installation. Reviewing the project, Charrett (2009) concluded: 'The real success of Edible Estates, I feel, is that it is not just in private back gardens that gardening can give pleasure and satisfaction, and that, in the public realm, horticulture as a healing tool can be incredibly potent.'

The intricate design of the garden draws its inspiration from the curved, ornate beds of flowers in many of Britain's royal parks. Two, large arabesque planting beds (Figure 5.14) [e5.2.1.1] are situated amongst trees to create several gathering spots where residents can sit on benches and tree stumps (Figure 5.15) [e5.2.1.2]. The park-like setting [e5.2.1.3] includes fruit trees, berry bushes, and edible ornamentals, in addition to customary vegetables and herbs [e5.2.1.4], and is designed to be best viewed from the surrounding balconies. The idea is that beauty and art can be used to motivate apartment dwellers to participate in the gardening process. Bags of potting soil, composters, and farming implements are stored alongside the building [e5.2.1.5].

Figure 5.14.

Figure 5.15.

It is a drizzly day in London when, on my way to the Borough Market (see Chapter 3.2), I take a detour to visit the Edible Estates. Expecting to pause there only briefly, I find myself surprised at how much I enjoy the modest site. When looking at the gentle, sweeping curves of planting beds, I struggle to imagine why it is that 99% of such elsewhere are always constructed in the same old, boring shape of square boxes or rectangles. These, as everyone knows, are forms antithetical to the lines that occur in nature. I sit down on a bench and reflect on how long it took stormwater management to make the change from square treatment ponds to more naturalistic designs (France 2003). In that case, it was only when the once dominant engineering mindset was supplanted by one based in landscape architecture that things improved. 'Supplanted' I scribble in my field notebook while huddled underneath an umbrella; how appropriate a word to use when referring to the transformation that is now taking place in urban agriculture in consequence as projects like the one in front of me. Happy at my little revelation of the parallelisms existing in the evolution of natural design in urban ecohydrology and urban agriculture, I spring to my feet to walk the few blocks to the Borough Market for a visit, lunch, and, this being England, a good pint.

5.2.2 The Edible Bus Stop (London, UK)

Reliance on public transport is an essential component in planning for sustainability, and depends on providing waiting spaces which can be integrated into frameworks for urban regeneration. London has about twenty thousand bus stops, whose design can have far reaching consequences since small changes in transportation spaces create impressions that, by definition, are carried far afield. One imaginative project which addresses these topics and has attracted the attention of scholars (Adhitya and Tyler 2015; Brovarone 2018) is the 'Edible Bus Stop' in South London. Designed as a series of pocket parks along the 'No. 322' bus route from Clapham Common to Crystal Palace, the project is motivated by a desire to change the urban landscape in tiny steps, one bus stop at a time (www.theediblebusstop.org). A collective of landscape architects, garden designers, horticulturalists, artists, and activists came together in a guerrilla gardening initiative to appropriate derelict patches of land situated beside bus stops and to transform them into gardens. The original project location was a streetside sliver of undeveloped land that had been a bomb site in the Second World War.

Each garden was developed via a process referred to as 'E.A.T.,' based on Engaging the community, Animating the forgotten space, and Transforming the site through design. The overarching goal was to 'encourage pride in the community and a sense of place,' with volunteers assuming responsibility for garden maintenance while learning and sharing knowledge about the seasonality of food. Plants were selected based on their colour, fragrancy, and attractiveness for wildlife, with some 'possibly even [being] edible.' A video on the website begins with the statements: 'galvanising the community, creating a sense of ownership, and engendering responsibility,' before moving on to state that throughout London there are forgotten corners lying in wait to be transformed through the dramatic improvement of their physical space. An effective film clip shows a progression of time-lapse photography from a ground level perspective of plants emerging from their planters while thousands of vehicles and people race past.

The goals of this grassroots initiative in creating edible, green infrastructure are as holistic as they are ambitious (Marshalls 2018):

> The core of our ethos is the belief that a brutal landscape makes for a brutal outlook and that by taking responsibility for our forgotten public spaces we can collectively improve upon the experience of inner city living. By gardening together and enjoying these community led growing spaces, conversation is encouraged, barriers are broken down, skills are shared and sense of local empowerment and belonging in a neighbourhood is nurtured alongside the gardens. The gardens provide a place for people to come together for the good of the neighbourhood and get to know each other. They will become urban village greens for people to find space to sit and relax in amongst the hustle and bustle. They will be a resource for education, providing opportunities for people to learn about where their food comes from and how to grow and maintain a garden.

From its guerrilla roots (Purvis 2014), the gardens have become respectable, well-loved places that have received municipal awards, including an accolade from the onetime mayor of London, Boris Johnson. Openings of new bus stop gardens are popular events attended by reporters and dozens of people (Hope-Urwin 2013). One critic of the urban form rhapsodised about the gardens, wondering why it took so long to think of the 'brilliant' concept, since it is a given that bus stops are 'ugly, dirty, neglected places where people are embarrassed to be seen' and look unhappy while they wait (www.theviewfromadrawbridge.com). The critic lists the benefits of bus stop gardens for providing beauty, food education, and wildlife habitat, at the same time as increasing community engagement and pride in safer neighbourhoods. The review concluded with the expressed wish that the idea would catch on around the world, something which has been achieved in places as far removed from London as the city of Thunder Bay, north-western Ontario (www.thewalleye.ca).

The Kerb Garden is the original Edible Bus Stop pocket park along the No. 322 line. When approached from a distance of a block away or on the other side of the busy street, it is immediately apparent that here is something that is different from the norm: a playful intervention of greenery and edgy architecture that is in contrast to the brick and blank sidewalls of the nearby houses (Figure 5.16) [e5.2.2.1]. Closer inspection reveals that it is possible to walk on the normal sidewalk paralleling the garden installation [e5.2.2.2]. However, a slight detour of only a few steps allows one to enter and then to navigate through the garden along several different flow-paths that are bordered by raised beds whose retaining walls are made of concrete blocks reclaimed from demolition sites around Lambeth (Figure 5.17) [e5.2.2.3]. Or, if not in a particular hurry and looking for a convenient spot in which to consume one's bagged lunch, or of course if waiting to catch the next bus, then it is possible to linger by sitting upon the welcoming benches. The planting beds overflow with effusive verdancy, including an attractive mix of apple and crab apple trees, various bushes, sunflowers, flowers such as lavender, and 'where possible, edibles', such as maize, chard, rhubarb, and strawberries [e5.2.2.4]. Several black plastic tubs presumably contain compost [e5.2.2.5]. A sign requests visitors to refrain from using the gardens as either a litter bin or toilet [e5.2.2.6], while another encourages members of the public to participate in periodic gardening maintenance [e5.2.2.7]. Attention has been paid to planting flowers around the base of one of the few large trees in the neighbourhood, beside a bench that is popular among locals [e5.2.2.8]. The open accessibility of the Kerb Garden is in stark contrast with greenery on the other side of the street from which the public is excluded by a high metal fence [e5.2.2.9; e5.2.2.3].

Figure 5.16.

Figure 5.17.

A bus ride transports me to another Edible Bus Stop pocket park: the Hoopla Garden. Here, a streetside widening had previously been occupied by a row of bollards used to prevent ad hoc parking. For the new design, the bollards were left in place [e5.2.2.10] and are now artistically reimagined as being part of a mock, giant game of hoopla; i.e. the bollards symbolise the pegs with the round gardens representing the concrete rings (Figure 5.18) [e5.2.2.11], some of which double as seating areas (Figure 5.19) [e5.2.2.12]. Gardens are planted with fruit and nut trees and wildflowers to attract pollinators [e5.2.2.13]. Garden visitations complete, it is but a short distance to see another impressive example of modern architecture: The Shard [e5.2.2.14].

Figure 5.18.

Figure 5.19.

It is hard not to be impressed by the Edible Bus Stop. The use of reclaimed concrete blocks assembled as a jagged row, one leaning against the other like dominos, in the Kerb Garden, and the rings and pegs design of the Hoopla Garden, are as imaginative as they are beautiful. The presence of lavender at the former site is a nice addition to the streetside scentscape. Several times I bend down to enjoy the fragrance of the flowers, which is almost strong enough to mask the smell of diesel and petroleum from the street. For several minutes I spy on a young lad decked out in active sportswear who is chatting with a scruffily dressed fellow (a street person, perhaps) while they sit upon a bench, thereby demonstrating that the pocket park is admirably fulfilling the social component of its design. However, the emphasis overall seems to be weighted more toward artistic form rather than either urban agriculture or environmental function; in other words, more on the park-like aesthetics and less on edibles. I find it hard to reconcile the well-meaning but possibly over-hyped goals with the reality. An artistically beautiful social space, certainly; but a place to really learn about urban agriculture? I'm not so sure. Besides, as an environmental scientist who has studied the uptake of trace contaminants by plants, I would be cautious about eating too much that is grown so close to traffic. The Hoopla Garden also leaves me torn. Whereas I admire the cleverness of its overall design, I cannot but help to think of all the work I've done in trying to educate about water sensitive design and low impact development, wherein raised planting beds are to be avoided at all costs, to be replaced by gardens which are depressed below grade so to absorb and treat surface runoff. I spend some time trying to envision how to go about conjoining the site's admittedly clever design with the idea of 'functional art' with respect to stormwater runoff (France 2005). But then not every site can incorporate every positive landscape aspiration; nor perhaps should it, especially given the limited space available in these tiny bus stop parks. Overall, I remain impressed by both the spirit and execution of these acupuncture-like garden interventions into the ailing urban body.

5.2.3 Abbey Gardens (London, UK)

'Abbey Gardens' is another remarkable urban agriculture project that exists in an otherwise nondescript location in the East End of London. During the twelfth century, the site was home to the twelfth-century Cistercian abbey of St Mary Stratford Langthorne, one of the wealthiest in England at the time of the Dissolution, and one whose impressive gatehouse remained standing until 1825 (Figure 5.20) [e5.2.3.1]. The area gradually became industrialised, with construction of extensive rail yards and terraced row-houses in the mid-nineteenth century that obliterated almost all traces of the abbey. Another intervention served as the inspiration for transforming the derelict site into the designed gardens that exist today. In 1906, a group of local unemployed men cleared a piece of land to establish a squatters' camp and series of triangular-shaped vegetable gardens. The idea of the squatters was to show that wasteland could be put into productive use and that the unemployed were willing to undertake the work. After this initial heyday, the site was neglected and became overgrown with brambles, eventually being used as a dumping ground for all manner of discarded trash, including burned-out cars. Spurred on by knowledge that town-wide revitalisation would be taking place in preparation for hosting the 2012 Olympics, the Newham borough council undertook a feasibility study of the site's potential. During 2007, locals aided archaeologists from English Heritage and the Museum of London to excavate the scant remains of the abbey (Figure 5.21) [e5.2.3.2] and to erect interpretive signs of its history [e5.2.3.3]. The site was cleared of bramble and seeded with grass and wildflowers, and an adjacent transit station was constructed for a new line of the Docklands Light Railway. The following year several artists (www.somewhere.org.uk) won a competition for a public art installation on the site. The idea was to create an artistic garden that was inspired by the name of the 1906 squatters' camp: 'The Triangle Camp' (so-named for the shape of the parcel of land that it had occupied). The site's occupational history and development is admirably described on the official website (www.abbeygardens.org).

Today much is made of the site's intriguing history concerning working class food sustenance and grassroots landscape regeneration. The original squatters' inspiring slogan of 'What will the harvest be?' has become the leitmotif around which the new Abbey Gardens were developed. As such, this slogan is posted on the side of the onsite office trailer [e5.2.3.4] as well as being prominently displayed in a huge sign affixed to a boundary wall (Figure 5.22). The face of the office trailer is covered with an enormous photograph of the squatters standing in bold defiance in front of a wall upon which the slogan is emblazoned (Figure 5.23) in a style of script [e5.2.3.5] that was obviously copied for the new sign. Thus, the past is physically linked to the present.

Figure 5.20.

Figure 5.21.

Figure 5.22.

Figure 5.23.

Several years of fundraising and work were involved in remediating the contaminated soil on the site and in growing plants in window boxes and balconies in the neighbourhood before transplanting them to the site. The resulting 2,000-square-metre community garden occupies a series of raised beds constructed of untreated oak and steel by local volunteers. It is the design of these beds in the shape of triangles in homage to the name of the squatters' camp and original garden layout (Figure 5.24) [e5.2.3.6] which is what makes The Abbey Gardens so interesting. The 34 raised beds contain vegetables and flowers laid out in a form reminiscent of formal Edwardian civic parks. These are interspersed with a dense network of paths for easy access (Figure 5.25) [e5.2.3.7]. As an added bonus, the garden's website includes an interactive site plan over which one moves the mouse to reveal the planting palette of each raised bed. In a move breaking from traditional allotments, individual plots do not exist. Instead, cultivation, maintenance, harvesting, and produce are shared amongst the regular gardeners [e5.2.3.8]. An autumn Herb Festival is held with music, baked snacks, and teas, yoga and massage sessions, and herb-themed face painting. The overall effect, easily observable from the adjacent street, is of an inviting public place that is both productive and beautiful. This is so different from the traditional allotment gardens which are characterised by higgledy-piggledy layouts and where uninvited visitors are often regarded as suspicious intruders.

Figure 5.24.

Figure 5.25.

Once again I find myself venturing into an area of London far removed from the normal tourist haunts. Bland apartment towers rub against sprawling industrial buildings situated along nondescript suburban streets. It is only when I reach the metal fence and peer through it to the gardens beyond that my interest is piqued. I stroll through the open gate and have the site to myself for the next hour. The sharply angled, triangular planting beds are great to behold, once again leading me to marvel at how such a simple act of simply reconfiguring the shape of such can completely transform one's experience of an urban garden. It is early spring and some of the planters have yet to reveal their botanical contents. For others, I bend down to examine the tiny sprouts of vegetables emerging from their sustaining blanket of rich topsoil. I reach down and rub the dark earth through my fingers before lifting it to inhale its fecund aroma. Later, in a remote corner of the site, I use my foot to disturb the ground in order to examine the contaminated soil beneath, again bending down to feel it with my fingers in order to contrast it with that from the planting beds. It is immediately understandable how, notwithstanding contamination issues, the physical makeup of the site's damaged soil necessitated the import of more healthy material. My little experiment in haptic soil science over, I replace my 'divots' and move on.

It is easy to admire the attention paid to the agrarian history as displayed through the giant sign and photograph of the Edwardian squatters. However, given the garden's very name, I feel that more could have been made of interpreting the onsite foundations of the original abbey. Later, when walking to the new transportation station, I am able to find the extremely informative panels about the abbey, thereby placating some of my criticism. But I still think there is a missed opportunity to provide onsite information about the medieval monks' use of medicinal plants, à la Brother Cadfael, through didactic plantings (perhaps things have changed in this regard since the time of my visit). Such a fusion of monastic ruins with an educational herb garden is something that is accomplished very effectively at Vauclair Abbey, a site I have visited while on a long-distance walk through the French countryside.

5.2.4 Carrot Green Roof and Garden (Toronto, ON, Canada)

One of the best examples intermeshing urban agriculture with attractively designed public space is the 'Carrot Green Roof and Garden' in the vibrant Danforth area of Toronto (www.carrotgreenroof.ca). The project occupies a fenced-in area of roof [e5.2.4.1] atop the 'Carrot Common,' a building that houses a co-operative grocery store, a bookshop, and a suite of offices for wellness practitioners, all located beside an inviting courtyard that is open to the street and popular with neighbourhood residents [e5.2.4.2]. The rooftop was previously a restaurant patio before being converted to a place to grow food in plastic containers in a similar fashion to the 'Food in the Sky' garden in London, which also sits atop a grocery store (see Chapter 3.1). In 2008, during renovations to repair the leaking roof, a decision was made to redesign the site with a focus on sustainable practices.

A series of design charrettes included a firm specialising in green roof installation, landscape architecture students, and community residents. The resulting 300-square-metre area was transformed into an intensive green roof containing planters filled with vegetables and medicinal herbs (Figure 5.26) [e5.2.4.3] as well as a mesic or, in their parlance, 'wetland' garden [e5.2.4.4]. These are interspersed with an attractively laid out extensive green roof (Figure 5.27) [e5.2.4.5] and living wall containing sedums and similar hardy plants that are the staple in such green eco-roofs (Figure 5.28) [35.2.4.6]. Rainwater is harvested (presumably to be used for irrigation) and several beehives occupy a secluded area of the upper roof. Academic studies have examined the success of the extensive plantings of native species and their role in improving microclimatic conditions (Anderson and Gough 2021).

Figure 5.26.

Figure 5.27.

Figure 5.28.

It is the blending of these spaces for plants with other places for people that makes the Carrot Green Roof such a successful agrourbanism project. The use of beautifully stained wood as space dividers [e5.2.4.7] to create a deck terrace (Figure 5.29) [e5.2.4.8] and outdoor kitchen area (Figure 5.30) [e5.2.4.9] shows how visual amenities can be incorporated to improve the utility of such place-making. In this case, the rooftop is designed for hosting events and workshops that promote food security, sustainability, and the sharing of urban gardening advice and experience. That all of this occurs in the midst of plants functioning as either edible produce or as green infrastructure is a good demonstration of outdoor classrooms. The overarching purpose of the Carrot Green Roof (Redekop 2012) is to be 'a learning hub, an eco-educational centre, a showcase of best practices for sustainability and green roof projects ... and most of all a community sharing space'. As Gorgolewski *et al.* (2011: 173) conclude, the project 'shows how urban spaces like rooftops can seamlessly integrate production, education, and community activities'.

As an educational space, the outdoor kitchen enables cooking demonstrations, and an area is set aside for ongoing experimentation to study the viability of various green roof plantings. Interpretive signs inform about rainwater harvesting, horticulture, and composting, including one [e5.2.4.10] that reads:

> The large wooden planters which frame the outer roof are filled with colourful ethnocultural crops which are fruits and vegetables particular to certain cultural cuisines. These crops are part of a greater initiative both here in Toronto and on the farmlands of the Vineland Research and Innovation Centre. The World Crop project in Vineland ... is working to determine which world crops can successfully be grown in Ontario, what is required for optimal production and how can pests be controlled biologically. To give the project context, each year $800,000,000 of ethnocultural crops are imported into Ontario alone. It has been suggested that if we can divert 10% of these imported crops, by growing them here in Ontario, it could result upwards of $80,000,000 remaining in Ontario horticulture. In addition, it would mean fresh, local produce for people in our city ... This pilot involves introducing world crops into various community spaces, or 'learning gardens' in Toronto. The purpose is to gather interest and create connections between urban consumers, both familiar and unfamiliar with these crops, and our farmers, the producers of our local food. The Carrot Green Roof is one such 'learning garden' site.

The idea of designing attractive productive landscapes to function as incubi for growing ideas as much as for growing food is an important lesson from this project. This theme is taken up in the organisation's newsletter, 'Top of the Carrot News.' The roof garden has partnered with a local group, 'Seeds of Hope,' to provide work for individuals recovering from isolation, mental illness, addiction, and poverty, through allowing them to participate in growing and harvesting medicinal plants (see Chapter 7.1 for similar projects).

Figure 5.29.

Figure 5.30.

Even before my arrival to the Carrot Green Roof I am excited by the prospect of my visit. For this is a homecoming of sorts, since I was once resident of this particular neighbourhood. Normally the area is a-buzz with activity. Today, however, it is a Sunday morning the day after Toronto's Gay Pride Parade (one of the largest such in the world), and a post-party hush seems to have descended upon the entire metropolis. My good mood soon disappears when, after waiting well past the pre-arranged time to be given access to the roof by the director, it seems that I've been stood up. Luckily, I am able to locate someone who works in one of the building's offices, and who, after my plea of saying that I am an academic who has travelled across the country to see the site, is sympathetic and unlocks the entrance door to the roof. For the next hour I have the luxury of the garden to myself.

The rooftop space is wonderfully inviting. It is easy to imagine relaxing under a terrace awning while sipping a drink and listening to music, or, alternatively, busily using a large table to prepare foods while attending cooking lessons in the outdoor kitchen. That said, the site seems to have much more to say about the use of plants in green infrastructure than it does about them in terms of edible landscapes. Striking the appropriate balance when creating such didactic landscapes is an obvious challenge. My overall impression of the present site is that it is primarily a green eco-roof with a nod toward urban agriculture, rather than being an urban agriculture project that includes some interesting environmental elements. For, as its very slogan states, the garden is 'a unique place for nature in the city;' i.e. that is 'nature,' not 'food'. On the other hand, the project, with its wonderful blending of plants, place, and people, is a singular example of the more holistic concept of agrourbanism. And it is for this reason that Carrot Green Roof should be both celebrated and emulated.

5.2.5 Lafayette Greens (Detroit, MI, USA)

With its complex history of economic problems, racial tensions, and property abandonment, Detroit is becoming synonymous with the modern resurgence in urban agriculture (see Chapters 4.1 and 7.1). One project that stands out and which has attracted both praise and prejudice, is 'Lafayette Greens' (Figure 5.31). This is one of the best examples anywhere of the inspired blending of artistic form with agrarian function, bringing 'healthy food, green space and an urban park together in a unique space,' as its publicity factsheet intones. It is not without reason then, that Philips (2013) begins her book – one of the few which focusses on the aesthetics of urban agriculture – with a brief overview of the Layfayette Greens.

The project is located in the heart of downtown Detroit on a property once occupied by the 14-story Lafayette Building which, following its demolition in 2009, had left a half-hectare derelict lot. The Compuware Corporation, whose offices are situated nearby, provided the financial support, planning, and construction expertise, in addition to volunteer hours from its staff in order to create a public space to be enjoyed by their employees as well as neighbouring office workers and visitors [e5.2.5.1]. Lafayette Greens is a good example of a partnership between the private and public sectors. As Philips (2013: 1) writes: 'By making the space an edible landscape, instead of just a plaza, the company is helping to educate the public about health, environmental responsibility, and how to grow food'. An overarching aim is to 'encourage other downtown-based businesses and residents to reach out and beautify the spaces that surround them.' The project won an award from the American Society of Landscape Architects, who stated that the designers 'broke the mould on how to create for the long term with a contemporary language. It [the project] raises the issue of the importance of accessibility to fresh produce in the inner city. This strikes such a great balance of design, scale, and economy that it makes a good model. And it's great that it's so visible right downtown' (ASLA 2012).

The overall concept is based on creating a productive vegetable garden that also functions as a public space in which to attract and engage the community. Attention was paid to the surrounding urban context in terms of creating a sophisticated, multi-functional space. The structural order of the site and the strongly angled geometry of its raised beds mimics the hard lines of the surrounding high-rise buildings. The resulting garden is composed of an ordered series of symmetrical raised beds in a linear layout that pays homage to the row cropping tradition of Michigan farming [e5.2.5.2]. Planting beds are designed as urban equivalents of the French *potager* kitchen garden. Attractively constructed of metal and wood [e5.2.5.3], these are filled with an artful arrangement of more than two hundred types of vegetables and herbs (Figure 5.32) [e5.2.5.4]. Signs inform about the seasonal availability of the plants [e5.2.5.5]. Beds are also maintained at a constant elevation across the metre-sloping grade of the site (Figure 5.33) [e5.2.5.6], thereby allowing gardeners, some of whom might have mobility issues, to select their own preferred height-location from which to operate. The beds incorporate seating benches for relaxation (Figure 5.34) [e5.2.5.7], and several are reserved for decorative plantings of flowers [e5.2.5.8].

Figure 5.31.

Figure 5.32.

Figure 5.33.

Figure 5.34.

Produce is harvested by volunteers (many of whom are local office workers) and donated to a community food bank. Onsite staff offer instruction about cultivation and nutrition as well as information about local food systems in Detroit. A 'Children's Garden' – located in the sunniest portion of the site so as to encourage lingering (Philips 2013) – includes reused steel drums of various sizes (Figure 5.35) [e5.2.5.9] that are 'filled with plants to discover, smell, touch and taste, [thereby] fostering a sense of wonder and connection to the natural world.' The garden is now managed by a local organisation which uses it as an outdoor classroom and community meeting place. Evening courses are taught on gardening for pollinators, as well as other non-botanical pursuits such as yoga, belly dancing, and tai chi. The place serves as a venue for concerts, weddings, and creative art installations. A small weekly farmers' market is run, and a programme of summer internships is available for school children.

Careful attention has been paid to the design aesthetics of the site by the landscape architects. Indeed, Philips (2013) notes that the garden's aesthetic geometry is such that it can be appreciated by those who look down on it from the surrounding office towers. A promenade is bordered with fragrant lavender beds and comfortable benches (Figure 5.36) [e5.2.5.10] to enable visitors to 'stop and smell the lavender' as an accompanying sign instructs [e5.2.5.11]. As stated by the ASLA (2012): 'The use of fragrant lavender, with its documented ability to induce calm, to line the public passage through the site reinforces the regenerative and restorative function of the garden in a bustling city environment.' Art installations take the form of playful sculptures in the Children's Garden [e5.2.5.12] and paintings cover the boarded-up windows of the abutting building [e5.2.5.13]. Colourful and creative birdhouses are arranged about the garden. Public meeting space for organised workshops or casual lunch breaks is provided by a patio [e.5.2.5.14]. Finally, there are two artistic additions to the gardens that are truly exceptional in their execution. An installation of more than a dozen steel girders recalls the verticality of the original building that once occupied the site (Figure 5.37) [e5.2.5.15]. But it is the three whimsically designed storage sheds (Figure 5.38) [e5.2.5.16], constructed from salvaged pallet wood and doors and named after one of the architects of the original Lafayette Building, that are most memorable.

Figure 5.35.

Figure 5.36.

Figure 5.37.

Figure 5.38.

And if all this were not enough, Lafayette Greens' attention to environmental design is exemplary. Philips (2013) makes frequent reference to the site's 'intricately detailed' attention to 'regenerative operations'. Recycled materials are used whenever possible. In several planting beds, master gardeners grow species that are attractive to butterflies (Figure 5.39) [e5.2.5.17] (since my visit, pollinator gardens have also been created for bees and birds). A beautiful heirloom orchard and meadow contains species selected to promote urban biodiversity [e5.2.5.18]. The lawn is composed of drought-tolerant fescue and the raised beds use drip irrigation. Finally, stormwater runoff is treated in a bordering bioswale (Figure 5.40), as explained in an accompanying sign (Figure 5.41).

Figure 5.39.

Figure 5.40.

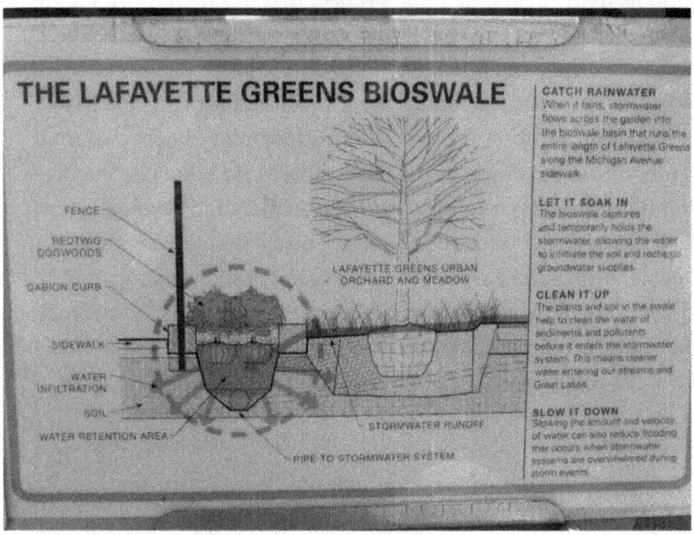

Figure 5.41.

Lafayette Greens represents the apogee of bringing urban agriculture out from the residential back-lots and into high-visibility downtown cores. And just as stormwater treatment wetlands have transitioned into becoming stormwater treatment wetland *parks* (France 2003), so will Lafayette Greens and other similarly inspired projects (France 2016) be important beacons along the road to the widespread creation of urban agriculture garden *parks*. In accomplishing this, they will be following a landscape architecture tradition that extends back to the royal ornamental gardens of Versailles and even distant Babylon. The accolades that have been received are many: 'the success of Lafayette Greens illustrates how much of an impact a well-built green space can impact a blighted community' (www.livable.org), while emphasising that 'urban farming is more than just about growing vegetables' (Smith 2015), such that it is 'an incredible asset to the city of Detroit ... a place of respite and solitude; a place where people reconnect with beauty and experience nature; a place of education and discovery' (www.greeningofdetroit.com).

After several days of touring all manner of urban agriculture sites in the troubled and depressing metropolis of Detroit, I am simply blown away by my visit to Lafayette Greens. And I am not alone in this regard. The same praise is offered by a group of university students and their professors whom I meet when visiting the site. They have travelled from out of state to specifically visit the site, and express their admiration for its successful mix of form and function. Are there any gardening tool sheds located anywhere that are as attractive as the ones here? As well as a feast for the eyes, I enjoy – once the tour group of students departs – smelling the aroma of the lavender while sitting on several benches in the promenade. Then, on top of all this, I'm happy to discover the most informative and well laid out demonstration bioswale for management of stormwater runoff that I've encountered at any urban agriculture location.

As I sit on the bench and scribble in my field notebook, I reflect that it is therefore a shame that the project has raised the ire of some of Detroit's urban agriculture activists. These individuals seem unable to get past their prejudicial belief that any corporation's largesse must be regarded with suspicion. For them, urban agriculture should remain exclusively about grassroots initiatives in producing food for the downtrodden and underprivileged. In no way should the movement be sullied by creating pretty spaces for affluent urbanites through such top-down commercial interventions. What these critics fail to realise is that it is the exposure of the public to such projects as Lafayette Greens, with its emphasis on aesthetic form, that will educate, inspire, and serve to raise interest and discussion among concerned and better-off citizens about urban agriculture in general; also this in turn may help to raise donations and volunteer support from these individuals for other projects whose emphasis is on functional food production.

5.2.6 Case study lessons

- The time is right for urban agriculture to be brought out from backyards and other sequestered spaces into high-visibility locations such as front-yards (Edible Estates), sidewalks (The Edible Bust Stop), rooftops (Carrot Green Roof), and street-side public parks (Lafayette Gardens), where sometimes the gardens can provide the background for the community placemaking (The Edible Bus Stop, Carrot Green Roof, Lafayette Greens).
- Attention to the aesthetics of garden design by knowledgeable landscape architects is essential for creating pleasing fusions of beauty and productivity that are important for raising awareness of urban agriculture (Edible Estates, The Edible Bus Stop, Abbey Gardens, Carrot Green Roof, Lafayette Greens).
- Something as remarkably basic as simply altering the geometric shape of planters from their traditional square box form to, for example, curves (Edible Estates), circles (The Edible Bus Stop), or triangles (Abbey Gardens), goes a long way to improving site aesthetics and the visitation experience.
- The park-like ambiance of regenerative artful gardens is also due to attention paid to the architecture of various site amenities and utilities, including seating areas (Edible Estates, The Edible Bus Stop, Carrot Green Roof, Lafayette Greens) and equipment sheds and children's play areas (Lafayette Greens).
- Whimsical installations are a welcomed element in artful gardens, including symbols of games (The Edible Bus Stop), homages to past conditions (Lafayette Greens), and creation of play areas for children (Lafayette Greens).
- Heritage and history can be remembered through imaginative means, such as the use of large photograph installations (Abbey Gardens) or the presence of steel girders of phantom buildings (Lafayette Greens).
- Artful gardens benefit from inclusion of both edible vegetables and ornamental flowers (Edible Estates, The Edible Bus Stop, Lafayette Greens).
- Through careful attention to plantings, artful gardens provide opportunities to educate about environmental issues, sustainable design, and green infrastructure (Carrot Green Roof, Lafayette Greens).
- Corporate sponsorship may be necessary to support high-end, expensive designs (Lafayette Greens).

References

Ableman, M. 2016. Street farm: growing food, jobs, and hope on the urban frontier. Chelsea Green Publishing, White River Junction, VT, USA.

Adhitya, S. and N. Tyler. 2015. It's the little things that count. In: Dloan, T. and B. Colhos (eds.) Infrastructure practice and social needs. International Institute of Applied Systems, Laxenburg, Austria.

Anderson, V. and W.A. Gough. 2021. Nature-based cooling potential: a multi-type green infrastructure evaluation in Toronto, Ontario, Canada. International Journal of Biometeorology 65: 1-12.

Arthur, B. 2017. The foodscape revolution: finding a better way to make space for food and beauty in your garden. St. Lynn's Press, Pittsburgh, PA, USA.

ASLA. 2012. Honor award: Lafayette Greens: urban agriculture, urban fabric, urban sustainability. American Society of Landscape Architects. Available at: www.asla.org.

Bieler, A. 2017. Contemporary art exhibitions as places of learning about reflexive food systems.

Breitbart, M.M. 2013. Creative economies in post-industrial cities: manufacturing a (different) scene. Routledge, London, UK.

Brovarone, E.V. 2018. When transport infrastructure shares the public space: public transport stops as urban spaces. In: Pezzagno, M. and M. Tica (eds.) Town and infrastructure planning for safety and urban quality. CRC Press, Boca Raton, FL, USA.

Charrett, S. 2009. Edible estates: conceptual art meets guerilla gardening in Southwark. Garden, January Issue: 45-46.

Crouch, D. 2003. The art of allotments: culture and cultivation. Five Leaves, Nottingham, UK.

France, R.L. 2003. Wetland design: principles and practices for landscape architects and land-use planners. W.W. Norton, New York, NY, USA.

France, R.L. 2005. Facilitating watershed management: fostering awareness and stewardship. Rowman & Littlefield, Lanham, MD, USA.

France, R.L. 2008. Healing natures, repairing relationships: new perspectives on restoring ecological spaces and consciousness. Green Frigate Books/Libri, Farringdon, UK.

France, R.L. (ed.) 2016. Integrated urban agriculture: precedents, practices, prospects. Libri, Farringdon, UK.

France, R.L. 2020. Waymarking Italy's influence on American environmental imagination while on pilgrimage to Assisi. Cambridge Scholars Publishing, Newcastle upon Tyne, UK.

Gorgolewski, M., J. Komisar and J. Nasr. 2011. Carrot city: creating places for urban agriculture. Monacelli Press, New York, NY, USA.

Haeg, F. (ed.) 2010. Edible Estates: Attack on the front lawn. A project by Fritz Haeg. Metropolis Books, New York, NY, USA.

Hope-Urwin, I. 2013. The Edible Bus Stop: from guerrilla garden to Lambeth 'pocket park'. Available at: www.brixtonblog.com.

Lawson, L. 2005. City bountiful: a century of community gardening in America. University of California Press, Berkeley, CA, USA.

Marshalls PLC. 2018. The Edible Bus Stop – changing the world in tiny steps. Available at: www.marshalls.co.uk.

Nordahl, D. 2009. Public produce: the new urban agriculture. Island Press, Washington, DC, USA.

Pearson, C. and J. Nasby. 2008. The cultivated landscape: an exploration of art and agriculture. McGill-Queen's University Press, Montreal, Canada.

Philips, A. 2013. Designing urban agriculture: a complete guide to the design, construction, maintenance and management of edible landscapes. Wiley, Hoboken, NY, USA.

Polinna, C. 2013. The governance of regeneration – London and Berlin. Regions Magazine 290: 19.

Purvis, K. 2014. The Edible Bus Stop: promising patch of possibility to blooming success. The Guardian, 24 April.

Redekop, S. 2012. Top of the carrot news. Volume 6.

Sereni, E. 1997. History of the Italian agricultural landscape. Princeton University Press, Princeton, NJ, USA.

Smith, K. 2015. Lafayette Greens spreads gospel of farming in Detroit. The Detroit News, 12 July.

Vega, K.A., J. Schlapfer-Miller and C. Kueffer. 2021. Discovering the wild side of urban plants through public engagement. Plants, people, planet. Independent Publication.

Viljoen, A. (ed.) 2005. CPULs: Continuous productive urban landscapes: designing urban agriculture for sustainable cities. Elsevier Architectural Press, Amsterdam, the Netherlands.

Viljoen, A. and K. Bohn. 2015. Second nature urban agriculture: designing productive cities. Routledge, London, UK.

Wilde, A. and A. Meier. 2016. Community food: the role of artists. In: CITIES (ed.) Farming the city: food as a tool for today's urbanisation. Trancity, Haarlem, the Netherlands.

Chapter 6
Healing heritage: restoring agri-culture

6.1 Heritage and homage

Anecdote

Personal experiences of visiting an urban site of agrarian memory. My first trip without being accompanied by my parents was to visit my paternal grandmother who lived in Toronto, Ontario. More than half a century later I still have fond and vivid memories of her taking me to see two sites that helped to instill my abiding interest in history. One was Casa Loma, a mock castle in the form of a Gothic Revival mansion; and the other was Black Creek Pioneer Village, an open-air heritage museum designed to recreate what life would have been like in a mid-nineteenth-century farming community of the region. Even as a twelve-year-old I was struck by the incongruity of visiting a farm surrounded by a grid network of major thoroughfares in this region of north Toronto. Only once before had I been to an open-air museum with costumed interpreters and educators.

Working with the motto that 'the past is all around you,' the museum provided introduction to dozens of different types of heritage breed animals, of which I remember being most impressed by seeing Clydesdales for the first time. Of the many gardens that could be viewed, it was the kitchen garden with its unrecognised heirloom vegetables and rich diversity of medicinal and culinary herbs that I found particularly interesting, especially when my grandmother shared stories of her childhood in England and how her own grandmother would tell her stories about growing the same plants in Yorkshire in the nineteenth century. Never before had I experienced the ability of plants to invoke such ancestral memory. The biggest attraction, however, were the dozens of historic buildings scattered around the site, cobbled together as they were from numerous locations about southern Ontario, and reassembled and restored here at the museum. The lesson, even to my young comprehension, was that heritage needs to be actively saved and preserved. And so, in addition to characteristic village buildings such as shops, homes, a schoolhouse, an inn, workshops, and a church, we visited those with agricultural purpose. From the timbered grain barn I learned that the large central area was for threshing and winnowing, two activities about which I had not known; from the rabbit hutch I learned that the absence of refrigeration meant that pioneer farmers needed to regularly butcher

DOI 10.3920/978-90-8686-938-1_6, © Wageningen Academic Publishers 2022

animals; from the fieldstone apple storage cellar I learned about climate control for food storage in the form of multiple doors for ventilation and straw bedding for insulation; and from the cider mill and its press iron I learned that apples are useful for much more than simply snacking upon.

Book

Plan Al Gharbia 2030 (Briefing book, Charrette proceedings, Regional structure framework plan) by the Abu Dhabi Planning Council (ADPC) presents the results of a regional plan for the western section of the United Arab Emirates, 'Al Gharbia,' based on the 2008/2009 work of an international group of leading design professionals (of which I was a participant, serving as an ecological, environmental engineering, and ecotourism expert). All aspects of comprehensive planning were addressed: infrastructure, health care and services delivery, housing, settlement town plans, transportation, tourism, and the environmental framework. Key directions of the plan were environmental respect, social health, cultural identity, and economic development. And as part of its sustainability foundation, the plan sought to protect and preserve the region's cultural memory through an entwined network of heritage landscapes and archaeological reserves. Included as one component of the overall plan was the desire to develop a tourism industry that was respectful of the unique values and fragility of the local culture and environment by focusing on traditional desert agriculture and cuisine.

Of the six locations specifically targeted in the plan, one was the Liwa Oasis, a marvellous 180-kilometre long crescent of foliage, farms, and interspersed settlements supported by an artisanal aquifer. Located inland and beside the Empty Quarter of Saudi Arabia, the oasis' valleys are filled with beautiful date palm plantations and dotted with camel farms. Significantly, the oasis holds iconic role in Emirati history and identity and is considered by many to be the spiritual heart of the entire country. For many locals the Liwa Oasis is regarded as a focal point for their heritage – a place of cherished and sentimental memory of a simpler time, where today retreat and sanctuary can be sought in face of rapid modernisation. As recently as the middle of the last century, before the discovery of oil transformed the region, women, children, and the elderly would move from the coast during the hot and humid summer months to the temperate oasis. There they would raise camels and farm dates and other vegetables for sustenance. The date palms were crucial to the daily lives of the nomadic Bedu tribes through providing both an essential food source as well as construction material. With the only fresh groundwater in Al Gharbia occurring as an ancient aquifer, the Liwa Oasis has become intensively farmed over the last three decades. Today agriculture, subsidised by government policies, is the largest employer and the primary economic driver in the region. At the same time, there is extensive tourism interest in the oasis due to its lush greenery and proximity to the largest sand dunes in the world. In addition to addressing pressing problems concerning agriculture, such as unsustainable cultivation and overgrazing, the regional plan offers suggestions on how to regulate development in the Liwa Crescent with respect to preserving its important cultural and spiritual place within Emirati culture.

From the start of the charrette, the group of international experts recognised there to be opportunities to tie agriculture to visitor experience in the Liwa Oasis through providing key ecotourism and cultural opportunities. Draft and refined strategic principles independently generated by different breakout groups were similar in their aspirations in this regard: 'Innovation of agricultural practices (date cultivation, cattle herding) to revive and enhance traditional practices and provide employment' (ADPC 2009: 49), with tourism to support enhancement of local values; 'Recognising the continued need to preserve the agricultural heritage of the region, Al Gharbia should become a world model in developing water conservation and supply technologies' (ADPC 2009: 49); and 'The agriculture and food sector in Al Gharbia will provide a strong economy of food, including processing and distribution, built on a supply of sustainably produced local and international agricultural products. Special attention will be given to the development of a 'new desert cuisine' that incorporates traditional food practices in order to meet the expectations of tourism and the Emirate population' (ASDPC 2009: 63).

Tourism is an important part of the government strategy to diversify the Emirate's economy away from a reliance on the petroleum industry. The shared opinion of the international charrette team was that it is difficult to imagine a tourism strategy without a vital link to local cuisine: 'There is a large opportunity to create a new desert cuisine in Al Gharbia. Desert-oriented food that could, at least in part, be grown locally and be processed in the region. Additionally, there are opportunities for chef/cooking schools within new resorts in Al Gharbia' (ADPC 2009: 138). Desert ecocultural tourism should build on existing farming activities. Agrotourism in this light could provide amenities such as driving tour routes and other cultural activities associated with celebrating the rich heritage of date farming and camel ranching. The overarching goal should be the cultural preservation of the Liwa Crescent as a living landscape that is protected and managed through the traditional Arab system of ethical resource conservation and use – *Harim* and *Hima*.

As part of the overall regional plan, a sustainable food strategy was proposed for Al Gharbia, with a focus on strategic agriculture in the Liwa Oasis Crescent:

- Grow and raise what is optimised for the region using intensive and advanced arid agriculture methods including glass houses and re-adapted historical agricultural methods.
- Pursue opportunities to promote legacy agricultural practices for both Emirati and tourist experiences.
- Enhance local food economic multipliers.
- Take a position on the world stage of great and creative food traditions by emphasising desert cuisine.
- Develop a sustainable food research and education system (farming, chefs, etc.).
- Celebrate food by building it into local settlements and the regional economy.

The final goal to be worked toward is for the Liwa Oasis to be nominated and then to achieve official designation by UNESCO as a 'World Heritage Cultural Landscape.' A successful example of this for another agrarian landscape of heritage significance are the Acadian dykelands of Grand-Pré in Nova Scotia, Canada.

Sites

6.1.1 Spryfield Kidston Urban Farm (Halifax, NS, Canada)

The concept of heritage in urban agriculture, if considered at all, has often been associated with preserving cultural traditions and identifies of Indigenous Peoples (as, for example, the case studies 'Restoring balance through traditional agriculture;' 'A new generation of Navajo farmers;' 'Learning and teaching heritage farming;' 'Land, corn, people: Preserving cultural identity;' 'Preserving culture through native seeds;' and 'Native foods define a people', all in Cobb 2012). However, as Scazzosi and Branduini (2020) show in an important recent contribution to the burgeoning field of urban agriculture scholarship, these topics also pertain to European agrourbanism. Unlike the educational 'city farms' in greater London (see Chapter 4.2), which were created *de novo* upon developed ground, all but one of the sites in the present chapter section occupy land that had itself been farmed until quite recently. Therefore, instead of being new introductions of farming into the urban fabric, these sites represent agrarian vestiges that have managed to survive in the face of surrounding urbanisation.

Spryfield, a suburban community in the greater Halifax Regional Municipality of Nova Scotia, is currently experiencing sprawl. A sign announces 'Kidston Estates' [e6.1.1.1], a new housing development comprised of uninspiring modern homes [e6.1.1.2]. Spryfield was established in the late eighteenth century by Captain Spry as a farming community to supply the naval port of Halifax. The Kidston family farm was a large property that was used as a woodlot before being converted to growing hay, grains, and vegetables, and raising livestock, all of which would be taken to the farmers' market in Halifax, the oldest such in North America (see Chapter 3.2). Spryfield's last farm closed in the 1960s, and the community is, as are so many, now dependent on food grown elsewhere. In 1996, a field that had laid fallow for decades was leased from the Kidston family and, following ploughing, was sown with oats. The organisation that leased the land was the 'Urban Farm Museum of Spryfield' (www.urbanfarmspryfield.com), who describe their role as follows: 'By promoting local and seasonal eating in Spryfield, the Society is celebrating the area's agricultural heritage as well as helping to produce a healthy, safe food supply with care for the environment. Fortunately, the rural heritage of Spry's Field is still vibrant in local memory and has left many traditions and skills of self-sufficiency which are still of great practical value today'. Their goal is achieved by establishing a working farm museum which aims to 'commemorate the agricultural heritage of Spryfield, to facilitate food production in the urban area, to serve as an adjunct teaching venue for local schools in the natural and social sciences, the arts and family economics, and to strengthen the traditional social fabric of Spryfield'.

Located adjacent to the new housing development, signs marked with the symbol of a key – used in the province for denoting museums and other historic sites – direct those interested to the urban farm [e6.1.1.3]. Entrance is through a large swing-gate [e6.1.1.4] that gives access to a lovely path through a forest glade, a stark contrast to the suburban sprawl left behind [e6.1.1.5]. Following a walk of five minutes, a sign announces arrival to the municipal historic site of 'Kidston Farm,' thereby providing the link to the name of the new housing development. The sign also informs that the site is run by the 'Urban Farm Museum Society' [e6.1.1.6].

Although there is no signage explicitly stating such, an assumption can be made that the former farm site was once located in the large open expanse that is presently surrounded by woods, beside which are situated picnic tables and a minor art installation [e6.1.1.7]. An attractive area of ornamental flowers and shrubs provides a buffer to the housing development [e6.1.1.8]. The centre of the site is a large community garden bounded by fences to exclude deer [e6.1.1.9]. Inside is the typical jumble of plots of various vegetables that one finds at such locations [e6.1.1.10]. Two attractive design elements include the network of brick walkways (Figure 6.1) [e6.1.1.11] and a large wooden bench (Figure 6.2) [e6.1.1.12]. Another nice touch is the use of a canopy of streamers to provide colour to the garden and whose rustling in the wind might also serve to dissuade visitation by grazing birds [e6.1.1.13]. A tiny shed is used to store equipment [e6.1.1.14] and hosts a chalkboard listing prices of vegetables that are available for sale [e6.1.1.15]. A sign mentions an evening programme focused on family gardening [e6.1.1.16]. The Urban Farm Museum produces a handful of instructional brochures and runs a number of courses, including one offering children instruction in growing vegetables. Other activities and events include workshops on tree planting, oxen ploughing, and restoring wagons, various seasonal celebrations including a harvest fair, and a weekly market, annual plant sale, and periodic heritage walks.

Figure 6.1.

Figure 6.2.

I smile at the contrast of the housing development's sign depicting a lush forest in peak autumn colours with the reality of what I see in front of my eyes: a few sad and spindly trees planted on manicured suburban lawns and along the sidewalk. On the mid-summer day of my visit I have the historic farm to myself. I spend several enjoyable minutes sitting at a table while listening to the twitter of birds. I swivel around to admire how something as simple as stringing up radiating streamers (in a form reminiscent of a medieval maypole) can do much toward aesthetically enlivening a site. I find, however, one aspect that is wanting. Despite much ado being made on the Society's website about the agricultural heritage of the site and their role in reviving and fostering 'a respect for and the pleasure in the heritage,' I cannot find replication of that information at the site itself. This seems like a missed opportunity. But then, perhaps, the only people who visit the site are those who tend the garden plots and therefore are likely to already know about its history.

6.1.2 Zorbis Office and Apartment Building (Athens, Greece)

Many middle-aged and elderly dwellers in cities around the world are but a generation or two removed from their familial rural roots. Athens, with a population of over three million, is the second largest city in south-east Europe. Much of its population lives within a dense urban fabric of high-rise apartments [e6.1.2.1]. That said, many well-off urbanites retain a connection to the countryside by returning to their ancestral homes during summer vacations, where they tend their olive orchards. For one Athenian, this annual sojourn was not enough. Based on 'a deeper need to bring nature into his city life' (Anonymous 2014), he decided to transplant Greece's agrarian traditions to the centre of its capital city. Doing so would allow him to promote principles of green architecture and urban ecology. What is so unusual about this project is that the farming takes place on the roof of a high-rise building.

An otherwise nondescript modern building located in the Ambelokipi neighbourhood of downtown Athens houses the offices of the Zorbis architecture and development firm in addition to sixteen floors of apartment residences (Figure 6.3), including one occupied by the owner of the firm and building. From the street, it is only with the aid of magnification or a keen pair of eyes that one can discern the presence of latticing and vegetation on the building's roof [e6.1.2.2]. Ascending to the penthouse suite and roof through use of a private elevator enables access to the urban farm. Here vegetables and herbs are grown in wooden boxes (Figure 6.4) [e6.1.2.3], and orange, fig, and olive trees in large containers [e6.1.2.4]. The dense vegetation is such that one can almost believe from certain vantage points that the garden is on the ground rather than being elevated seventeen storeys in the air [e6.1.2.4]. At other times, there is no disguising the fact that the farm is on a roof in the middle of a city [e6.1.2.5]. A small patio provides an intimate space for relaxation and entertaining [e6.1.2.6].

Figure 6.3.

Figure 6.4.

If this was all there was to this particular rooftop urban farm, it would not be different from the hundreds of such that have been built around the world in response to the recent increase in popularity of urban agriculture. What makes the present farm so distinctive is what goes on indoors on the top floor. There, in an array of cages [e6.1.2.7], are kept chickens, guinea fowl, and rabbits (Figure 6.5) [e6.1.2.8], and, most remarkably, several goats (Figure 6.6) [e6.1.2.9]. Animals are feed on vegetables that are discarded daily from nearby supermarkets. The daily chores of watering, pruning, milking, cleaning, and collecting produce are undertaken by an employee of the firm. Inside facilities include those related to the production of goat cheese: antiseptic metal working counters (Figure 6.7) and curing areas (Figure 6.8), in addition to fermenting equipment and refrigerated storage [e6.1.2.10]. All this goes toward the stated aim of not just planting an urban garden and raising animals, but managing the farm as an entire ecosystem based on a desire to know everything about the production of one's own food. The site has become popular with friends and visitors, and its owner hopes it will become part of a new movement.

Figure 6.5.

Figure 6.6.

Figure 6.7.

Figure 6.8.

After several days spent visiting more traditional urban agriculture sites in Athens it is with anticipation that I arrive at the building and am soon whisked up to its roof. As interesting as the garden plantings are, there is no denying that it is the livestock which makes the site so special. The incongruity of goats being housed seventeen floors in the air in the middle of a city is hard to wrap one's mind around. My immediate concerns about animal welfare are assuaged (largely but not completely) by the obvious loving care which the owner bestows upon the goats, in addition to learning that they join him and his family on the summer return to the ancestral farm in rural Greece. My admiration for the undertaking increases when I am given a fresh pepper to bite into and a section of goat cheese to nibble on. It is easy to agree with the owner's statement that no store-bought products could be their match for taste. As impressive as the site is, however, there is no escaping the realisation that it is brought about by special circumstances that might make it difficult to be easily replicated. These include the rooftop farm being in a building that is designed, built, operated, and lived in by the owners; a private service elevator that can be commandeered for specific use to maintain the farm; and the daily chores being undertaken by an employee of the family firm in lieu of him being temporarily laid off due to the economic crisis that was occurring in Greece at the time when of my visit. However, it takes but a second bite of cheese for me to refocus my attention on the enjoyment of the immediate experience à la Heidegger's *Dasein.*

6.1.3 Chiswick House Kitchen Garden (London, UK)

There is a long-established tradition of land being set aside in English estates to grow vegetables within protected, walled 'kitchen gardens' (France 2016; Wilson 2010). 'Chiswick House' is an eighteenth-century neo-Palladian villa (Figure 6.9) located in west London. It has been in public ownership since 1929 and is now managed and maintained by 'English Heritage.' Its walled gardens date back hundreds of years, and during the twentieth century were used as a plant nursery before being abandoned in the 1980s. In 2005, a group of volunteers came together to revive the gardens from their state of overgrown neglect. Today they are used to teach local residents about horticulture and the benefits of growing food (www.chiswickhouseandgardens.org.uk).

A sign near the entrance informs about the property's naturalistic design [e6.1.3.1]. The 30-hectare landscape incorporates all the features one expects to see in such estates (as gleaned from watching BBC costume dramas), including wide gravel pathways that bisect expansive lawns [e6.1.3.2] and sculpted water features [e6.1.3.3], some featuring a beautiful bridge or folly (Figure 6.10) [e6.1.3.3]. Also present is a famous ornamental Italian garden (Harris 1994; Woodstra 2013), modelled on those of Roman antiquity [e6.1.3.4]. From here a path leads to the conservatory [e6.1.3.5], a site used by the Beatles for filming a promotion video for their song 'Paperback Writer.'

Figure 6.9.

Figure 6.10.

Tucked into the side of the property, and actually inherited from a neighbouring estate whose house was removed, is a brick-walled enclosure (Figure 6.11) that contains the 'Kitchen Garden' (Figure 6.12) [e6.1.3.6]. The half-hectare garden is filled with abundant plantings of a wide variety of edibles [e6.1.3.7], including berry bushes protected by grazing screens [e6.1.3.8], fruit trees [e6.1.3.9], and a few plots given over to flowers [e6.1.3.10]. A meeting room in a nearby building is used to plan the garden's maintenance programme [e6.1.3.11]. Here, several pieces of antique equipment, including a wheeled table barrow [e6.1.3.12], recall the estate's long tradition of gardening. A single sign provides a brief overview of this history [e6.1.3.13]. Mention is also made of workshops for school children, horticultural therapy sessions, and that harvested produce from the gardens is available for purchase in the café, where it can be enjoyed in the outdoor seating area [e6.1.3.14].

Figure 6.11.

Figure 6.12.

I finally arrive at Cheswick House and Gardens after negotiating a complicated series of linked public transport. The beauty of the Regency Period villa and its landscaped grounds immediately distracts from the intended purpose of the visit. For more than an hour I dreamily stroll about the landscaped grounds as if walking through the pages of an Austen novel. Doing so, I am aware of following in the footsteps of such illustrious visitors to the estate as Voltaire, Rousseau, Ben Franklin, Garibaldi, Queen Victoria, and Gladstone. It is only the sight of the formal Italian garden that brings me back to task. I eventually find my way to the walled Kitchen Garden and seek out a worker from whom to ask whether admission is possible. Once inside, I wander along the paths taking photographs while trying to imagine how busy a place it must have been in its heyday when the garden would have been filled with grounds-staff tending the plots while rubbing shoulders with domestics from the kitchen who were harvesting the produce. Today, it is just me and the ghostly memory of those who worked diligently to sustain the noblesse. Once again, I find myself disappointed in regard to the absence of more information being posted about the history of the place. After all, it's not as if there is a dearth of space along the extensive brick walls upon which to affix several interpretive panels. Soon, however, my reveries are interrupted by a pair of squirrels who, oblivious to my presence, scamper about the garden in a mad flurry of spring courtship. My immediate smile is replaced by a frown when I sadly realise that these are in fact grey squirrels, an invasive species imported by Victorians from North America, and a species which has contributed to the demise of the red squirrels native to 'this sceptred isle'.

6.1.4 Riverdale Farm (Toronto, ON, Canada)

Perched on the scenic west bank of the Don River Valley in the Cabbagetown district of Toronto, 'Riverdale Farm' occupies the site of a former farm that later became a zoo which operated here for more than eighty years before its closure in 1974. Opened in 1978, the Riverdale Farm is meant to be representation of a rural Ontario farm circa the turn of the twentieth century. This is very much a working urban farm where visitors can interact with real farmers doing daily chores. Stressing that it is not a petting zoo, visitors are asked to refrain from hand-feeding the animals (www.riverdalefarm.ca). The farm is home to many heritage breeds that have been carefully selected to be as close as possible to the original farm stock of southern Ontario between 1880 and 1920. These are general-purpose breeds which would have originally accompanied their immigrant owners from England, Ireland, and Scotland: cows suitable for beef and milking, sheep for producing both wool and meat, and horses for riding and timber hauling. Visitors therefore learn how these animals have become rare as a result of decades of industrial agriculture based on breeding specialists. Visitors also have the opportunity to view and learn about the farm's collection of antique farm equipment.

Located but a short distance from the downtown centre of Canada's largest metropolis, and proudly announced by an entrance sign cheekily proclaiming it to be 'A CITY WITHIN A PARK' [e6.1.4.1], Riverdale Farm is a bucolic oasis of three kilometres of rustic lanes and trails that snake their way through wooded areas, alongside fenced fields and paddocks, around ponds, and between historic farm buildings [e6.1.4.2]. Waymarking posts help guide visitors to the farm's attractions [e6.1.4.3]. Many of the onsite outbuildings are classed as heritage structures, originating from regional farms which have been collated and reassembled at the present location (Figure 6.13) [e6.1.4.4]. One building is a careful reproduction of a Victorian-style Ontario farmhouse [e6.1.4.5]. Two towering windvanes, prominent wooden fences, and presence of rusted farm equipment lying about evoke an aura of rusticity (Figure 6.14) [e6.1.4.6]. Cows and horses mill about in pens and paddocks (Figure 6.15) [e6.1.4.7], while goats eat and pigs laze on hay [e6.1.4.8], the latter accompanied by cautionary signs [e6.1.4.9]. Inside a barn [e6.1.4.10] chickens gather on perches or in front of their egg-laying nooks [e6.1.4.11], with breeds identified through hand-held info-paddles [e6.1.4.12]. Vegetables and herbs are grown in raised beds (Figure 6.16) and sold along with fresh eggs and bread and pizzas made in a heritage baking oven at the midweek farmers' market hosted on the site [e6.1.4.13].

Figure 6.13.

Figure 6.14.

Figure 6.15.

Figure 6.16.

Ivy-draped relics of the old zoo add to the historical ambience [e6.1.4.14]. The old zookeeper's residence, which has had a fascinating history over the last century (including serving as a temporary morgue for the adjacent necropolis), houses a museum and is used for community events. An atmospheric old stone bridge provides access to several wildlife ponds [e6.1.4.15], which, along with a butterfly garden [e6.1.4.16], are preserved as a nature reserve in the heart of the city and are used as part of a stewardship programme that educates about conservation and biodiversity. Educational activities for children about food literacy are run throughout the year. The Farm Kitchen provides beverages and homemade snacks for visitors.

The farm, run by the municipal Parks, Forestry and Recreation Division, has become a vital part of the social and economic fabric of the city. 'Thousands of people from the local community and across the City of Toronto to tourists from afar' enjoy the place. Interpretive activities and lessons about agriculture in addition to farm chores of making pottery, weaving, and spinning are offered. The farm has become integrated into the community and was voted as being one of the top reasons to live in the neighbourhood. It is not surprising then that locals volunteer to help with garden planting, clean-ups, and routine maintenance. The farm faces the challenge of securing long-term funding, and has been threatened with closure before being saved through a generous legacy donation.

For a period of four years while a graduate student at the University of Toronto, I cycled past, jogged by, and took public transit within half a kilometre of the Riverdale Farm without knowing of its existence. And so it is during my agrourbanism visit that I find my present-day self repeatedly chastising my past-self for egregious ignorance of such a wonderful urban greenspace. Had I but known of this peaceful retreat located so near my one-time home I would have eagerly sought respite here from those seemingly overpowering trials and tribulations that can feature in one's youth. As a result, I find my time during the present visit to be occupied alternatively between that spent as a scholar actively engaging in the elements of the site, and that spent as a *flâneur* simply lollygagging along while enjoying the tranquillity of the place. It is while wearing the hat of the former that I appreciate how the weathered buildings, livestock, wooden fences, and windvanes all contribute to conveying a feeling of historical authenticity that is so palpable.

6.1.5 Cole Harbour Heritage Farm Museum (Halifax, NS, Canada)

Due to the presence of a climatically sheltered saltmarsh and well-drained drumlins, the Halifax Regional Community of Cole Harbour's potential for agriculture was recognised early in the region's period of European colonisation. As such, it became established as the premier location for supplying the market gardens of the nearby city of Halifax, a location where farming was difficult due to the rocky coastline. As a result, the actual site occupied by the 'Cole Harbour Farm Museum' has been cultivated continuously for more than two centuries (www.coleharbourfarmmuseum.ca). Even well into the twentieth century, local farmers were regionally known for their use of wood and glazed sash frames for germinating seeds in warm beds of manure and soil, a strategy for early spring replanting. The property was saved from development in 1970s through actions of an historian and land-use planner who worked to mobilise the community to preserve the site as a living example of a Nova Scotian family farm. Today, the farm museum is managed by the 'Cole Harbour Rural Heritage Society' and is designated as a 'Provincial Heritage Property' that represents 'a way of life that has all but disappeared in this area.' Government tourism websites refer to the farm museum as 'a rural oasis amid a modern suburban area,' providing 'an authentic piece of living history [that] connects community and visitors to their rural roots and develops understanding of plants, animals and farming today.' The farm museum's director cautions, however, about viewing the farm entirely from a perspective of tourism: 'I'd like to avoid '*The little house on the prairies*' image,' she states, instead focusing on the many educational opportunities provided for schools to learn about such things as composting, animal husbandry, market gardening, heritage construction, and local nature and history (Langan 2021).

The contrast between the Cole Harbour's busy streetscape of gas stations, banks, parking lots, and retails stores, and that of the adjacent farm museum's wooden fences, weathered buildings, and antique machinery is striking [e6.1.5.1]. Access to the farm museum is from a typical suburban development of bungalows [e6.1.5.2]. Signage identifies the site, its sponsor, notable features, and the need for donations [e6.1.5.3]. Immediately upon entrance, the most dramatic observable feature are heritage buildings – dating from 1780-1938 – which have been moved to this location from elsewhere in the region (Figure 6.17) [e6.1.5.4]. It is while standing in the central courtyard bedside the array of picnic tables, each brightly painted in a colour of the rainbow, where one can pirouette and truly appreciate the diversity of the lovely, wood-shingled buildings (Figure 6.18) before wandering off, information pamphlet in hand, to explore each in turn.

Figure 6.17.

Figure 6.18.

A side-by-side couplet of small buildings are the blacksmith shop and milk shed [e6.1.5.5], the interior of each displaying an impressive assemblage of equipment [e6.1.5.6]. Other nineteenth-century buildings include hand-hewn barns whose open doors beckon visitors to venture inside [e6.1.5.7]. It is there where many of the farm museum's two thousand artifacts are displayed, including carriages and ploughs (Figure 6.19), ice harvesting and wood cutting equipment, and an intriguing potato digger [e6.1.5.8]. Interpretive panels inform about the various implements as well as the site's history and operations, such as 'hot bed' growing, weekly trips to market, logging, and ice harvesting [e6.1.5.9].

The farm museum partners with area farms to house and display real farm animals, while emphasising that they are not a petting zoo. Guided by directional waymarking [e6.1.5.10], visitors encounter ducks in water basins, and sheep, calves, and amiable goats in paddocks [e6.1.5.11]. Proximal signs inform of the costs to feed the animals and ask for donations [e6.1.5.12]. Visitors can partake in old-fashioned outdoor games of throwing horseshoes and rolling barrel rings [e6.1.5.13].

Constructed in the 1780s and dismantled and brought to this location in the 1970s, the 'Giles House', with its vernacular saltbox and central chimney design, is one of the oldest buildings in Cole Harbour (Figure 6.20) [e6.1.5.14]. Its backyard contains a kitchen garden that includes an assortment of attractive flowers [e6.1.5.15]. Interior furnishing reveals glimpses of farm domesticity: the icebox in the kitchen, the portrait of Queen Victoria over the mantel in the sitting room, and the spinning loom in the 'housewives corner' (Figure 6.21) [e6.1.5.16]. It is here where one visitor recounted with fondness her experience of volunteering at a spinning wheel for the living history exhibit, mentioning how the rhythmic sounds and feel of fibres from her fingers to the bobbin allowed her to 'escape from the urban worries and responsibilities' of her hectic modern life (Langan 2021).

Figure 6.19.

Figure 6.20.

Figure 6.21.

Fronted by a lovely veranda, the mid-nineteenth-century the 'Harris House' contains the tearoom and gift shop as well as the museum administrative offices and archives [e6.1.5.17]. At its rear is a heritage garden brimming with edible and ornamental plants that has received a national award and is included among the province's top ten 'garden road trip' destinations [e6.1.5.18]. Signs inform about the ecological benefits of companion planting, how the plants derive from others 'rescued' in the area, how the harvested bounty can be enjoyed in the tearoom, and for the need for volunteer gardeners [e6.1.5.19]. Here, too, is a tiny 1930's pumphouse that has been repurposed as a garden tool-shed [e6.1.5.20]. The tearoom (Figure 6.22) offers salads, daily soups, freshly baked biscuits, and most notably, its popular English 'creamed tea' consisting of a cup of tea with scones, strawberries, and clotted cream. The small gift-shop sells a combination of interesting craft items and uninspiring knickknacks (Figure 6.23) [e6.1.5.21]. Upstairs, and accessible by appointment, are the community archives. Here, around a working table, are shelves of books and filing cabinets and cardboard boxes that are stuffed with documents, photographs, and oral history tapes, metal slide-boxes, and magazines (Figure 6.24) [e6.1.5.22].

Figure 6.22.

Figure 6.23.

Figure 6.24.

As is often the case when I visit such agrourbanism projects, it is the contrasting juxtaposition of the rusticity of the farm with the modernity of the city that is most memorable. Nevertheless, at other times when looking out over the collection of wood-shingled buildings, I find it possible, when there is a lull in the noise of the traffic, to imagine myself to have somehow slipped through a wormhole and be transported back in time. This is particularly the case when I observe several children rolling barrel hoops. The combined sight of the laughing children at play in front of the antique barns and livestock in the paddocks really does its best to try to convince me, against all reason, that I am indeed a time traveller. Later, as I circulate around the site, examining and reading about the various artifacts inside the barns and Giles House, I cannot agree more with the farm's claim of being a bona fide 'museum,' an impression that is reinforced when I visit the historical archives.

While partaking in the guilty pleasure of an English creamed tea, I eavesdrop on a conversation occurring at the nearby till. A local has arrived to pick up an order of scones for preparing an afternoon tea at his home to celebrate the arrival of his relatives from England. When it is my turn to pay at the till, I inquire and learn that the tearoom does a brisk take-away business in this regard with people from the community.

6.1.6 Norsk Folkmuseum (Oslo, Norway)

Situated on a large peninsula across from the downtown core of Oslo is the world's first, and one of Europe's largest, open-air museums. As part of a group of national museums located in the area, the 'Norwegian Folk Museum of Cultural History' was founded in 1894 to preserve cultural memories, artefacts, and archives about everyday Norwegian life through the centuries (www.norskfolkemuseum.no). The developmental history of the folk museum is well described on the website, as is its role in contributing to establishing the Norwegian cultural identity. Today, the museum is a research institution that publishes a journal and books, and employs historians, conservators, ethnologists, and other scholars. Exhibitions occur on a wide range of topics, such as folk costumes, emigration to North America, weapons, knitting, folk art, priests, Sami (Lapp) culture, and farm work. It is one of the most popular attractions in Oslo and is visited by more than a quarter million people a year. In addition to the spectacular Gol Stave Church, the folk museum contains 160 buildings from the Middle Ages to today which have been relocated to the site from around country. Many of these are rural farm buildings which are now scattered throughout the museum grounds, whereas others are gathered together in the Old Town' Guided tours and educational programmes are offered to children, youths, and adults. An historic holiday school is run during the summer in which school children dress in costumes and assume old-fashioned names while participating in and learning about farm chores. Seasonal markets, banquets, and performances take place.

It is but a brisk walk of half-an-hour from the city centre to reach the museum district. There, beside buildings housing famous ships of Norwegian seafarers – a Viking long boat, the *Kon-Tiki*, and the *Fram* – is the folk museum. Once inside, admission fees paid, visitors wander along gravel paths winding through the woods. Fields, a few with livestock (Figure 6.25), segue into clusters of farm buildings huddling together at the edge of woodland copses (Figure 6.26). In several locations, buildings are arranged along the central streets of artificial hamlets (Figure 6.27). Although a few buildings sport slate or shingled roofs, most are covered in the traditional manner with turf (Figure 6.28). Most of the homes are constructed of roughly hewn logs or cut planks. Farmhouses are either elongated *loans* or multi-floored 'attics' and stables. Other structures include summer kitchens, timber and grain mills, outhouses, and raised storehouses (*stabbur*). The folk museum website provides detailed information about many of the individual structures and their former occupants. Several buildings are open to enable visitation of their antiques (Figure 6.29), facilitated with explanations provided by costumed interpreters. Another area of the site is dedicated to showcasing a working farm from the 1950s.

Figure 6.25.

Figure 6.26.

Figure 6.27.

Figure 6.28.

Figure 6.29.

My visits to Oslo's boat museums leave me in a state of exploration bliss. It is several decades since I pulled a sled along the complete length of Ellesmere Island (France 2010) following in the footsteps of those who had arrived there aboard the *Fram*, and it will be another decade until my interest and research will lead me to visit both Easter Island, from where the *Kon-Tiki* sailed, and L'Anse aux Meadows, the spot to where Vikings first arrived to North America aboard their long-ships. And so, as I walk toward the folk museum, I know it to have a very tough act to follow in terms of capturing my interest and fostering my enthusiasm. However, it takes but a few moments of wandering from the entrance buildings and into the landscape of faux farm fields and historic farm buildings for me to realise that I need not have been apprehensive. For the folk museum is simply marvellous.

It is a weekday morning early in the tourist season when I visit. As a result, there are times when I am alone but for the lingering presence of imagined ghosts. In such instances it is almost impossible to believe that the bucolic scene in which I am immersed is but a short walk from the high-rise, downtown core of the capital city. The tracks through meadows to clusters of rough-hewn homesteads are so evocative of Norwegian farming heritage. I stop several times to admire how well the atmospheric site has been laid out. It is at such moments when I can also appreciate the folk museum's other gifts of enveloping stillness and quietness. It is really as if the very nature of Time itself has paused, and that there is no such thing as a booming capital city located a few kilometres away.

It is the sight of the turf roofs on the farm buildings that wrenches me back to the present. For much of my recent professional life has been involved in educating architects and landscape architects about the environmental benefits provided by green eco-roofs: their use in mitigating urban runoff and the heat-island effect, and in providing habitat to sustain urban ecology. It is while examining the planted roofs on the historic buildings that I reflect, as I often do when visiting archaeological sites around the world, on how much of what is regarded as 'innovative' sustainable design and green infrastructure is really just reinventing the wheel from the time of our ancestors. This was when humans lived more in balance with nature's constraints and possibilities, thereby avoiding what the Indigenous Hopi refer to as an existence of '*Koyaanisqatsi*'.

6.1.7 Case study lessons

- Agricultural heritage in urban settings can be celebrated on sites of original farms (Spryfield Kidston Urban Farm, Riverdale Farm, Cole Harbour Heritage Farm Museum) and estates (Chiswick House Kitchen Garden) or in novel settings (Norsk Folkmuseum), including private properties (Zorbis Office and Apartment Building).
- Heritage urban farms can exist in bucolic settings (Spryfield Kidston Urban Farm, Riverdale Farm) or adjacent to busy streetscapes (Cole Harbour Heritage Farm Museum, Norsk Folkmuseum).
- Historical authenticity is enhanced through presence of heritage buildings, be they those originally existing on the site (Chiswick House Kitchen Garden, Riverdale Farm) or those that have been relocated there (Cole Harbour Heritage Farm Museum, Norsk Folkmuseum).
- The agricultural heritage displayed can represent a particular period (Chiswick House Kitchen Garden, Riverdale Farm, Cole Harbour Heritage Farm Museum) or cover a range of time (Norsk Folkmuseum).
- Education can occur through establishment of hands-on learning programmes (Chiswick House Kitchen Garden, Riverdale Farm) and the use of 'living history' costumed docents accompanied by displays of farming artifacts (Cole Harbour Heritage Farm Museum, Norsk Folkmuseum).
- Sites for showcasing agricultural heritage for the public can also function as resources for historians through archiving local records (Cole Harbour Heritage Farm Museum) or conducting of scholarly research (Norsk Folkmuseum).
- Produce grown in gardens can be sold in onsite cafés (Chiswick House Kitchen Garden, Cole Harbour Heritage Farm Museum) and markets (Riverdale Farm).

6.2 Civitas

Anecdote

Personal experiences of visiting a suburban neighbourhood remembering its giving trees. Shel Silverstein's best-selling classic, *The giving tree*, deals with the ways in which a tree unselfishly sustains an ungrateful boy. Fully three-quarters of the land area of Massachusetts in the nineteenth century was composed of open farm fields, with only one-quarter remaining in its original forested state. Today, due to the natural process of rewilding that has occurred following abandonment of thousands of pioneer farms, the areal percentages of the two-land use are reversed. When walking through the woods in rural parts of the Commonwealth one is often forewarned when approaching the foundations of a former homestead by the presence of apple and maple trees, a legacy of their planting for sustenance rather than ornamentation. Nearby my home in Nova Scotia, several apple trees in a large field also mark the location where an original farmhouse once stood. My sister lives in a modern subdivision elsewhere in the province that was entirely created on a former apple orchard, such that the properties each contain a vestigial tree. Then there is the Sheldrake Lake subdivision located within the Halifax Regional Municipality. It is a visual treat to visit the leafy neighbourhood of affluent homes in the autumn when the maple trees are at their peak red colour. It is more interesting, however, to time a visit during the early spring when one can observe the activities of the Sheldrake Lake Community Maple Project in full operation. More than thirty trees are regularly tapped by the homeowner's association for harvesting maple syrup. A community sugar shack is set up and a maple education programme has been developed. Technical workshops are run on tapping and boiling tree sap, and the press is often present to film and interview crowds from the neighbourhood. Similar programmes based on rediscovering the original *raison d'être* for planting maple trees and using 'sugaring' to build community identity have spread to other suburban and peri-urban neighbourhoods in both Nova Scotia and Ontario.

Book

Restorative redevelopment of devastated ecocultural landscapes, by myself and contributors, is based on a conference I hosted at Harvard University to provide guidance on how to approach the long-term recovery of the fabled marshlands of southern Iraq – both the wetlands and agricultural plots – following their ecocide by Saddam Hussein after the first Gulf War. Based on a fusion of ecological restoration and sustainable development, 'restorative redevelopment' is advanced as an emerging paradigm for regenerating large-scale landscapes. Rather than merely fixing the broken elements of nature, restorative redevelopment advocates the reuse of devastated landscapes to improve the value and liveability of locations for humans at the same time as effectively reinstating natural processes and functions. Case studies from more than fifteen countries that were presented at the conference by leading authorities, are examined in terms of their insight for the recovery of the Iraqi marshlands as well as other ecocultural landscapes devastated by conflict or natural disaster. Many of the chapters focus specifically on addressing cultural heritage in terms of recognising, remembering, and, where possible, regenerating the agricultural legacy. The overarching aim of the book is to present practical approaches that can be employed for sustaining the process of restoration efforts, both during and after the reparation work has been completed.

The important points are made that restoration can be technically successful but socially unsuccessful, and that it is easier to restore objects than it is relationships. The full potential of restorative redevelopment is not simply the technical restoration of nature but is also the restoration of a cultural relationship with nature. General lessons can be learned from the complexities involved in the post-war rebuilding of Warsaw, and from projects of the International Center for Agricultural Research in Dry Areas which are based on rebuilding agriculture in countries affected by war and climate change. Included with the latter are such aspects as rebuilding seed and food systems, safeguarding and restoring agrodiversity, rebuilding human and institutional capacities, reducing future vulnerabilities, making relief more effective and efficient, and generating returns on investments.

In addition to case studies that focus on technical issues such as transportation, wetland and biodiversity restoration, water treatment and reuse, and salinity management and water conservation in arid agriculture, other chapters focus on developing mutually-supportive linkages between tourism and agriculture. Topics considered here include: community-based management models that are respective of conservation and indigenous culture, the importance of self-development and capacity-building for locals, and compensating farmers for land taken out of cultivation by establishing new markets for other products.

The book closes with an epilogue that showcases three examples of Cambodia's institutional frameworks for ecocultural restorative redevelopment which offer optimistic management models that are useful for Iraq. One of these is the 'Sustainable Livelihoods Project' by the Asia Development Bank, which is based on giving equal credence to all of human, social, natural, physical, and financial capital. The management strategy is described as follows:

> The livelihoods approach is a way of thinking about the objectives, scope, and priorities for development. It seeks to develop an understanding of the factors that lie behind people's choice of livelihood strategy and then to reinforce the positive aspects and mitigate against the constraints or negative influences. Its core principles are that poverty-focused development activity should be people-centred, responsive and participatory, multi-level, conducted in partnership, sustainable, and dynamic (p. 491).

Clearly, such an ecocultural approach that supports environmental integrity as well as cultural heritage in the form of agriculture is a much more complex and challenging undertaking than are approaches concerned with merely fixing the broken pieces of nature as if humans were not inhabitants of complex social-ecological systems. Once again, this encapsulates the difference between regarding landscapes as embodied places rather than as mere spaces.

Sites

6.2.1 Woodlands Farm Trust (London, UK)

The 'Woodlands Farm Trust' is a 40-hectare working farm located south of the Thames on the border between the London boroughs of Greenwich and Bexley. The peri-urban area contains a number of forest patches, and the farm is situated along Watling Street, the historic roadway linking London to Canterbury built by the Romans and later traversed by Chaucer's pilgrims. The farm is a registered charitable trust and is the only farm in the greater London region that is managed in accordance with a national countryside stewardship scheme to foster biodiversity through preserving natural habitat, in this case based on showing how farming and land conservation can work cooperatively through holistic management (www.thewoodlandsfarmtrust.org). The site was cleared of its woodland in the early nineteenth century and has operated as a farm ever since, initially growing cereals, vegetables, and fruit, and raising livestock, and later specialising as a piggery with fields given over to growing fodder. In the 1990s the farm was saved from being destroyed for a motorway and later a housing development. In 1997 the Trust was founded 'to keep access to the farm for community use on the principles of sustainability, conservation, education and community involvement'. Today the site is run as a livestock farm as an educational and recreational resource to the local community.

Signs at the front of the property along the road announce arrival to the farm and inform of its activities [e6.2.1.1]. Situated in a yard filled with wildflowers behind a wrought-iron fence, lies a somewhat foreboding building which was once the farmhouse (Figure 6.30) [e6.2.1.2]. What is obvious from the first steps into the site is that this is a different experience from visiting city farms in the core of London (see Chapter 4.1). Here, the presence of tractors and the jumbled ensemble of outbuildings [e6.2.1.3] and overall scruffy look to the place indicates that this is very much a real working farm, not a simulacra of such. Pens hold rare breeds of cattle [e6.2.1.4], including impressive English longhorns. Friendly pigs scamper about in their mud pen while checking out visitors (Figure 6.31) [e6.2.1.5]. Other animals include lambs, a Shetland pony, guinea pigs, tamworths, chickens, ducks, and bantams, some of whom are located inside a barn [e6.2.1.6]. Nearby, large cut-out signs denote the demonstration milking station (Figure 6.32).

The farm's role in conservation becomes apparent as one moves beyond the cluster of buildings to enter the natural landscape that spreads out behind. Here, using seeds collected from the local area, a tree nursery grows shrubs for replanting in nearby schools and parks [e6.2.1.7]. Rolling hayfields and wildflower meadows are managed by a flock of sheep [e6.2.1.8]. No fertilisers are used, and cutting occurs late in the season to allow flowers to produce seeds for the next year.

Figure 6.30.

Figure 6.31.

Wetlands have been created to sustain wildlife and to separate visitors from an apiary. More than four kilometres of hedgerows comprised of hazel, blackthorn, field maple, hawthorn, and crab apple are maintained as biodiversity corridors for birds, small mammals, and invertebrates. These are managed in the winter to avoid disturbing breeding animals, and contain multiple bird feeders (Figure 6.33). Because of this, the area is popular with birdwatchers [e6.2.1.9], with organised walks and annual biodiversity surveys by volunteers that feed into the national network of data collection. The farm is bordered by an oak woodland [e6.2.1.8; e6.2.1.10] that is linked with others in the area.

Figure 6.32.

Figure 6.33.

Educational programmes [e6.2.1.11] are provided with a focus on farming literacy and skills development for different age groups. These follow the national curriculum or can be tailored to the needs of individual schools. A young farmer's club and a weekly club for tweens and teens allow them to learn about what goes into running a real farm. A self-guided activity trail is based on a character in a popular children's picture book. Tours also exist for community groups, and corporate days are arranged for volunteers to engage in maintenance projects. An unfortunate wrinkle in the fostering of neighbourhood goodwill, especially given the farm's claims of supporting the community, is a conflict that has developed with a local ramblers group concerning right-of-way access to an established footpath that runs through the property.

By the time I arrive at the farm I have been walking for many hours, having taken a ten kilometre detour from the Watling Street pilgrim way to visit the ruins of Lesnes Abbey along the Green Chain Walk. Despite another five kilometres to go until I reach Greenwich Park, the day's destination, I take advantage of the opportunity of being in the vicinity to visit the farm. My interest is immediately taken by the understated ambiance of the site. The straightforward and utilitarian impression that is evoked echoes the famous 'stay calm and carry on' spirit of this nation. It is the integration of the working farm with nature conservation that makes the place special. As my visit is in October, it is not the best time in which to observe wildlife. Nevertheless, I spend several minutes watching birds and enjoying their chatter as they flitter back and forth between the feeders and their hideouts in the dense jungle of the hedgerows. Upon spotting the wetland pond, I leave the trail and bushwhack through the riparian plants to have a look at the water, provoking some waterfowl to take flight. How marvellous the place must be at its springtime peak of biological activity. As I re-shoulder my pack to continue my pilgrimage walk to Greenwich, I reflect on how the Woodland Farm represents a fine example of the multi-functionality of agricultural land in terms of the ability of such places to sustain both humans and nature.

6.2.2 Farm Centre's Legacy Garden (Charlottetown, PEI, Canada)

Prince Edward Island (PEI) is referred to as Canada's 'garden province' as it produces a quarter of the country's potatoes and because farming remains the backbone of the island's economy. The 'PEI Farm Centre' is located in the provincial capital of Charlottetown, where the 'Confederation of Canada' was established in the nineteenth century. The Farm Centre is a non-profit organisation begun in 1975 to provide a central commodity marketing board to promote sales and to build cooperation among agricultural sectors (www.peifarmcentre.com). Its 'Legacy Garden,' created in 2014 only a few minutes from the centre of the city, is the largest urban farm in Canada. The property contains the oldest experimental farm in the country and is unique for being the only research station operated by Agriculture and Agri-Food Canada (AAFC) that permits public access. The Legacy Garden is designed to be a place for rural and urban 'PEIslanders' to come together to share a common history based on remembering past, and celebrating present, connections to agriculture. A section of the farm is run by the federal agency to showcase new farming approaches and to reignite excitement in food production.

Situated near a supermarket and beside the Farm Centre's office building [e6.2.2.1] (whose other tenants include various farm organisations and environmental groups), an entrance sign and attractive wooden promenade (Figure 6.34) [e6.2.2.2] announce arrival to the Legacy Garden. Here, on a 4-hectare plot of land [e6.2.2.3] beside the busy main road through the city [e6.2.2.4], are a series of boxed and unboxed plots filled with a bounty of easily recognised plants (Figure 6.35) [e6.2.2.5]. More than one hundred and fifty plots are rented out at a nominal cost as community gardens, for which there is a waiting list. Circulation about the plots is facilitated via grass or wood chip walkways [e6.2.2.6]. Plants are sustained by drip irrigation and tomatoes grown in a hoop-house [e6.2.2.7]. A fun garden is provided for children [e6.2.2.8]. Other productive areas include a memorial garden, family pastime garden, and urban beehive. A goodwill garden donates thousands of kilograms of produce to local charities that support the most vulnerable through food banks, soup kitchens, and homeless shelters. The lush agricultural fields and accompanying remnant trees [e6.2.2.9] provide a contrast with the adjacent strip development of offices, warehouses, retail outlets, and restaurants [e6.2.2.10]. A newly planted 'food forest' [e6.2.2.11] is based on bio-orchard principles involving mixing fruit and nut trees with large varieties of berries, herbs, and flowers.

The garden's attention to site amenities and decorative elements is commendable. An equipment shed boasts a mural that depicts an old agricultural fair that one imagines the likes of Anne of Green Gables might have attended [e6.2.2.12], while another posted sign informs about the Legacy Garden rules [e6.2.2.13]. Spaces for relaxation are provided through an elegantly designed outdoor patio that features benches and flower-filled planters (Figure 6.36) and a picnic table under a rustic wooden shelter [e6.2.2.14]. Children can climb about a nearby model boat [e6.2.2.15]. A cubist art installation adds a bit of whimsy to the site (Figure 6.37).

Figure 6.34.

Figure 6.35.

Figure 6.36.

Figure 6.37.

Two hectares are run by the research arm of the AAFC [e6.2.2.16]. This is where one aspect concerning 'legacy' comes in with respect to providing a 'celebratory and educational landscape' wherein contemporary farming activities are shown to have had a long history of use on the property itself (historic photographs of these operations are shown on the organisation's website). The area is divided into three zones to showcase PEI's agricultural history, with food being grown in the same proportion as it was on the island during those time periods. A fourth section is dedicated to a vision of the future by emphasising new species that have recently been introduced to the province. The Legacy Garden therefore 'honours the past, recognises the present, and looks to the future in terms of new crops that haven't yet been commercialised in PEI'.

Another way 'legacy' is invoked is through the organisation's motivational slogan that 'from seeds of community spring a legacy of hope'. As mentioned, the site is a former experimental farm that has been repurposed through partnership with local communities. Special outreach programmes encourage recent immigrants (particularly refugees) to share their farming experiences from their birth countries as a way to ease their integration into the multi-cultural melting pot of Canada. Educational programmes are organised with schools concerning healthy eating and sustainable farming. Workshops and seminars provide instruction about the storage, cooking, and preservation of fresh food. A series of social activities brings farmers and city dwellers together. More than a thousand volunteers take part in growing food in the goodwill garden each year, thereby contributing to increasing food security for the city's most vulnerable residents. Due to being located near the centre of Charlottetown, the gardens provide a venue for cultural and community events such as concerts, art exhibitions, seed expos, and farming trade shows.

Although most of the food is donated to charities, some is sold to help cover operating costs. The site can be rented as a beautiful space for private functions such as weddings, family reunions, picnics, campfire gatherings, and other outdoor social occasions (kitchens in the adjacent Farm Centre building can also be rented). Fundraising is underway to buy a freezer to prevent food from going to waste and to build a four-season greenhouse with room to store field equipment. Future plans are to create a tree nursery to educate about carbon sequestration, biodiversity restoration, and soil regeneration.

The Legacy Garden is almost impossible to miss when visiting Charlottetown as it is located on the main thoroughfare that links both city and island to the mainland. It is the large size of the garden that is perhaps its most distinguishing feature. My visit is unexceptional except for a conversation had with the single person I encounter while there: a university summer student involved in the maintenance of the site. I follow him about for an hour while watching him work through a series of routine tasks, each of which he is happy to elaborate on in detail once he learns I am a university professor with a professional interest in agrourbanism and wider concerns related to sustainability. His youthful enthusiasm for his work, combined with his passionate interest in environmental issues, is a reassuring salve to assuage feelings of pessimism about the next generation. I ask him about what the 'legacy' in the title of the garden means in terms of its layout and to his particular job in maintaining operations. He admits to being confused about this. I wander over to the area where he says the experimental plots are located, those that represent the legacy of the Island's history of different farming periods. But I am unable to locate any interpretive signage to help distinguish what is in front of me. I wander into the grocery store located right beside the farm and am disappointed to be unable to find any vegetables for sale from the neighbouring garden; i.e. different from the situations with respect to Food from the Sky in London (Chapter 3.1) or the Mill Creek Urban Farm in Philadelphia (Chapter 7.1). This, I expect, says more about the buying patterns of large stores than anything with regard to the good intentions of the Farm Centre's garden. It is a telling example of how difficult it is to convince retailers to become responsible in terms of reducing food miles through supporting local food initiatives. I leave the site saddened by this realisation, thinking of both the quixotic act of jousting at establishment windmills and of the fact that often the efforts of environmentalists can seem to mimic the labours of Sisyphus. Back outside, I look to the young summer student working away in the distance and silently wish him well in maintaining his optimism about societal change.

6.2.3 Mikveh Israel Agriculture School (Tel Aviv, Israel)

Few agricultural sites are so emblematic of the instrumental role played by farming in nation-building as the 'Mikveh Israel Agriculture School'. Established a few kilometres from the ancient port of Jaffa in 1870, when Palestine was part of the Ottoman Empire, the youth village and agriculture school was the first rural Jewish community in modern times. Funded by European Jewry, the school's mandate was rooted in the Zionist ideal of educating young people about agriculture and having them disperse and build settlements (including *kibbutzim* and *moshavim*) throughout 'Eretz Yisrael' in order to 'make the desert bloom' (Aaronsohn 1995; Fischer 1942). Today the site continues to function as an agricultural school that instructs both day pupils and resident boarders (www.mikvehisrael.org.il). As an 'educational-cultural space,' the school champions its core values of 'heritage, humanism, work, knowledge, sustainability, and responsibility' as a way to 'connect man to the earth in the heart of Israel's urban reality [through] nature conservation and love of country'. This is accomplished by both its focused education programmes as well as tours provided for the general public and school groups.

The school occupies an area of 200 hectares of open space that, given the choking sprawl of surrounding housing development, is referred to as a 'green lung in the heart of the city.' The most impressive building is the synagogue (Figure 6.38), which is famous for being where Zionist leader Theodor Herzel was photographed meeting German Emperor Wilhelm II in 1898, an event that is commemorated by a statue. From here, a grand boulevard of palm trees traces the original access road (Figure 6.39) back to the impressive entrance gate. The site is occupied by a jumble of brick buildings, including a series of barns housing dairy cows and chickens, classrooms, residences, and a somewhat lacklustre visitor centre. Interesting historic features include a blacksmith shop, wine cellars, antique farming equipment displayed out-of-doors, and an impressive Ficus tree from India that is more than a century old. The landscape in the central area is beautifully laid out with groves of trees, lawns, and gardens. Most notably among the latter is the experimental botanical garden that contains more than a thousand species. It is here, from 1930 onwards, that the school planted and studied the adaptability of different species of trees which had imported from locations around the world (Figure 6.40). Located elsewhere on the extensive property are fields of grain crops (Figure 6.41), vegetable plots, fruit tree orchards, and greenhouses, all of which are used as a living laboratory for educating the country's next generation of farmers.

I am in Israel as part of an international team of landscape professionals brought together to participate in a design charrette for transforming the Hiriya landfill into a world class park. To provide us with a familiarity with the immediate area, we have been given a private tour of the neighbouring agricultural school. While on the tour bus driving to the site we hear a large blast in the distance and are told that it can only mean one thing. Indeed, later that evening we are to learn the sad details of the bombing. And so it is in a sombre mood that we are met by our hosts and given a tour of the agricultural school.

Figure 6.38.

Figure 6.39.

Figure 6.40.

Figure 6.41.

The school is an impressive location and the historical details recounted by the guide are interesting. Those members of our group with religious or secular links to the country are much moved by the foundational stories we are told. Later, I wander off by myself to have a look at the crops in order to examine the pumps used to raise the groundwater for irrigating the plants. I stare transfixed as a sudden gust of wind temporarily animates the stalks of wheat, and find myself translocated in time and space to previous visits made to the shores of Lake Kinneret (the Sea of Galilee) as well as toother locations in neighbouring Syria and Jordan. It is during this reverie that I reflect on the archaeological ruins and grain fields that I've seen, all part of the Fertile Crescent, the place where Neolithic agriculture was born. Thinking about the thick history of this particular part of the world and its palimpsest of pain through millennia of conflict makes me remember the sound of the explosion. Checking my watch to confirm that there is enough time before we are due to depart, I walk over to the botanical garden and lie down beneath a massive eucalyptus tree, seeking the solace of horticultural therapy before it is time to re-join the troubling world.

6.2.4 Les Hortillonnages (Amiens, France)

The Somme River flows through the city of Amiens in northern France. Located near the famous cathedral in an area of riparian wetlands are a series of what are referred to as 'floating gardens' but which are actually built atop alluvial islands, just like those of Xochimilco in Mexico City (France 2011). These market gardens, *hortillons,* have existed since Middle Ages and were created when vast beds of peat were excavated for use as fuel, as shown in a carving in the nearby cathedral (Figure 6.42). Although regarded as a heritage landscape in Picardy (Anonymous 2017; Chauzel 2014; Leleux 2012), the gardens have only rarely been discussed in the international literature on urban agriculture (France 2016; Mancebo 2018). Today, 300 hectares remain of the original wetland area of 10,000 hectares. And of that vestigial amount, only 25 hectares are still being actively farmed. At one time up to a thousand people lived on the islands; now little more than a dozen do so full time. Many of the plots have been converted to locations for pleasure gardens and vacation homes. The Association pour la Protection et la Sauvegarde du Site et de l'Environnment des Hortillonnages was established in 1975 to safeguard the site from destruction by a proposed highway bypass. Their ongoing work involves cleaning canals, stabilising banks, and organising boat tours for the general public. The *hortillonnages* are regarded as 'an exceptional tourist site' that offers a 'return to the past'. More than two hundred thousand people visit the urban wetland gardens each year. A small museum opened in 2017 to provide additional information. During the annual Art, Cities & Landscape Festival, art installations throughout the site 'take an offbeat or poetic, funny or critical, but always new, look at this complex environment, on its history and its future, on the link between nature, culture, agriculture and heritage'.

Figure 6.42.

The gardens are composed of plots of land (*aires*) separated by canals (*rieux*) that are navigated by flat-bottomed boats with raised bows designed so as to enable access to the plots without causing damage to the encasing banks. Three locks are used to regulate water levels and to ensure circulation throughout the wetlands. Banks are composed of sediment extracted from the canals and are shaped with use of special shovels. The agro-ecosystem is very productive, with three crop yields per year. In the past, produce was boated to a central floating market where residents of Amiens would lean down to buy directly from farmers in their watercraft. In homage to this tradition, one such market has been held annually since 1993. For the rest of the year, produce from the farms is sold at traditional, terrestrial markets on Saturdays.

Where the Somme leisurely flows through Amiens past riverside dwellings and its stunning cathedral [e6.2.4.1], citizens take advantage of the beautiful setting for their recreational pursuits [e6.2.4.2]. When the city-side sections of the *hortillonnages* is reached, signage informs about the site's history and biodiversity [e6.2.4.3]. Here, also accompanied by interpretive signage [e6.2.4.4], area small bridges, some being simple constructions and others ornately designed, that cross over the perimeter canal (Figure 6.43) [e6.2.4.5], giving access to properties on the edge of the wetlands. Some of these vacation or weekend properties are simple affairs, whereas others show considerable attention to landscaping [e6.2.4.6]. In order to see the area of the wetlands occupied by the 'floating gardens' it is necessary to take an organised tour.

Boats powered by silent electric motors [e6.2.4.7] guide tourists through the maze of canals (Figure 6.44), each of which, according to the map in the provided pamphlet, bears a somewhat grandiose name. The landscape is one of particular beauty [e6.2.4.8]. The canals are lined with willows, alders, and poplars, and contain aquatic macrophytes such as water lilies. All are populated by swans and other waterfowl. That such a thriving urban wild is juxtaposed to the high-rise towers of the modern city [e6.2.4.9] is remarkable. Soon the farms are reached. Here, vegetables are grown in planted rows (Figure 6.45) or on lattices and in hoop-houses [e6.2.4.10]. The banks of the plots are formed of wood and metal, some displaying signs of erosion (Figure 6.46) [e6.2.4.11]. The tilled soil of some plots indicates their dedication to vegetable production [e6.2.4.12]. The presence of flowers and stately trees on other properties denote their use as holiday or weekend cottage retreats [e6.2.4.13]. None of the shelters have electricity and some properties seem to be used only for camping.

Figure 6.43.

Figure 6.44.

Figure 6.45.

Figure 6.46.

I am in Amiens following completing another leg of a walking the complete length of the Western Front. After weeks of traversing the battle-scarred landscapes of loss, seeing the famous statue of the Weeping Angel in the cathedral is a poignant capstone to the sombre pilgrimage. It is the sight of the carving in the narthex of a peasant using a spade to build the *hortillonnages* and of another in the choir stalls of a *paysan* carrying a produce basket to market that helps to shake the gloom away. I exit the cathedral to finish my trip by immersing myself in the vibrant landscape of Amiens' historic urban farms.

The walk along the edge of the wetlands facing the city provides a wonderful tease of what is to come. I jockey for position in the bow of the boat in order to be able to take unfettered photographs during the tour. Soon we are moving along through the maze of canals. Being a scholar of human inhabitation around water, and in wetlands in particular (France 2007), I find myself like a child in a candy shop. I savour the present experience while finding it impossible not to reminisce about similar excursions taken over the years in such places as Mexico, Cambodia, and Peru, all part of researching how agriculture and tourism in these locations might serve as models for restoring the famous Mesopotamian marshes of Iraq (France 2011). The present landscape is so beautiful that it hard to maintain the level of concentration that I normally need to comprehend the rapidly spoken French spoken by the guide. Once, when the boast is paused for a moment at a location, I completely block out the larger environs and focus on the particulars of a single white waterlily in the water in front of me. This the plant that many, including Monet, consider to be the most beautiful of all aquatic flowers. Moving forward again, I am captivated by the unfolding waterscape of properties devoted to urban farming intermixed with those dedicated to being urban refugia. Tour over, I take my leave of the *hortillonnges*, thinking about how the area as an exemplar of multifunctional agrourbanism.

6.2.5 Yedikule Gardens (Istanbul, Turkey)

Istanbul is the only ancient Mediterranean city that still retains a localised concentration of agriculture within its urban framework. On the side of the city opposite to the Bosporus, walls were built in the sixth century by Emperor Theodosius to protect from a land-based invasion. Today, more than five kilometres of the Byzantine walls have received UNESCO World Heritage designation. In the Yedikule ('Castle of the Seven Towers') district, a series of market gardens, called *bostanlari*, are situated in a narrow strip along the outer side of the walls. This is an area that has been farmed for 1,500 years according to descriptions in a sixth-century Byzantine text (Durusoy and Cihanger 2016; Kaldjian 2004). As shown on historic maps and in antique drawings and old photographs, the present structural form of the gardens have exited here since late eighteenth century when the moats were filled in. Records show that the gardens played an important role in producing vegetables for the palaces during the Ottoman era.

Today, municipal officials delineate the borders of hundreds of plots and rent them to individuals deemed capable of tending the land. The livelihoods of dozens of families and hundreds of people depend on produce sold to wholesale agents for distribution and resale at markets and bazars throughout Istanbul. Crops include tomatoes, mint, potatoes, corn, cabbage, beets, carrots, onions, parsley, and the renowned Yedikule *marul* lettuce. This romaine variety, notable for retaining its crispness without wilting, is so famous that it is the focus of an annual lettuce festival. The landscape is described as being 'intensively, expertly and sustainably cultivated to maximise harvests through clever and efficient management of space and resources' (Durusoy and Cihanger 2016). Cultivation methods have been passed down through generations, including techniques in using manure and collecting seeds. Crops are irrigated from seventeenth-century wells, with water pumped into a network of narrow channels that supply wood-framed planting beds, each of which can be opened or closed depending on the growing cycle needs of specific plants. The Yedikule gardens are therefore very much a heritage landscape that mixes old traditions with a modern living culture. By offering glimpses into the city's past, they are an ongoing, self-sustaining way in which 'to reconnect the people of Istanbul with the soil, tradition, and history' (Durusoy and Cihanger 2016).

The most impressive feature of this urban agriculture site is its enormous size. Even at a brisk pace, it takes more than an hour to walk along the full extent of the Byzantine walls and their accompanying heritage gardens. Some sections of the walls have been completely rebuilt whereas other parts exhibit broken towers and piles of crumbled brick. The linear strip of bright green farm plots provides a stunning contrast to the white and pink bricks of the walls (Figure 6.47). From the vantage of the parallel sidewalk, one looks down on clusters of trees and areas of urban wild that alternate with well laid-out rows of vegetables within the bed of the former moat (Figure 6.48). As this is a bona fide agricultural landscape, the scene is one of bustling activity. Farmers can be seen either turning soil with hoes and shovels, picking their produce, or regulating water flow through the numerous ditches. Many plots have ramshackle equipment sheds constructed of metal.

Figure 6.47.

Figure 6.48.

By existing at the interface of the city and countryside, peri-urban gardens are always vulnerable due to the omnipresent threat of urban expansion. In this case, notwithstanding the global rise in popularity of urban agriculture, Istanbul's drive for modernisation has resulted in farmers being evicted from land cultivated by their families for centuries (Anonymous 2016; Hattam 2016; White *et al.* 2015). As one farmer bemoaned: 'I was hoping to be able to pass down this profession to my son in time and establish our own family legacy of sorts beside these walls, but now it seems I will have to look for a new profession myself. There is no place to farm in this city'.

Despite claims that 'what these urban farmers do is very specific and unique to Istanbul's history and world heritage,' the mayor has dismissed the gardens as having any historical value. Although several (Durusoy and Cihanger 2016; Kaldjian 2004; Turan 2018) have argued that the gardens can contribute to the sustainability of the city in terms of ensuring food security, the municipality has stated that growing food in a polluted city is a dangerous activity. The City claims that it is public land and that the farmers are illegally squatting, ignoring the fact that the farmers are regularly charged rent. The erasure of centuries of farming heritage began when tool sheds were dismantled by an excavator accompanied by armed police. The justification given was that the makeshift structures posed a threat to the integrity of the historic walls. The reality is that it was the concomitant removal of soil (Figure 6.49) that damaged the walls of the UNESCO World Heritage site. The problem is that the gardens are viewed by the municipality as locations where undesirables congregate. Farmers, however, argue that they are in fact providing a public service by occupying the land, their presence being a deterrent to the drug addicts and homeless peoples who customarily occupy the run-down sections.

Istanbul's plan is to replace the historic gardens with a green recreational zone that will protect the UNESCO walls at the same time as enable visitor access. This was begun in 2013 when two gardens were bulldozed to create a highly manicured park (Figure 6.50) based on the belief that the farmed plots were visual pollution counter to the image of Istanbul as a modern city pleasing to tourists. Due to outrage from around the world, including a well-publicised visit by noted farming activist Vandana Shiva, the original plan for a recreation park with ornate gardens in addition to playgrounds, shops, restaurants, and parking lots, has been temporarily suspended. But the site remains a contested space whose future use is far from being settled.

Figure 6.49.

Figure 6.50.

From the impressive cisterns, to the Topkapi Palace, to the Blue Mosque, to the breathtaking Hagia Sophia – certainly one of the most impressive buildings ever constructed – it is hard to believe that this wonderful city can offer further architectural wonders. But as soon as the taxi drops me off at the Theodosius walls, I realise that Istanbul has indeed much more to showcase in terms of its heritage. My emotions rollercoaster in but a heartbeat from enthralled admiration to abject disgust as I drop my eyes from the structure of the walls to observe the kaleidoscope of colours of the flower beds at their base. My visit is occurring at a time coincident to criticism being levelled at the municipal government for commencing its plan to destroy the historic gardens and replace them with manicured parks like the one in front of me. The grounds-workers are happily mowing the lawns and tending the flowers, and seem generally pleased that someone who is obviously a tourist is photographing the fruits of their labours. I nod my head and smile at them but quickly move away lest I be unable to contain my disappointment at their actions. In contrast, I spend the next several hours in admiration of the farming activities while I traverse the complete length of the gardens. Seeing the farmers working in their plots flush against the ancient walls is like looking at an illustration in a medieval psalter that has come to life (see Chapter 5.1). It is as if I have suddenly slipped through a temporal wormhole.

About halfway along the walk, I feel limited through being restricted to looking down on the pastoral tableau from the height of the sidewalk. I therefore seek out and find a place of access to descend into the former moat. After strolling for a few minutes, I approach a middle-aged man who is picking vegetables. With an *As-Salaam Alaikum* greeting, followed quickly with an English explanation that I do not speak Turkish, I gesture at some nearby plants and ask if those are Yedikule lettuce. He nods affirmatively and gestures if I'd like to taste one. I agree wholeheartedly and munch away while listening incomprehensively as he talks, so I assume, about the lettuce. Expressing my delight at the taste of the plant, I reach into my daypack and reciprocate by offering him a drink from my bottle of cold water. While he is drinking, I bend down to take a close-up photo of the plant. Then, as I replace both camera and water bottle into my pack, I am offered a partial head of lettuce as a parting gift. And with a vigorous handshake and a heartfelt *tesekkürler* (thank you) I am once again on my way, this time while enjoying the taste of the famous lettuce.

6.2.6 Parc Agrai de Baix Llobregat (Barcelona, Spain)

Situated in the southwestern peri-urban fringe of Barcelona is a remarkable landscape that is the apogee of how agrarian tradition and productivity can complement one another. To understand and appreciate the site, one must realise that urban 'parks' are not just pleasant greenspace set aside for recreation or relaxation. The word can also refer to areas of specified purpose, such as technical or business parks. The park in this case is a working landscape of agricultural fields and natural spaces that covers an area of almost three thousand hectares within the Llobregat River delta, a region of 14 communities and home to 730,000 people. Due to the Mediterranean climate and rich soil of the alluvial plain, the history of farming in the area extends to the Neolithic era. Construction of irrigation canals in the nineteenth century allowed vegetables and fruits to replace wine as the primary crop. An architectural heritage of dozens of traditional buildings is distributed throughout the delta farmlands.

The 'Llobregat Agrarian Park' was established in 1998 as a bulwark against the rapid urbanisation and industrial expansion that threatened to erase the farming tradition of the delta (Dorda and Berenguer 2008; Serafina 2010). Today the park is managed through a consortium of government officials, technical experts, and local farmers' groups dedicated to preserving competitively sustainable agriculture in the area. The site's basic mission is to 'consolidate and develop the territory and to foster the continuity of agrarian activity, [while] promoting specific programmes to conserve production, ecological and cultural values in the agrarian space and [to] developing its economic, environmental and social roles in a framework of sustainable agriculture' (www.parcs.diba.cat/baixllobregat). Five objectives (Dorda and Berenguer 2008) are used to achieve the mission, as described in the management and development plan (Anonymous 2009). These are targeted toward the goals of conserving agricultural land, promoting agricultural enterprise, and preserving and defending the ecological integrity and cultural heritage of the area:

1. Achieve *infrastructural efficiency* (roads, irrigation networks, improving water quality, re-using regenerated water, etc.) and *general services* (rural vigilance, supplies of diesel oil, etc.).
2. Promote *efficient production systems* that respect the environment (integrated pest management and ecological production, zero residues, efficient irrigation systems, etc.) and *commercialisation* (quality brands, Protected Geographical Indication that boosts the increase in farm income).

3. Encourage the *establishment of services* and the *modernisation of farms* in order to improve their viability (standardisation centre for products, agreements with cooperatives and crop plants protection associations ...).
4. Achieve a quality area integrated into the land and in harmony with the natural environment (centre for collecting plastics related to the greenhouse effect, monitoring environmental quality parameters, town planning and environmental disciplines, etc.).
5. Consolidate and promote the natural and cultural heritage of the Agricultural Park, without interfering in agricultural activity, (educational programme for schools, agro-territorial and environmental interpretation centre, Arboretum of traditional varieties of fruits, etc.).

Implicit in all endeavours is the underlying recognition that agricultural land cannot be conserved unless the agriculture taking place is economically viable (Anonymous 2009).

The agrarian park contains a diversity of natural habitats, including cultivated fields, riparian forests, hedgerows, wetlands, and coastal pine-barrens and dunes; this despite the strangulation of motorways, railways, aqueducts, and pipelines, on top of a sprawlscape of factories, waste dumps, and the regional airport. The dominant feature is the presence of 600 operational farms, some of which have been in same families for centuries. More than a thousand fulltime workers are employed, with most of the produce being sold at Barcelona's wholesale market. One quarter of the arable land is planted with artichokes, for a yield of 7,000 tonnes per year. In a strategy to promote regional fare, the park management consortium has begun to brand the produce, a trend that is supported by some of the city's restaurants. Given the climate, plants in this region are thirsty. Historic irrigation channels distribute water throughout the fields. Use of recycled wastewater proved problematic when it was found to be too saline, necessitating other interventions such as a shift to drip irrigation. Tours are organised to allow city dwellers the opportunity to learn about the history of the agroecosystem while observing farmers at work in their fields.

Signage marks the entrance to the park and arrival to the administration offices and information centre [e6.2.6.1] at the Masia de Can Comas, a repurposed farm villa (Figure 6.51). Inside, the exhibition hall contains a series of information panels (Figure 6.52) [e6.2.6.2] that do an admirable job of reviewing the agricultural heritage of the landscape through textual descriptions and the display of old black and white photographs. The images [e6.2.6.3] offer a glimpse into an easily romanticised pastoralism: farmers in their fields stacking hay, picking olives, or simply standing proudly beside their vegetables; smiling families sorting fruit or posing in front of baskets laden with harvest; men guiding oxen-pulled carts and women in an abattoir decked with hanging carcasses; early machinery being used, as well as other scenes of domestic life (Figure 6.53) [e6.2.6.4]. Other panels depict colour photographs of modern tractors and irrigation channels [e6.2.6.5]. A table offers visitors a selection of brochures on the topics of fresh produce branding and its availability at participating restaurants in Barcelona, bicycling routes along the river corridor through the park, cautions about encountering onsite wild boar, a guide to the biology and identification of raptors, and various maps that identify places of interest.

Figure 6.51.

Figure 6.52.

Immediately upon leaving the park offices, visitors are immersed into a fascinating peri-urban landscape where fields are juxtaposed with anthropogenic structures such as power lines and elevated motorways (Figure 6.54) [e6.2.6.6]. Visitors are guided by directional signage [e6.2.6.7] with passage along a network of farm lanes [e6.2.6.8] which occasionally have to be shared with tractors (Figure 6.55). Expansive fields of vegetables, some with stacked bins at the ready to receive the collected harvest, spread out in all directions [e6.2.6.9]. It is obvious that artichokes are the preferred crop [e6.2.6.10]. Irrigation channels of various dimensions border and bisect the fields (Figure 6.56) [e6.2.6.11]. Olive orchards are fed with drip irrigation [e6.2.6.7]. Although modern barns dot the working landscape [e6.2.6.12], it is the encounter of the old farm villas (Figure 6.57) [e6.2.6.13] that makes the visit memorable. Whereas these somewhat dilapidated structures are no longer occupied by humans, their enclosed yards serve as pens for sheep [e6.2.6.14].

Figure 6.53.

Figure 6.54.

Figure 6.55.

Figure 6.56.

Sheep and goats are let out of their pens (Figure 6.58) [e6.2.6.15] to graze on grasses and bushes in the riparian zone [e6.2.6.16]. This latter is the recreational area of the Llobregat Agrarian Park [e6.2.6.17]. Parallel paths exist for cyclists and for pedestrians, the latter on top of the flood protective embankment [e6.2.6.18]. Trailside amenities include resting benches, signs informing about wildlife, and a map showing locations of interest [e6.2.6.19]. Views are provided to the river and its riparian forest and wetlands in the foreground, and an enormous stadium and its flashing sign in the distance [e6.2.6.20]. Exiting the park by crossing over the pedestrian bridge [e6.2.6.21] provides a similar experience of nature colliding with development that characterises this peri-urban landscape. Here, the placidly flowing river [e6.2.6.22] is in sharp contrast to the speeding vehicles on the parallel motorway [e6.2.6.23].

Figure 6.57.

Figure 6.58.

I find myself completely befuddled from the very start. I've taken public transport from the centre of Barcelona to the Llobregat district hospital with the intent of walking from there to the agrarian park's information centre located on the other side of the river. But, in the absence of having a map, I find myself thwarted by the presence of not one but two enormous interchanges that look as if they have been magically transported here from Los Angeles. With no easily discernible way in which to navigate a path through the pretzels of multiple roads, I return to the hospital to hire a taxi from the rank.

The exhibition on the area's agricultural heritage in the visitor centre is informative and useful for experiencing the site. In this regard, each time I encounter one of the historic farmhouses while walking around the park, I reflect back to the collage of photographs of local pastoral life. This helps to give faces to the ghostly presences felt lingering about the now-vacant properties.

Except for the occasional farmer on his tractor, I see no one during my walk through the fields until I join the river corridor trails and encounter several strollers and joggers. I am fascinated by this side-by-side pairing of agriculture and recreation as it reminds me of the UNESCO World Heritage landscape of dykeland farming and hiking trails of Grand-Pré in my present home of Nova Scotia. But it is when the released sheep and goats cross over the farm road to graze in the riparian zone that I really appreciate how the two landscapes can be integrated through holistic management. Above all, it is the not-so-gentle collision of sprawl with greenspace that I find most memorable during my visit to the Parc Agrai de Baix Llobregat: the way the vegetable fields flow around the electrical pylons or underneath the elevated motorways; and the glaring intrusion of the tennis court-sized electronic sign on side of the stadium that flashes it's banal messaging across the verdant river valley.

While riding public transport back into the city centre I examine the maps I'd picked up in the information centre and am struck by two things. First, what is remarkable is that despite several hours spent walking through the landscape, I realise that I've visited less than 10% of the total area of the agrarian park. Second, given that the section I walked around is completely surrounded by busy motorways – designated in alarming red on the map – I cannot really see how pedestrians can move from one section of the immense park to another along the walking tracks marked in yellow. Perhaps such interconnecting through-routes do exist and they have simply not been illustrated by the cartographer. I make a mental note to endeavour to sort out this quandary should I have an opportunity to return to Catalonia in the future.

6.2.7 Case study lessons

- Heritage farms provide opportunities to showcase conservation strategies (Woodlands Farm Trust) and legacy crops (Farm Centre's Legacy Garden, Mikveh Israel Agricultural School).
- People farm and reside in agricultural heritage locations (Les Hortillonnages, Yedikule Gardens, Parc Agrai de Baix Llobregat), demonstrating that these are true working landscapes, not mere museums.
- Agricultural heritage landscapes can be integrated into regional recreation planning (Parc Agrai de Baix Llobregat) and can become tourist destinations (Les Hortillonnages).
- Education occurs through programmes of classes and workshops (Woodlands Farm Trust, Farm Centre's Legacy Garden), crop research and dedicated schooling (Farm Centre's Legacy Garden, Mikveh Israel Agricultural School), and via visitor centres with exhibitions and outreach publications (Parc Agrai de Baix Llobregat).
- Agricultural heritage landscapes often require deliberate action to protect them from being destroyed through urbanisation and modernisation (Woodlands Farm Trust, Les Hortillonnages, Yedikule Gardens, Parc Agrai de Baix Llobregat).
- Onsite preservation of old buildings and other human structures fosters connectivity to the agrarian past (Woodlands Farm Trust, Mikveh Israel Agricultural School, Yedikule Gardens, Parc Agrai de Baix Llobregat).
- Inclusion of community gardens and attention to site aesthetics, including art installations, goes toward creating community places where the past and present can be simultaneously celebrated (Farm Centre's Legacy Garden).

References

Aaronsohn, R. 1995. The beginnings of modern Jewish agriculture in Palestine: 'indigenous' versus 'imported.' Agricultural History 69: 438-453.

Abu Dhabi Planning Council. 2008. Plan Al Gharbia 2030. Briefing book. Government of Abu Dhabi.

Abu Dhabi Planning Council. 2009. Plan Al Gharbia 2030. Charrette proceedings. Government of Abu Dhabi.

Abu Dhabi Planning Council. 2011. Plan Al Gharbia 2030. Regional structure framework plan. Government of Abu Dhabi.

Anonymous. 2009. The management and development plan of the Park Agrari del Baix Lloregat. Edicions La Terra, Barcelona, Spain.

Anonymous. 2014. Greek man turn city roof into farmland. Available at: www.en.enkos.gr.

Anonymous. 2016. Istanbul's urban farmers face eviction from ancient gardens. www.middleeasteye.net.

Anonymous. 2017. The Hortillonnages. Association for the Protection and Backup the Site and the Environment of Hortillonnages. Amiens, France.

Chauzel, L. 2014. Les Hortillonnages: Une tradition maraîchère. Association Pour la Protection et al Sauvegarde des Hortillonnages, Amiens, France.

Cobb, T.D. 2012. Reclaiming our food: how the grassroots food movement is changing the way we eat. Storey Publishing, North Adams, MA, USA.

Dorda, J. M and S. C. Berenguer. 2008. The Baix Lloregat Agricultural Park (Barcelona): an instrument for preserving developing and managing a periurban agricultural area. Available at: www.Ruralitynearthecity.com.

Durusoy, E. and D. Cihanger. 2016. Historic landscape vs. urban commodity?: the case of Yedikule urban gardens, Istanbul. Megaron 20: 125-136.

Fischer, E. 1942. The Mikveh Israel School during the war years, 1914-1918. Jewish Social Studies 4: 269-274.

France, R.L. 2007. Wetlands of mass destruction: ancient presage for contemporary ecocide in southern Iraq. Green Frigate Books/Libri, Farringdon, UK.

France, R.L. 2010. High Arctic extreme science: environmental research for the Trans-Ellesmere Island Ski Expedition. Green Frigate Books/Libri, Farringdon, UK.

France, R.L. 2011a. Restorative redevelopment of devastated ecocultural landscapes. CRC Press, Boca Raton, FL, USA.

France, R.L. 2016. Designing urban agricultural forms: history, education, proposals and projects. In: France, R.L. (ed.) Integrated urban agriculture: precedents, practices, prospects. Libri, Farringdon, UK.

Harris, J. 1994. The Palladian revival: Lord Burlington, his villa and garden at Chiswick. Yale University Press, London, UK.

Hattam, J. 2016. In Istanbul's gardens, a battle for future harvests. Yale Environment 360, 25 February.

Kaldjian, P.J. 2004. Istanbul's *bosans*: a millennium of market gardens. The Geographical Review 94: 284-304.

Langan, K. 2021. Farm country: Cole Harbour heritage farm is a precious little jewel of history and culture in the hectic commotion of the modern world. www.saltscapes.com.

Leleux, P. 2012. Hortillonnages et hortillons. Editions de la Librairie du Labyrinthe, Amiens, France.

Mancebo, F. 2018. Urban agriculture for urban regeneration in the sustainable city. In: Grifoni, R., R. D'Onofrio and M. Sargolini (eds.) Quality of Life in urban landscapes. Springer, Chem, Switzerland.

Scazzosi, L. and P. Branduini 2020. Agricultra: urban agriculture and the heritage potential of agrarian landscape. Springer, Berlin, Germany.

Serafina, S. 2010. Farming at Parc Agrari: Barcelona's backyard. Barcelona Metropolitan, 26 August.

Turan, E. 2018. Gardens of resistance: urban agriculture in the Yedikule market gardens, Istanbul. M.A. Thesis Cultural Studies, Istanbul Bilg University, Turkey.

White, C., A. Shpov and M. Ostovich. 2015. An archaeology of sustenance: the endangered market gardens of Istanbul. In: Cherry, J. and F. Rojas (eds.). Archaeology for the people. Oxbow Books, Oxford, UK.

Wilson, C.A. (ed.) 2010. The country house kitchen garden 1600-1950. The History Press, Cheltenham, UK.

Woodstra, C.A. 2013. The Italian garden at Chiswick House. English Heritage Historical Review 5: 110-133.

Chapter 7

Healing sites and societies: regenerative landscape design

7.1 Structures and sites

Anecdote

Personal experiences of visiting urban farms in Venice. Venice, as everyone knows, or perhaps more accurately, as they believe they know, is drowning. Realistically, this is more in consequence of the incessant tsunami of eighteen million or so 'bight-and-run' tourists that inundate the city every year than it is to the occasional but increasing number of *aqua altae* floods that pour in from the Adriatic. The decaying brickwork palazzos, declining local population, and desultory tourist kitsch are enough to send the sensitive visitor fleeing. For three summers I lived and taught in Venice while conducting research on how to 'save' the city from rising sea levels, tourist hordes, and its own governmental greed and mismanagement (France 2008). A popularist version of that story (France 2011) included a question posed to me by the publisher: 'what is the best thing you can recommend doing in Venice?' My answer was to suggest visiting the nearby island of San Erasmo because it functions as the peri-urban garden for Venice. Here one can walk unimpeded by crowds while observing and interacting with locals working in the fields. The growers take great pride in their canal-side plots and are quick to offer, as is the nature of farmers everywhere, multitudinous suggestions on how the status quo could be improved. One solution they offer towards preserving the economic viability and wellbeing of Venice's ever-diminishing local population is to convert some of its greenspace and relic buildings to gardens that would serve both the community as well as the multitude of hotel restaurants. Regardless of whether or not such a scheme would work, there was an undeniable optimism that exudes from San Erasmo's residents that is so lacking in their neighbours who struggle on in ever decreasing numbers in the historical centre of Venice. Many are the strategies that have been proposed to prevent, or at least slow down, the continued devolution of Venice into that Disneyfied version of its former self: 'Veniceland'. Could not the regeneration of the certain regions of the city, perhaps some of its sporadically visited parks or the under-utilised Arsenale complex, be repurposed in the form of urban gardens?

DOI 10.3920/978-90-8686-938-1_7, © Wageningen Academic Publishers 2022

Book

Socially restorative urbanism: the theory, process and practice of experiemics, by Kevin Thwaites, Aice Mathers and Ian Simkins, addresses the issue that urban regeneration has become increasingly de-humanised, despite warnings made by Jane Jacobs more than half a century ago. To counter this trend, the authors call for placing 'human experience at the forefront of how the urban places that people occupy are understood and made' (p. x). They believe the problem is a consequence of urban planning wrongly focusing on a narrow perception of what constitutes 'place,' by emphasising the physical form of locations while failing to capture their lived experiences. In contrast, a holistic integration of material, spatial, and social attributes is called for in order to build a better form of urban social sustainability. Such an approach is dependent on integrating the activities of architects, landscape architects, and urban designers in order to 'embrace wider sociological and psychological aspects of urban human-environmental relations' (p. 20). Doing so recognises the cardinal need to develop 'a more social conception of restoration' (p. 5). Socially restorative urbanism operates through an experiential landscape lens that envisions place as a fusion of spatial and experiential dimensions (both morphology and participation), and is rooted in phenomenology. By considering the social aspects of sites, the methodology, which the authors refer to as 'experiemics,' forces urban planners to deal with the inhabitants of their spaces based on the qualities of liveability and localism.

Much of the book dives deep into socio-spatial domains, including territorial relationships, particularly the importance of transitional edges (something ecologists have known for decades vis-à-vis 'ecotones'), top-down versus bottom-up design, delivery, and management of sites, the specific case of citizens with special needs (referred to herein as the 'disabled'), and the need to move the focus from lifestyles to communities. Interestingly, the first pair of photographs used in the book to highlight differences between conventional and socially restorative urban design, is from China. One photograph shows a phalanx of sterile apartment towers and the other a traditional street food market. The latter therefore demonstrates Jacob's clarion call for regenerative planners to acknowledge the inherent 'intricate ballet' that is urban life.

Taking a page from the field of environmental psychology, the authors insist that outdoor urban places provide restorative experiences by providing sensations of revival and renewal that reduces stress and mitigates mental fatigue. And a key methodology in which to engage in the experiemic development of socially restorative urbanism is in assessing the direct experience obtained through site visits; in other words, paying attention to the emergent patterns concerning the perceptions of place, such as place- and object-specific experiences, feelings of emotional significance, aspects of imagination and recollection, as well as all manner of social interactions. In short, these are similar to the research tools employed in the present book (see Chapter 2).

Sites

7.1.1 60 Richmond Street Housing (Toronto, ON, Canada)

In 2011, following four years of construction, the Toronto Community Housing Authority opened the city's first co-operative housing project in two decades on a busy street in the downtown core. The 11-storey, mid-rise building contains 85 units of subsidised social housing for low-income workers. This urban in-fill project on the site of a former emergency shelter is intended for people in the hospitality industry who were displaced from another social housing project that was undergoing extensive renovations. The building is popular among its tenants and is well regarded by architects, as attested to by dozens of published articles and the receipt of numerous awards, including ones for design excellence and for best residential building (Novakovics 2019; www.canadianarchitect.com). The building achieved LEED gold standard certification for sustainable design, materials, and operations, due its architecture 'act[ing] as medium to cultivate greenery, cool and cleanse the air, and absorb stormwater' (www.teeplearch.com). For example, in addition to various structural niches that allow daylight to penetrate into the structure, a major feature is the full-building interior atrium that funnels air into the apartments through natural circulation.

One aspect of the innovative sustainable design strategy was to incorporate urban agriculture directly into the building's structure (Gorgolewski *et al.* 2011). The intended occupancy by hospitality workers became the inspiration for the ensuing design plans which included a resident-owned and operated restaurant and training kitchen at street level. A community/kitchen garden occupies the open-air terrace on the sixth floor which is intended to provide the restaurant with herbs and vegetables. The plan is for organic waste from the restaurant to be used in the gardens which are irrigated by rainwater collected from the roof and stored in a large cistern. As the designers state, 'the building is viewed as a living, growing form that responds equally to the urban form of the city and the environmental condition of the site.' The idea was to create a full-cycle ecosystem at a very small scale based on the self-sustaining condition of what they refer to as 'urban permaculture'. And like so many agrourbanism projects, the terrace garden is also meant to serve as a social space to help residents build a sense of community while they build their horticultural skillset.

It is not difficult to spot the award-winning building when walking along the street. For its irregular arrangement of stacked and projecting modules, with brightly coloured balcony niches separated by recessed areas, is distinct from the other mundane structures in the neighbourhood [e7.1.1.1]. Most notable is the large notch on the sixth floor that provides a glimpse into the large interior atrium (Figure 7.1). An airy walkway crosses the open-air atrium to gives access to the terrace garden [e7.1.1.2]. Here, several planter beds contain soil and a few plants (Figure 7.2). Black garbage bags are filled with weeds that have recently been removed [e7.1.1.3]. A system of metal struts and wires affixed to the wall provide a climbing surface for creeping vines [e7.1.1.4]. Several warning notices are posted, one about not exceeding the terrace's load capacity of 60 people, and another that surveillance cameras will record anyone continuing to let their pets defecate or urinate in the gardens [e7.1.1.5].

Figure 7.1.

Figure 7.2.

My first reaction on seeing 60 Richmond Street is to smile. For its intriguing external shape is, as others have noted, certainly redolent of the sort of edgy, modern architecture that is common in the Netherlands but, regrettably rare in Canada. Once inside, however, I find myself singularly disappointed by what I see... or actually, what I don't see in terms of urban agriculture. Even though it is several years since the building was completed and a year following its occupation, there is little that I see in front of me to match the hyperbole I have read about the place. The nearly vacant planters and scraggly dead vine clinging desperately to the mostly barren wall make it hard to believe the necessity for posting a sign cautioning against having more than sixty people on the terrace at any one time. For there is nothing here to make it an attractive place to socialise, except perhaps for waste-depositing dogs. Back outside on the street, I stare into the window of where the restaurant is supposed to be, but find it empty.

In retrospect, this is another example of a site that was visited at an inopportune time. Recent online photos show a verdant sixth-floor garden brimming with a rich diversity of flowers and other plants, and a green wall covered with a creeping vine. And a restaurant has been opened as a lunch spot catering to the many nearby businesses. A lesson here is that successful urban agriculture projects can take time to reach fruition.

7.1.2 King's Cross Central Skip Garden (London, UK)

One of the defining traits of many urban agriculture projects is their ephemeral existence. The temporality of land vacancy therefore encourages innovative approaches to be taken, as for example, in the form of mobile gardens that can be shifted around a site or transported between sites as development progresses. The 27-hectare area around the King's Cross and St. Pancras train stations in central London is undergoing major redevelopment (Cockrall-King

2012) to make it 'one of the biggest regeneration schemes in Europe' (Kamvasinou 2017). In 2009, *The Guardian* newspaper and media group moved into the area and was looking for ways to establish links with the local community. They partnered with Global Generation, a local sustainable education charity that 'strives to develop an understanding of the importance of the environment among the younger generation and promotes enterprise between young people and business' (www.kingscrosscentral.com). As they claim: 'Our gardens provoke curiosity. They are sustainable and creative, giving individuals the opportunity to step beyond the familiar 'business as usual' world, encouraging them to join together with others to grow a new story in which the seemingly impossible is continuously in a process of becoming possible' (www.globalgeneration.org.uk). The fit was ideal as the editors of the progressive thinking newspaper wanted their staff to move beyond simply writing about sustainability to be actually doing something participatory.

The idea was to create a 'productive, bio-diverse and social ecology' for the area; something that was built of abandoned materials recycled from the nearby construction sites, such as wood scaffolding, old water pipes, and 'skips' – the British term for mobile industrial waste containers designed to be loaded onto a special type of truck. Given the ubiquity of skips at all construction sites, their re-purposed use as mini-gardens seemed most appropriate. Significantly, the gardens were completely mobile such that, as different sites in King's Cross Central were being redeveloped, the skips would be shuffled to new locations, all the while continuing to produce their harvest uninterrupted. Different skips were therefore transformed for growing herbs, fruit trees, rotated crops of vegetables, as well as for a worm farm.

An important element of the project is its creation of social enterprise for neighbourhood youth. Toward this end, school groups were involved in the construction and maintenance of the gardens Harvest is sold to local restaurants and to the canteen in *The Guardian* newspaper building. A YouTube video shows speeded up footage of youths harvesting lettuce and spinach, walking down the street and into the newspaper building, where they ascend the escalator to hand over the fresh produce to the canteen chef, all within five minutes of the plants being picked. As one youthful volunteer stated (www.kingscross.co.uk/skip-garden):

> It was the construction of the Skip Garden, where I realised all the magic. Businesses, students, builders, those that were so close yet so far away all became a community, working thoroughly on creating an ecological solution – a garden in a skip! I had then attended a week's training programme, it was a rite of passage to becoming a 'Generator' [i.e. in reference to the affiliated education charity]. I was educated in public speaking, developing my carpentry and gardening skills, and we all transformed into budding young event planners and hosts whilst excelling in our ability to haggle 'pro bonos'.

In turn, food waste from the local restaurants to whom vegetables are provided is routinely collected and goes to power the 'worm factories' that produce the soil used in the skip gardens.

Between 2009 and 2011, the 'Skip Garden' was situated opposite St Pancras Station, where it became a landmark for locals and tourists alike. At the time of my visit in 2012, the skips had moved several blocks away and operations had expanded to include a handful of small hoop-house gardens. An onsite kitchen and workspace are present, and since 2013, visitors are allowed access.

It is a short walk from the King's Cross Station to the construction site marked by the presence of a surrounding barrier fence depicting green foliage [e7.1.2.1]. Here, sky-cranes stand among the hulks of rising apartment towers [e7.1.2.2]. All is shiny glass and metal, with only a few vestiges of the previous brick buildings left in place. A design illustration shows that the future for the regenerated site will bode more of the same [e7.1.2.3]. An information sign welcomes visitors to the mobile Skip Garden, described as a 'self-sustaining vegetable garden with a twist' [e7.1.2.4]:

> As different areas in King's Cross Central are developed, the skips will be moved to new locations across the development site and continue to produce their harvest of vegetables. A new part of London is being created here – and we want it to be the greenest too. This does not just mean building eco-friendly homes and offices. It also means investing in environmental projects like this one. Here in the Skip Garden, we hope to produce enough herbs and vegetables to sell and re-invest in other community projects.
>
> The Skip Garden is the work of many hands. It is being created and gardened by volunteers from the Guardian and the Observer, young people from the area and construction apprentices. The site and materials have been provided by King's Cross Central and its contractor partners ... Global Generation, part-funded by the Big Lottery, are managing the project and directing the gardening.
>
> The project is part of Global Generation's Network of rooftop and ground level food growing sites that are adapting to create a productive, bio-diverse and social ecology in King's Cross. Global Generation is supporting young people to play a pivotal role in transforming cities and the way we live in them. They are being trained to become more social innovators who are building gardens as well as inspiring others to imagine and create a new and sustainable future.

The accompanying map highlights the diversity of existing and proposed edible landscapes in the area [e7.1.2.5]. Inside the construction site, a concrete lot adjacent to several old brick buildings has been given over to the urban agriculture project. The site is filled with a clutter of objects, such as trailers and bicycles [e7.1.2.6], as well as hoop-houses and small crate planters that are brimming with vegetables [e7.1.2.7]. Here too are half a dozen construction skips whose bounty of plants are accessible by wooden steps [e7.1.2.8]. One sports a wire trellis as a surface for climbing plants (Figure 7.3).

In 2015, the skips were on the move again to another nearby location, where they existed until finally closing down in 2019. Global Generation now runs a permanent garden behind the British Library and has hopes to re-establish the Skip Garden at a future date.

Figure 7.3.

I have difficulty in finding the garden skips. They are not where I had thought they would be in front of the train station, and no official I talk to seems to know where they went. On a whim, I ask a newsagent if he might know what happened to them. Happily, he is able to direct me to their new location. The gate is locked and no one is present. My visit, therefore, consists of viewing the site from the sidewalk. Even such a distance of separation, however, does not take away from my favourable impression in the ingenuity of the project. The use of skips repurposed as gardens on an active construction site is perfect. The fact that they are filled with edible plants and not with their customary building waste is a powerful metaphor for the healing and regeneration that can arise through agrourbanism.

7.1.3 Hellinikon Airport Community Garden (Athens, Greece)

In preparation for hosting the 2004 Olympic Games, Athens closed its old Hellinikon Airport and opened a new one in 2001. The initial plan was for the relinquished land of the 600-hectare site to become a large, municipal park. However, there is a history of Greek politicians reneging on promises and commitments for building parks. And so, when rumours began to circulate that the land was about to be sold to an international developer, local residents 'bearing seeds and armed with shovels' moved in to create a communal and productive agricultural space (Yannacopoulou 2012). The country was then plunged into its economic crisis and the tensions between differing visions for repurposing the abandoned land were exacerbated. For politicians in a country facing bankruptcy, the lure of obtaining revenue through selling assets was hard to resist. For a segment of the population faced with increasing austerity measures, the use of community gardens to reclaim public land in the fight against its privatisation was an act of defiance. Residents worked with a university on a land use plan and solicited letters of support for their aims. They then staged a symbolic olive tree planting to assert their rights to the land. The result was that the municipality leased use of 2,600 square metres of land for agriculture.

The role of the 'Hellinikon Airport Community Garden' in the struggle against the controversial redevelopment of the former airport in particular, and as part of the larger movement concerning community gardens as political tools for societal transformation is detailed in Apostolopoulou and Kotsila (2021), where the garden's purpose is explained:

> The self-managed garden is governed by the principles of alternative, social and solidarity economics and sustainability. It believes in an alternative development model ... where self-management, environmental and social justice define a solidarity economy different from today's political, economic and social organisation characterised by greed, competition, individualism, and violence. Solidarity economy is a social movement ... that contributes to the establishment of genuine economic and political democracy. With continuous work, the abandoned space has been gradually transformed.

A considerable portion of the harvest is donated to social service and homeless organisations as part of the gardeners' demonstration that the benefits of their self-organised urban agriculture initiative ensue to a wider community than simply those individuals directly involved in tending the plots.

The portion of the garden site that I am taken to is modest in size and is surrounded by a variety of buildings, some of whose original purpose is difficult to fathom [e7.1.3.1]. Much of the site is a green carpet of a diversity of plants sown directly into the earth and provided with an array of wooden stakes upon which to climb [e7.1.3.2]. The ability of vegetables to flourish on what amounts to a rubble field of stones and loose, sandy soil (Figure 7.4) is astonishing and a testament to the recuperative, healing power of nature. A small working area has been set up with a shed, chairs, and potting trays underneath a shady tree [e7.1.3.3]. Nearby and behind a pile of mulch, a gardener is seen tending to a row of young stalks of corn (Figure 7.5).

The community garden was active between 2013 to 2018 when authorities cut the electricity and water supply. The final announcement from the gardeners in 2019, as relayed in Apostolopoulou and Kotsila (2021), put a positive spin on the outcome of their guerrilla struggle: 'The Hellinikon community garden by resisting the profit-driven privatisation of public land and by (re) claiming urban public space for all, has managed to create a space of coexistence and sharing, of environmentalism, health and food culture, and of solidarity and direct democracy.'

When the final master plan for redeveloping the former airport was released in 2017 it is clear that private interests had won out. Intended greenspace was reduced to a third of the surface area, with much of the new development being catered toward elite residents and tourists. The

Figure 7.4.

Figure 7.5.

plan is now to build a world-class metropolitan park that will be the largest in Europe (www.thehellinkton.com). Elements in the projected 8 billion euro project include a resort, high-rise apartment buildings, marina, shopping centre, business park, sports facilities, casino, and coastal settlement. Components in the park include a nature centre, athletics facility, sculpture park and exhibition area, plaza, some preserved woodlands, and a small piece of land set aside for urban agriculture which is to include something that is referred to as 'a prototype eco-cultivation park'. Conceptual renderings depict people milling about large fields of row plantings of cereal grains and vegetables, as well as orchards. Structural intentions are to repurpose a former terminal building as an exhibition centre, aircraft hangers as aviation museums, and parts of the runways as roads. Time will tell whether the urban agriculture component will respect the legacy of grassroots guerrilla gardening that existed on the site or whether it will be little more than a superficial example of gentrified green-tokenism.

It is as easy as it is sad to notice the big difference in the upkeep of Athenian parks between the times of my agrourbanism visit during economic crisis and that of a previous archaeological-themed visit made three decades before. Where once there had been a pride of place in the well-maintained oases of greenspace, now there is decrepitude, the smell of urine from the homeless and destitute squatters having replaced that of flowers. It is therefore reassuring to visit the airport garden and see how lovingly it is being tended. I watch the gardener as he makes several trips to the mulch pile to load up his wheelbarrow and cart it back to where he works it into the soil between the rows of maize. Later, a feeling of calm descends while I admire the sunflower heads as they sway gently in the breeze. And all, if but for a moment, seems well with the state of the world in terms of its resilience.

7.1.4 Greensgrow Farms (Philadelphia, PA, USA)

Philadelphia has more than four thousand vacant lots which many believe could and should be used for urban agriculture as a way to reduce food miles and food deserts. Urban agriculture is considered to be particularly valuable as an interim use for brownfields which cannot be immediately developed for permanent post-industrial purposes. Enter 'Greensgrow Farms', who believe that urban agriculture works best on such sites that have not yet been fully remediated, but whose contamination pathways are closed and stabilised. Operating with a motto of being growers of food, flowers, and neighbourhoods, the organisation astutely recognises that 'restoration of urban land through revitalisation efforts doesn't affect only the immediate site itself – it has the power to attract and maintain new and old businesses and residents' (www.greensgrow.org).

The organisation's first project addressed a Superfund site in North Philadelphia; an abandoned lot that had been a steel-galvanizing plant (Carroll 2016). In addition to the contaminated soil, open tanks were filled with a cocktail of heavy metals in which children would occasionally swim (Cobb 2012). Remediation of the industrial brownfield necessitated the original soil being capped with a metre-thick protective layer of concrete. Following this, the site became a receptacle for trash until it was transformed, in 1998, into 'a colourful oasis' (Cobb 2012: 96) to function as 'a centre of sustainable business for the local and regional community'. Greensgrow Farms calls itself a 'regenerative farming project,' and operates a nursery, a farm stand, and a CSA, as well as educating about urban agriculture and food systems. Greensgrow Farms considers itself to be 'the premier urban agriculture project in the country,' and operates through the following values:

> Caring for the planet – We value waste reduction and reuse, employ organic farming principles, and support farmers who share our commitment to environmental stewardship.
>
> Caring for people – We provide fair and supportive workplaces, are committed to the success of our customers, and carry products and services that improve quality of life.
>
> Educating by doing – We lead by example and inspire others to achieve their own visions for healthy and verdant communities.
>
> Staying local – We are dedicated to supporting regional farmers and producers within 150 miles of Philadelphia.
>
> Welcoming everyone – We create spaces that are open for all to learn, share, grow, and nurture themselves, their families, and their neighbourhoods.
>
> Maintaining transparency – We are open and honest in our business practices, and committed to sharing our successes and failures for public benefit.
>
> Being ground-breaking – We believe in pushing ourselves to lead cultivation of a healthier, greener city.

Hoop-houses are used for growing trials and to extend the growing season. All plantings are in raised beds with imported soil piled on top of a crushed rock drainage system lying above the concrete encasement. Vegetables grown include chard, kale, arugula, tomatoes, peppers, eggplants, beets, radishes, onions, and leeks. An integrated pest management system is used rather than pesticides. A hydroponics system demonstrates the utility of soilless growing atop concrete, and a vermi-composter converts organic matter into natural fertiliser. Several environmental features are notable additions to site management. For example, leftover fryer oil is obtained from restaurants for the biodiesel powering of their delivery vehicles and other farm machinery. And as the entire 'farm' sits on a slab of concrete capping, water management issues are of paramount importance, thereby necessitating the use of green eco-roofs, rain barrels, and closed-loop water circulation techniques to reduce the runoff. Similarly, sections of the surrounding sidewalks have been opened up and planted in order to facilitate stormwater infiltration.

A portion of the vegetable harvest is provided to help community garden start-ups. More than fifteen thousand people visit the site each year, of which over four hundred are enrolled members in the market-rate CSA program. A parallel, special CSA for low-income families has been established. Nursery and farm sales gross more than a million dollars a year, with much of the revenue going toward paying the salaries of the dozen full-time staff (Hanson and Marty 2012). Greensgrow Farms produces a limited number of its own products, including jams, sauces, veggie spreads, and it's 'Honey from the Hood' from its onsite beehives. Landscape designers and horticulturalists are available to provide advice to shoppers at the nursery. Trees and shrubs are raised for the city's parks, as are grasses and herbaceous plants for new green infrastructure projects. A successful community education programme (Travaline and Humold 2010) provides workshops and master classes in the central hoop-house on sustainability issues related to current environmental problems, food and nutrition, home gardening, and green construction. It is for all these reasons that Greensgrow Farms has received favourable mention in compilations of urban agriculture projects (Cobb 2012; Hanson and Marty 2012; Johnson 2010; Rich 2012).

The 'farm' is located in a mixed-use neighbourhood of two/three-storey working class row houses and occasional vacant lots [e7.1.4.1] in what was once a heavy manufacturing area of the city. The half-hectare city block containing the 'farm' is bordered by greenery along the sidewalks: small trees interspersed in linear strips of flowers that function to absorb runoff, and shrubs growing in lovely decorated pots (Figure 7.6) [e7.1.4.2]. The entrance is marked with banners that give a festival-like ambiance to the site [e.7.1.4.3]. A sign identifies the founders of the project, informs that a self-guided tour 'of the premier urban agriculture project in the country' can be made by reading similar signs about the site, and explains that Greensgrow Farms 'operates three programs: a nursery, a farm stand, and a City Supported Agriculture or CSA.' Furthermore, 'it doesn't just produce farm-fresh produce and nursery plants,' it also provides 'education about urban agriculture, food systems, and environmental issues ... [while being] dedicated to developing profitable green businesses that grow the best products, people and neighbourhoods' [e7.1.4.4].

Figure 7.6.

Once through the entrance, visitors are greeted by a small booth where sales are transacted (Figure 7.7). For it is immediately apparent that the so-called 'farm' is really all about revenue generation. A nearby hoop-house functions as the centre of the commercial enterprise. Here, visitors can browse and purchase such items as canning supplies, living wall planters, terraria, dried moss, glass decorations, hanging baskets, flowers, and cacti (Figure 7.8) [e7.1.4.5]. Outside, bags of soil are for sale, accompanied by useful signage explaining the benefits of various types of soil and mulch [e7.1.4.6]. The bulk of the site is occupied by the nursery which a sign proudly announces won an award from a local magazine [e7.1.4.7]. Here, spread over the concrete pavement that caps the site [e7.1.4.8], are arrays of shrubs and small plants (many of which are flowers) displayed in pots or atop wooden shelves (Figure 7.9) [e7.1.4.9].

Figure 7.7.

Figure 7.8.

Figure 7.9.

A signpost directs visitors to various features of the property [e7.1.4.10], while a notice board informs of events such as a workshop on terraria, classes and certificates in food safety, lectures about summer canning or drip irrigation, and a book signing [e7.1.4.11]. The presence of a brightly decorated truck and trailer indicate the 'farm's' activities extend beyond the bounds of the specific location [e7.1.4.12]. The handful of nestled hoop-houses are the most prominent structures on the site [e7.1.4.13], with associated signage explaining their microclimatic benefits and that raised beds are used [e7.1.4.14]. Inside, a variety of vegetables are grown in long rows [e7.1.4.15]. This, despite the braggadocio of the organisation, represents pretty much the extent of urban agriculture on the site, making the presence of an historic poster about growing food in victory gardens seem somewhat ironic [7.1.4.16]. Other agricultural traces include compost bins, a sleeping hog, and a sign describing how to manage unwanted pests [e7.1.4.17]. A tiny, token hydroponics system is set up outdoors [e7.1.4.18], and an apiary is situated on a green eco-roof atop several storage sheds [e7.1.4.19]. A chicken coop sports overt signage while its residents remain elusive [e7.1.4.20]. Both the farm stand and CSA area are vacant [e7.1.4.7.21]. A sign explains that the impressive diversity of produce available to members in the CSA program [e7.1.4.22] is actually grown elsewhere [e7.1.4.23], the present location functioning more as a distributor rather than a producer of food.

Other features of the site include the imaginative use of a wire roll repurposed as a handwashing fountain (Figure 7.10), and a composting toilet with a green eco-roof (Figure 7.11) and interpretive signage [e7.1.4.24]. What looks like it might perhaps be the touted biodiesel generator lies idle [e7.1.4.25].

Since the time of my visit, Greensgrow Farms has opened another location in Philadelphia and has expanded its garden design consulting service. But they have also faced a series of challenges along the way (Digacaomo 2017; Hanson and Marty 2012). This includes the death of the founder and the loss of her inspired vision, residual animosity of some neighbours leading to theft of gardening tools, increased competition from supermarkets that are stocking more local and organic produce, and the development of other urban farms in the city as well as various home delivery meal services. A further complication has arisen from the need to confront their role in contributing to the gentrification of the surrounding, working-class neighbourhood.

Figure 7.10.

Figure 7.11.

Many agrourbanism enterprises experience changes in their direction over time. Greensgrow Farms began as a commercial hydroponics farm supplying gourmet lettuce to high-end restaurants. Later, a shift in emphasis occurred from growing its own vegetables to being a retailer that positioned itself as a bridge between rural producers and urban consumers (Hurdle 2008). Subsequently, while operating as a stable, for-profit business (Rich 2012), Greensgrow turned to growing and selling flowers as a means to increase revenue (Cobb 2012). At the time of my visit, it was obvious that the nursery side of the operations had completely superseded those pertaining to onsite farming of edible plants. More recently, the organisation refers to itself as a 'non-profit educational urban farm with demonstration gardens that provide an essential connection to food and nature in an urban environment.'

I am in Philadelphia for only a few hours in between flying in from touring regenerative agrourbanism projects in London and before continuing on to visit other sites in Detroit and Toronto. I have high expectations when I arrive at the farm by taxi. After all, its webpage boldly claims that 'Greensgrow Farm is THE nationally recognised leader in urban farming.' Unfortunately, the reality comes nowhere near the hype, especially when compared to the Mill Creek Urban Farm which I have just visited (see Section 7.1.9). I learn a lesson with respect to agrourbanism projects: be wary of statements made in upper case caps. But as unimpressed as I am at finding the 'farm' to be basically a garden centre, there are still elements to admire about the site in terms of its attention to sustainable design. In particular,

I ascend the ladder to the eco-roof of the apiary and enjoy watching and listening to the bees buzzing about the grasses and wildflowers while on their way to and from their hives in search of pollen.

7.1.5 Evergreen Brick Works (Toronto, ON, Canada)

For much of the 20th century, Toronto's Don Valley was home to one of Canada's pre-eminent brickworks, which, at its peak, was producing about three million bricks per year, many of which were used in construction projects across the entire country (www.evergreen.ca). In 1989, the 17-hectare site was acquired by the Toronto and Region Conservation Authority in recognition of its architectural heritage and its role in the development of geological science. The location formally became part of the Toronto Parks System in 1994. Although the initial idea had been to use the space as a plant nursery for municipal re-greening projects, gradually other elements were added (Sengupta *et al.* 2008). Since 2010, the environmental organisation Evergreen has transformed the site through adaptive green design into 'a demonstration of how past and present can work together to create greener models for urban living.' Today, the $55 million project contains the brickworks buildings that have been imaginatively repurposed and the former brick quarry that has been regenerated as an urban oasis for biodiversity and ecotourism, complete with created wetlands, a wildflower meadow, and accompanying observation decks. The project has attracted the attention of scholars with interests as diverse as sustainability initiatives, technological innovation, landscape design, graffiti heritage, experiential tourism, and picturesque ruins (Beaton and Todd 2015; Dodds and Jolliffe 2016; Flannery and Smith 2015; Herrington and Lopes 2019; Irvine 2012; Sengupta *et al.* 2008).

With a motto that 'Cities Need Nature,' Evergreen works with local groups on greenspace planning and sustainable design in urban settings. This is evident in their transformation of the derelict industrial home-site into 'Canada's first large-scale environmental community centre [which] represents a striking new model for the future – a heritage destination where nature meets innovation in the heart of an urban centre.' Guided by the concept of filling old buildings with new ideas, their LEED Platinum-rated central campus building houses the 'Centre for Green Cities' and has been the recipient of numerous awards for sustainable design and construction. A former city mayor stated that the site is important for regenerating natural space and architectural heritage though the 'creative interweaving of economy, ecology, and community.' Revenue comes from venue space hire, sales from the café and gift shop, and rental of office space to progressively minded firms.

Education plays a major role in the vision of Evergreen and its design and operation of the regenerated brickworks site. With most Canadians now living in cities, it is important to show how people and nature can coexist in harmony, as well as to address the urgent need 'to heal our land with nature' (www.evergreen.ca). Citing Richard Louv about how lack of exposure to nature results in stunted sensory development among children (see Chapter 4.1), 'Evergreen Brick

Works' provides hands-on activities, interpretive tours, summer camps, school programmes, community festivals, lectures, walking tours, and family pizza nights. Their aim is to become a destination centre where visitors can 'become active participants in shaping a more sustainable future through education that inspires action' and 'have fun and learn while saving the world' (Sengupta *et al.* 2008: 18, 20).

From its inception, the project integrated urban agriculture into its plans and design as one of the core means for obtaining community engagement as well as a way to promote healthy eating and lifestyles (Gorgolewski *et al.* 2011). A farmers' market with a vendor-linked CSA program was opened onsite in 2006, and a gardening and eco-art program began in 2009. With its native plant nursery and ecological demonstration gardens, the site has become a retail destination for gardening products, a place where urbanites can purchase native and heritage plants, organic soil, and fertiliser, as well as wildlife houses and an assortment of eco-friendly objects. Central to the original initiative, the site plan calls for demonstration garden and nursery areas of more than ten thousand square metres dedicated to teach about urban agriculture. Other edible landscapes included in the site plan are a children's play area with fruit trees and berry bushes and a welcoming courtyard with a garden that includes produce intermixed with native trees and shrubs, wildflowers, and marsh plants. All this led urban agriculture scholars Gorgolewski *et al.* (2011), writing when the project was still in its design stage, to conclude that the Evergreen Brick Works will succeed in increasing food security and bringing people together.

The reality is that at the time of my visit, urban agriculture seems to have been air-brushed away in the subsequent development of the site. This can be shown from its invisibility in the brochures about the various self-guided tours that are offered. The geology tour focuses on the industrial excavation of the site, its significance in terms of fossils, interglacial periods, and the concept of deep time, as well as watershed landscapes. The art tour focuses on trash recycling, heritage photographs of the site, and eco-art interpretations of watersheds, geology, and mapping nature. Whereas it can be countenanced that urban agriculture might not figure prominently in these particular tours, its absence from mention in the next pair of tours is puzzling. The sustainability tour highlights the Centre for Green Cities, rainwater harvesting cisterns, the ice-rink heat recovery system, stormwater management pond, vegetated greenways, permeable pavement in parking lots, and other green energy concerns. And the ecology tour flags the stormwater management sites, wild meadow, erosion control plantings, managed coppice woodland, mesic greenways, wildlife plantings, and the propagation gardens that provide seeds for onsite rewilding plantings.

The size and complexity of the Evergreen Brick Works regeneration project is immediately apparent from the schematic site plan and building identification key shown on the entrance sign [e7.1.5.1]. The farmers' market is in full swing on the Saturday morning of my visit, part of it under a roof that allows natural light to flood into the space [e.7.1.5.2]. Here, urban foodies browse booths and tables filled with fresh produce, bagged greens, eggs, meat, fruit, bread, and pre-cooked snacks (Figure 7.12) [e7.1.5.3] while interacting with more than sixty regional farmers, producers, entrepreneurs, and chefs (Figure 7.13). In another building, an information panel and large banner denote the garden market [e7.1.5.4] which is stocked with a diversity of plants for sale in addition to raised planting box kits [e7.1.5.5]. One planter is used to demonstrate the utility of using fertiliser to enhance growth [e7.1.5.6]. At the rear of the building, a courtyard purposed for children contains edible plants in wooden or metal planters (Figure 7.14) [e7.1.5.7] and in a small glasshouse (Figure 7.15).

As scant as the urban agriculture offerings may be – it seems that the food programme was always to be based on selling rather than producing local and organic food (Sengupta *et al.* 2008) – it is its regenerative design and attention to sustainability issues where the Evergreen Brick Works both generally excels and occasionally disappoints. Portions of the site's 'water sensitive design' (*sensu* France 2002) are commendable, including best management practices and interpretive signage about porous pavement in the parking lot and rooftop rain harvested and stored in painted barrels [e7.1.5.8], street and parking lot runoff directed through notched curbs into bioretention swales (Figure 7.16) [e7.1.5.9], and raingardens and bioswales beside the buildings (Figure 7.17) [e7.1.5.10]. However, the design of other areas seems to have forgotten that water, unless in an

Figure 7.12.

Figure 7.13.

Figure 7.14.

Figure 7.15.

Escher drawing, runs downhill. Here, the raised areas of plants actually shed rather than absorb runoff. Whereas perhaps there might be a design constraint necessitating raised planting beds in the innovative ice-rink area [e7.1.5.11], there is no excuse for such mistakes in sustainable design in the other spaces such as the commons area [e7.1.5.12], and especially not the parking lot [e7.1.5.13] where inclusion of plug-in power source for electric cars does little to balance out the poor stormwater design [e7.1.5.13]. And then, rather than building stormwater treatment wetlands with sediment forebays, which have become an accepted practice in the landscape architecture of sustainable design (France 2003), the site features hard-edged detention basins [e7.1.5.14].

Other design elements of Evergreen Brick Works are to be praised, such as the innovative ice skating rink, use of natural materials in outdoor seating areas, and art installations [e7.1.5.15]. Of particular note, is the metal and living wall map depicting Toronto's system of ravines, including the Don Valley [e7.1.5.16]. The organisation has also done an excellent job at informing about the past use of the renovated brickworks buildings [e7.1.5.17] as well as in preserving remains of the site's industrial heritage [e7.1.5.18]. In a similar vein, whereas many regeneration projects tend to minimalize or even ignore their site's industrial heritage (France 2008), Evergreen Brick Works celebrates that heritage and uses it for educational purposes with photographs and interpretive signage (Figure 7.18) [e7.1.5.19].

The propagation gardens are located in a tiny parcel of land near the front of the building [e7.1.5.20] and are accompanied by a sign describing their purpose [e7.1.5.21]:

> Established during the early stages of Evergreen Brick Works' revitalisation, the propagation gardens are a great location to observe pollinators in action.
>
> Maintained by Evergreen staff and volunteers, the seed beds feature large groupings of native wildflower and grass species such as Pearly Everlasting, Wild Bergamont, Wild Strawberry, Canada Wild Rice, Switch Grass and Showy Tick Trefoil. When the plants

Figure 7.16.

Figure 7.17.

> produce seed, it is collected, processed and stored as required – providing educational opportunities, as well as an on-site seed source for planting throughout the GTA [Greater Toronto Area].
>
> Weekly from April to October, Evergreen leads an evening garden group comprised of dedicated volunteers who work to monitor and maintain this garden, as well as other areas on site.

Above all, it is the project's ecological 'restoration' of the former brickworks quarry that is to be lauded. Here, the created wetlands, trails, boardwalks, wildlife nests, wildlife meadows, quarry gardens, observation lookouts, and interpretive signs (Figure 7.19) [e7.1.5.22] contribute to Evergreen Brick Works being meritorious of receiving international recognition and acclaim, regardless of whether it has lived up to its original promise vis-à-vis urban agriculture. The views from the re-naturalised park back over the former factory nestled into the greenery of the Don Valley, with the downtown skyline of Toronto in the background and both the brickworks' chimney and the CN Tower reaching skyward, are truly memorable [e7.1.5.23].

The attention paid to urban agriculture in the Evergreen Brick Works has increased since the time of my visit. Small food gardens have been established to demonstrate innovative growing techniques with respect to balcony gardens, companion planting, heirloom plants, indoor growth tubs, vermicomposting, and closed-loops systems. Several cooking classrooms exist for workshops. The website contains a rich depository of information with reports, guides, videos, and online tools. Included here are a series of resource guides on the urban agriculture topics of backyard herb gardens, managing weeds and pests organically, composting, growing sprouts, container gardens, urban orchard maintenance, growing social programs, and outdoor edible classrooms.

Figure 7.18.

Figure 7.19.

My expectations are high when I visit the Evergreen Brick Works. After all, it advertises itself as being the 'international hub of sustainability' and the 'greenest facility in North America'. Once again, I come away feeling the hype is not necessarily matched by the reality on the ground. As I wander around the site, I cannot help but think about how a particular parcel of land here or there might have been devoted to showcasing different elements of urban agriculture (I am happy to learn that more effort has been made in this direction since the time of my visit). And as impressive as the re-naturalised quarry certainly is, with its wetlands and meadows, it seems a missed opportunity that a section of the large space could not have been put aside for community gardens. As I sit down on a bench to scribble my notes, I reflect on how in contrast to concerns related climate efficiency, the sustainable design profession has been slow to address issues related to food efficacy. I think, too, on the different meaning behind those two words, 'efficiency' and 'efficacy', the latter certainly being the gold standard toward which to strive.

But it is the strange bipolar way that stormwater runoff is treated on the site that most rankles. The raised beds of plants and hard-edge detention basin are in sharp contrast to the innovative water sensitive design exhibited elsewhere on the site. I seek out someone from the organisation for a chat, mentioning that for more than a decade I've taught and written books on the subject. Unfortunately, she is unable to answer any of my questions and tells me, upon receipt of my business card, that I can expect to receive communication from someone in the know in the very near future. But as that never comes, the lasting memory is of receiving a classic institutional brush-off.

7.1.6 Artspace Wychwood Barns (Toronto, ON, Canada)

Toronto, with a comprehensive programme of community gardens, numerous farmers' markets, food and garden tourism endeavours (e.g. *Ontario culinary adventure guide: discover what's on our tables* and *Exploring Toronto garden map*), and even its own food gardening magazine (*Edible Toronto and the Golden Horseshoe*), is recognised as a leading city for agrourbanism. It is not a surprise then to find a project such as 'Artspace Wychwood Barns' that is notable for being a success in terms of coupling regenerative design and urban agriculture. Indeed, Ladner (2011: 202) can't resist use of an exclamation point in his summary of the project: 'It houses a large greenhouse, a kitchen, a demonstration garden, after-school programs for elementary school children, seedling production for local community gardeners, vermicomposting demonstrations, cooking classes and catering. All this with accommodations for artists in the same building!'

From 1913 to the 1980s, the site's building functioned as a facility or 'barn' to maintain the city's fleet of streetcars. Today, as a brilliant example of adaptive reuse, the heritage building has been transformed for multipurpose use and the site functions as one of the most important food hubs in Toronto [e7.1.6.1]. The major tenant is the 'Green Barn' run by 'The Stop Community Food Centre' [e7.1.6.2], an organisation that has been addressing food concerns for Toronto's underprivileged for decades (www.thestop.com; see Chapter 4.1). A lively farmers' market (Campigotto 2010; Dodds and Jolliffe 2016) is hosted onsite every Saturday, with vendors situated under canopies in front of the building [e7.1.6.3]. Here, Torontonians can purchase food [e7.1.6.4] while interacting closely with producers from the region (Figure 7.20) [e7.1.6.5]. Opportunities also exist for acquiring other items such as antique photos or participating in storytelling sessions [e7.1.6.6]. Alongside the renovated building and underneath stencilled letters proclaiming the organisation's motto of 'the power of food', a series of small plots are interspersed with benches (Figure 7.21) [e7.1.6.7]. This is the 'Global Roots Garden,' a place where seniors and youth from the city's diverse cultural communities can grow food while sharing knowledge from around the world [e7.1.6.8]. Attached to the building is a high-tech, computer-controlled greenhouse which is designed to Gold LEED environmental standards and is accessible to interested visitors [e7.1.6.9]. The Stop uses the space to grow food (Figure 7.22) [e7.1.6.10] for their own programmes as well as for providing hands-on learning opportunities for children. A poster informs about a survey conducted in relation to The Stop's activities and challenges [e7.1.6.11]. A courtyard is dedicated to a sheltered garden [e7.1.6.12] which capitalises on the beneficial heat-radiating walls to demonstrate approaches for extending the growing season of various plants (Anonymous 2011).

Designed as a 'community space where people can mix and mingle while watching or participating in various activities,' it is ultimately the multiuse blending of spaces at the Artspace Wychwood Barns – the integrative celebration of food, fun, and art – that leaves such a strong impression upon visitors. Donations are acknowledged on a decoration showing sponsors' names emblazoned on metal leaves [e7.1.6.13]. Other attractive elements include the rainbow-coloured wrapping of the old smokestack [e7.1.6.14] which towers over a beautifully designed water sprinkler playground

Figure 7.20.

Figure 7.21.

that is as popular with young children during the day (Figure 7.23) [e7.1.6.14], as is the courtyard wood-fired oven [e7.1.6.15] with adults during make-your-own pizza evenings. The open space of the renovated former workplace for servicing streetcars (Figure 7.24) is used as a venue for many functions, one of which was the final reception at the 'Urban Agriculture Summit,' an international conference that was held in Toronto in 2012. The space is designed to allow the building's past history to be clearly viewed and easily understood (Gorgolewski *et al.* 2011). In homage to the past, the brick walls are decorated with large photos that highlight the previous streetcar-related history of the site [e7.1.6.16]. Upstairs, office space is rented to nonprofit environmental and arts organisations [e7.1.6.17], including a theatre group for youth [e7.1.6.18]. Outside are found artist's studios marked by attractive sidewalk plantings of trees and flowers [e7.1.6.19].

Figure 7.22.

Figure 7.23.

Figure 7.24.

Artspace Wychwood Barns is well plugged into the local community. Several abutting neighbours have replaced their front lawns with pocket gardens of edible plants, thereby showing how food can be grown in the smallest of urban spaces [e7.1.6.20]. The Stop Community Centre runs several nearby community gardens, with some of produce being donated to its food bank for distribution to the needy in the surrounding low-income neighbourhood (see Chapter 4.1). The site therefore serves as a high visibility base for The Stop to function as a 'thriving, holistic community centre where food is used to build health, skills and community'; in other words, as an agrourbanism 'food hub' (De la Salle and Holland 2010). The wide diversity of programmes include group cooking lessons, skills workshops, seed exchange, YIMBY ('yes in my back yard') yard sharing [e7.1.6.21], sustainable food system education for school groups, family support and education about nutrition, 'eat local, taste global' initiatives, community action, after-school sessions, and a summer camp. Testimonials from the benefiting individuals in the community are effusive (Anonymous 2011), motivating food celebrities as the likes of chef Jamie Oliver and writer Raj Patel to intone, respectively, that: 'I've travelled all around the world, and I've never seen anything like The Stop. Every city should have one,' and 'A truly revolutionary organisation. The Stop is at the forefront of a movement to create a more equitable, sustainable and healthy food system.' Nasr *et al.* (2015: 28) refer to the site as an exceptional example of 'a multifunctional community space where production of food is secondary to other functions: production of seedlings for distribution to various groups around the city, space for teaching about growing and preparing food, even space for special events and training meetings.'

It has been several months since my initial site visit to Artspace Wychwood Barns and I'm back at the location to enjoy the festivities marking the end of the Urban Agriculture Summit. The space is filled with practitioners, scholars, and urban farming/gardening enthusiasts from around the world, all of whom are milling about, exchanging banter while endeavouring to partake from their plates of food and containers of beverage balanced precariously in their hands. I spend time with several authors (Karen Landman, Joe Nasr, Jennifer Cockrall-King) whom I have solicited to contribute to my first book on the subject, *Integrated urban agriculture: precedents, practices, prospects*, while we share in celebrating an award being given to another contributor to that tome, noted expert in urban agriculture and international development, Luc Mougeot. There is a palpable feeling amongst those present that we could be participating in a cuspal moment in terms of recognition and attention being paid to urban agriculture.

As the evening of socializing goes on, I take a break from the reception to seek a moment of quietness in one of the outdoor gardens. There, oblivious to all the highfalutin experts waving their arms about inside, is an elderly woman on her knees harvesting plants from her section of the Global Roots Garden. Yes, the process of growing food in cities is something to be talked about and celebrated, but also, more importantly, the product that derives therefrom, the food itself, is something to be simply harvested and consumed without much ado. That, I think, while taking another sip from my drink, would mark the true achievement of success in the agrourbanism movement.

7.1.7 Allen Street Garden (Boston, MA, USA)

Somerville, part of greater metropolitan Boston, is New England's most densely developed city. In addition to three-storey family residences, the Allen Street neighbourhood contains industrial vestiges such as derelict buildings, wide expanses of broken concrete pavement, piles of discarded heating radiators, metal scrap containers, and abandoned rail cars (Figure 7.25) [e7.1.7.1]. A former residential property that had been, like many others in the area, a vacant eyesore through illegal dumping for more than four decades, was assessed, cleaned-up, and redeveloped under the city's brownfield redevelopment programmes as a community garden and pocket park (Anonymous 2007). Since then, the 450-square-metre site has attracted the interest of architects and brownfield experts (Wissinger 2016; www.brownsfieldsresearchlab.com), as it is looked upon as being a good example of success in regenerative landscape design. This results from its adding value to the surrounding neighbourhood through improving, rather than fixing, a space.

Sidewalk tree plantings [e7.1.7.2] and signage [e7.1.7.3] signify arrival to the community gardens. An attractive and inviting entrance [e7.1.7.4] gives access to an artfully designed small plaza [e7.1.7.5] from where pathways lead out to the plots (Figure 7.26) [e7.1.7.6]. Remediation of the site involved removing the top metre of soil which was determined to be polluted with coal and wood ash, trace organics, and metals. This was replaced with clean fill upon which the new surface features of the regenerated landscape were constructed to create what the designers refer to as a 'welcome oasis.' The community gardens are comprised by more than a dozen raised beds which have been filled with the imported soil. The beds are bordered by either brick or plastic and grow a variety of produce [e7.1.7.7]. A waiting list exists for acquiring a plot. The perimeter fence and an adjacent building are used to affix strings for climbing plants [e7.1.7.8]. Compost bins and a water tap complete the horticultural accessories [e7.1.7.9].

The site is particularly interesting in demonstrating that gardens can be seamlessly combined with park elements to create vibrant, multipurpose greenspace in inner cities. The park contains attractive seating benches (Figure 7.27) [e7.1.7.10], flower beds [e7.1.7.11], and a colourful

Figure 7.25.

community notice board [e7.1.7.12]. And in a region of the city with has so little greenspace, it is no surprise that the site also attracts a considerable diversity of birds and butterflies. That there has been a ripple effect of the project in the neighbourhood is shown by the mural on a grade school around the corner which depicts children gardening [e7.1.7.13].

Following my visit, a children's playground has been incorporated into the site. This was coincident with renovations to the community gardens, including building new plots, improving wheelchair accessibility, adding a larger storage shed, tables in the plaza, and a switch to permeable pavers for stormwater infiltration. The community garden is now one of eleven administered by the city.

Figure 7.26.

Figure 7.27.

It is good to be back in Somerville, only a ten minute walk from where I lived for a decade and a municipality whose working-class character I always admired as a healthy counterpoint to the hoity-toity airs of neighbouring Cambridge. I have paternalistic feelings toward these gardens as they were designed as a summer work project by former students of mine. My most memorable sensory impression of visiting the site is an auditory one. Indeed, as I walk through the neighbourhood and approach the gardens, I actually hear them before they even come into view. The cacophony of birdsong is remarkable, and is in such stark contrast to the silence that accompanies the biological sterility of adjacent city blocks. Tour and photography completed, I sit down on one of the attractive benches in the plaza and gaze up at the open windows in several of the adjacent homes. How pleasant it must be to wake up in a city to the sound of chirping birds greeting the morning rather than the din of cars. I settle back and relax while reflecting on how urban agriculture, even small pockets such as the present gardens, can truly provide benefits far in excess of its primary utilitarian role of growing food.

7.1.8 PERKA (Thessalonki, Greece)

The 'PERKA' – [PE]*riastikes* [KA]*liergies* or peri-urban agriculture in Greek – project is part of the great boom in urban agriculture that occurred in Greece in consequence of food security concerns related to the economic crisis. At same time, these projects have an acknowledged role in community building, horticultural healing therapy, reconnections with nature, city greening, and sustainability (Anthropoulou *et al.* 2018). The city of Thessaloniki is the second largest in Greece and is located in the northern part of the country where a onetime nervousness in relation to several of its neighbours led to a strong military presence. A 70-hectare former military encampment on the western edge of Thessaloniki was 'captured' in 2011 by a group of indignant citizens in a bottom-up political action against the authorities. This was coincident to the global 'Occupy Movement.' The greenspace guerrillas were opposed to the privatisation and redevelopment of the former base, and insisted on the importance of ensuring the public character of the area: 'Not only do we not want to take over the area, but by maintaining its open spaces and buildings, we function like a temporary bridge between today's official abandonment of the camp and the realisation of a complete and integrated plan for turning the camp into a free and open green space for the citizens' (Karagiorgas 2016). The squatters reactivated a well, laid new pipe, illegally connected into the water supply system, and divided a portion of the vast military base into a series of 35-square-metre parcels to be used for gardening (www.perka.org). A series of workshops attracted so many interested people that they decided to split into half a dozen 'nuclei,' each to be managed by its own members. Following a period of initial harassment, the military backed off once it became clear that there were questions about the actual ownership of the property. PERKA grew to more than one hundred and fifty members as the first self-organised communal garden in northern Greece (Gavrilidou *et al.* 2015). It operates on principles of organic, biodynamic farming, with a focus on the cultivation of plants from heirloom seeds. At the same time, it promotes the social benefits of 'working with and being in contact with the earth' and supports local biodiversity. Gardeners regard their enterprise as a sort of 'research lab' that bridges the gap between rural farmers and urban consumers while bringing city dwellers closer to nature. And in the spirit of optimism and a sense of collectiveness, donations of produce are given to support vulnerable social groups in the area.

There is an eerie, post-apocalyptic feel to the abandoned military base. Empty, tree-lined boulevards [e.7.1.8.1] connect derelict, graffiti-blazoned buildings that are missing windows, doors, and in some cases, even roofs (Figure 7.28) [e7.1.8.2]. Large sections of the ground surface are composed of broken rubble (Figure 7.29). Interspersed among the buildings are gates that give access to the garden plots for the various groups [e7.1.8.3]. Here, vegetables such as melons, tomatoes, eggplants, and sunflowers [e7.1.8.4] grow in lush profusion (Figure 7.30). Some sections have screening or elaborate assemblages of climbing stakes [e7.1.8.5]. The contrast between the verdant gardens brimming with life and the ghostly hulks and skeletal remains of the formerly occupied buildings is stark (Figure 7.31). Animals are limited to chickens in a large, screened coop and bees in a rooftop apiary [e7.1.8.6]. Elsewhere on the site, a small half-shell hoop-house nestles against a building [e7.1.8.7] from where a pipe harvests rainwater from the roof into a storage barrel [e7.1.8.8]. Amenities for the gardeners are few and are generally scavenged from materials found onsite [e7.1.8.9]. One building has been taken over to store gardening equipment [e7.1.8.10] and to function as the central location for socialising [e7.1.8.11].

Figure 7.28.

Figure 7.29.

Figure 7.30.

Figure 7.31.

My Saturday morning visit to PERKA becomes the most enjoyable time spent in any of the agrourbanism projects that I visit for this research initiative. I love the spooky *Walking Dead*-esque ambiance of abandonment that pervades the place, and how this contrasts with the life being wrought from the ground through the dedicated efforts of the gardeners. From 'dead' to 'dedication' I reflect; what better coupling of words is there to describe the transition from landscapes of desolation to mindscapes of optimism as characterised by regenerative agrourbanism?

Site photo-tour complete, I am taken over to the outdoor table to meet the gardeners. There I sample tomatoes and cucumbers from the gardens while engaging in more than an hour of cheerful, translated banter on all manner of topics, often comparing and contrasting various aspects of urban life and gardening in Greece and Canada. They are happy to learn that I had previously visited Greece during my first European backpacking trip while in my twenties. Toward the end of the visit – and even though it is still a few minutes shy of noon – I am not permitted to depart before sampling wine from the countryside family home of one of the men. And as often happens, one glass leads to another, accompanied by toasts of *yassou* and much laughter. As I take leave of the amiable group, with strong handshakes all around, I am reminded that the benefits to wellbeing accruing from participating in urban agriculture are just as likely to be social as they are gustatory. Not without reason are the words and concepts 'epicurean' and 'Dionysian' aspects of this country's manifold gifts to the Western World.

7.1.9 Mill Creek Farm (Philadelphia, PA, USA)

The Mill Creek neighbourhood of West Philadelphia is an impoverished area characterised by boarded-up rowhouses [e7.1.9.1], vacant, rubble-strewn lots, and few places to buy healthy food. It is a landscape that has been referred to as 'a catalogue of the failures of 20th-century urban policy, planning and design' (Spirn 2005) and has been studied by landscape architects for more than two decades (Spirn 1998). By the end of the nineteenth-century, Mill Creek itself was so severely polluted that it was buried in a combined stormwater/sewage pipe and had its floodplain filled in and built upon. Because the pipe diameter was too small to accommodate the loads of waste and runoff, this led to frequent ruptures. As a result, over the decades, buildings became undermined, and cave-ins occurred, occasionally resulting in deaths. In the 1960s, entire city blocks comprising more than one hundred homes were demolished due to the instability of their foundations. The low-lying abandoned blocks eventually came to be occupied by community gardens. In 2005, some of these gardens and an adjacent piece of litter-strewn land which had lain vacant for three decades were transformed into the 'Mill Creek Urban Farm' [e7.1.9.2].

Designed with the goal to provide the community with inexpensive produce and educational programmes (Walker 2012), an additional aim of the project is the use of the property to detain stormwater and reduce the long-standing problem of flooding in the area (Hodgson *et al.* 2011;

Walker and Rosen 2006/08). As announced on the entrance sign (Figure 7.32) and website (www.millcreekurbanfarm.org), the project is described as being 'an educational urban farm located in West Philadelphia that is dedicated to improving local access to fresh produce, building a healthy community and environment, and promoting a just and sustainable food system'. This is achieved by focusing on six core values:

> (1) The neighbourhood in which we farm. We work with our neighbours to strengthen our community. (2) Experiential education. We provide hands-on, farm-based learning opportunities. (3) Our environment. We demonstrate innovative ways to foster sustainable living. (4) Community and collaboration. We work with and exchange ideas with others. (5) Urban food production. We foster connection between people and the food they eat. (6) Equity and Justice. We believe healthy, affordable food is a basic human right.

The site occupies an entire block and is contained within a gated, chain-link fence [e7.1.9.3], portions of which, in a nice touch, are decorated with ribbons [e7.1.9.4]. A remarkable range of more than fifty types of produce [e7.1.9.5] is grown in the farm's fields [e7.1.9.6] in relation to consumer demand and seed availability. These include 'peas, potatoes, asparagus, garlic, okra, carrots, beets, leeks, pears, apples, peaches, berries, herbs (culinary and medicinal), flowers, beans, tomatoes, peppers, eggplant, cabbage, mustard greens, lettuce, scallions, broccoli, cucumbers, chard, kale, collard greens, spinach, turnips, radishes, sweet potatoes, summer squash, and zucchini'. A drip irrigation system of valves and plastic tubing is used throughout [e7.1.9.7]. During my springtime visit, many planted rows were covered in protective blankets of cotton gauze

Figure 7.32.

to thwart insects depositing their eggs on leaves [e7.1.9.8]. The farm also boasts an orchard strip of fruit trees along a sidewalk (Figure 7.33). An apiary is located in a sequestered area [e7.1.9.9], as is the compost pile [e7.1.9.10]. The farm uses natural weed and pest control and fertilises with plant compost, fish and seaweed extracts, worm casings, and compost tea. The farm is run by the non-profit agency, 'A Little Taste of Everything,' and is worked by volunteers [e7.1.9.11]. Sale of produce to the neighbourhood occurs at an onsite farmstand, at a nearby farmers' market, through an arrangement with a large co-op store in the vicinity, and occasionally via several restaurants. In addition to a small tool-shed [e7.1.9.12], another structure provides an expansive view of the farm from its open porch (Figure 7.34). The structure is constructed using cob (i.e. clay and sand) and showcases green building techniques, including solar panels, a living eco-roof (Figure 7.35) of sedum and other succulents [e7.1.9.13], a composting toilet [e7.1.9.14], and graywater collection and reuse. A mosaic composed of found objects decorates the rear wall above the sink with the motivational slogan: 'OUR WORLD OUR WORK OUR FOOD OUR POWER' (Figure 7.36) [e7.1.9.14]. Mill Creek Farm claims its primary purpose is to be an education centre based on community-building and the facilitation of inter-generational exchange (Travaline and Hunold 2010). It accomplishes this by giving tours, hosting school field trips and skill-share workshops, and employing a summer internship position in marketing as well as high-school students in a job training programme, while also using the site for community events such as BBQs. Nairn and Vitiello (2009) describe the farm being visited by a group of schoolchildren 'exclaiming their ambivalence through fear, disgust, and delight at the composting toilet, beehives, and worms. Few have ever seen how food is grown, so the revelation that carrots come from the dirt in the neighbourhood is startling to many. This is the joy of discovery, of learning how people and cities are sustained.'

Figure 7.33.

Figure 7.34.

Figure 7.35.

Figure 7.36.

The passion for urban agriculture as a tool for addressing social injustice radiates in the voice of the young woman who is giving me a tour of the farm. Again and again, while explaining the objectives and itemising the good deeds of the organisation, she stops mid-sentence when something botanical catches her attention. And for a moment, larger societal issues temporarily abandoned, she focuses on a particular type of plant and various aspects involved in its gardening. Although still early in the season, it is hard not to be impressed by the bounty emerging from the soil. In contrast to the hype exhibited by some other urban agriculture projects in Philadelphia, understated Mill Creek Farm is the real thing: a regenerative model of humility, hope, and healing. As I stroll about the site and gaze out over the rows of plants covered by the long, twisty swaths of white cloth, I cannot rid my mind of the thought that they resemble giant caterpillars. Might they, I wonder and simultaneously smile, rise up from the Philly soil and be as threatening as the giant sand worms of Arrakis?

7.1.10 Case study lessons

- Vacant lots in cities are attractive to real estate developers, sometimes necessitating long struggles and grass-roots initiatives to save the land for growing food (Hellinikon Airport Community Gardens, PERKA).
- Imaginative urban agriculture projects can stay one-step ahead of new development (King's Cross Central Skip Garden) or be incorporated into new buildings (60 Richmond Street Housing).
- Regenerative agrourbanism flourishes as a new use for derelict landscapes (Hellinikon Airport Community Gardens, PERKA), abandoned city blocks (Allen Street Community Gardens, Mill Creek Urban Farm), and post-industrial buildings (Evergreen Brick Works, Artspace Wychwood Barns).

- Workshops/programmes and design/build projects provide education about sustainable design (60 Richmond Street Housing, Greensgrow Farms, Evergreen Brick Works, Allen Street Community Gardens, Mill Creek Urban Farm).
- Community gardens, educational programmes, CSAs, hosted farmers' markets, and direct sales address issues related to food justice and security (Hellinikon Airport Community Gardens, Greensgrow Farms, Evergreen Brick Works, Artspace Wychwood Barns, PERKA, Allen Street Community Gardens, Mill Creek Urban Farm).
- Attention to the aesthetics of site amenities and the inclusion of art will enliven visitor experiences (Greensgrow Farms, Evergreen Brick Works, Artspace Wychwood Barns, Allen Street Community Gardens, Mill Creek Urban Farm).
- In addition to being locations for growing food, urban agriculture sites provide social capital in the form of space for community gatherings and local functions (Hellinikon Airport Community Gardens, Evergreen Brick Works, Artspace Wychwood Barns, PERKA, Allen Street Community Gardens, PERKA), while also being knowledge hubs for educating about sustainable urban living (Greensgrow Farms, Evergreen Brick Works, Artspace Wychwood Barns).

7.2 Souls and societies

Anecdote

Personal experience in learning of gardening as an escape from Nazi oppression. Due to geography and complicated familial circumstances, I did not meet my paternal grandfather until I was in my late twenties. It was during a subsequent visit to his London apartment at a time when I was a professor in the landscape architecture department at Harvard University that I told him of the interest of some of my students in garden design. My grandfather, when he learned of this, immediately took me outside to visit the small edible garden plot that was tended by himself and several other elderly residents of the retirement home. I was then informed that he had been a post-war editor of the British allotments' magazine at the time during the late 1940s when food rationing still existed and many were encouraged to grow their own vegetables. He then went on to recount his gardening experiences during the years when he was in Stalag Luft III, one of the most notorious Nazi prisoner of war camps, and famous for being the location from where *The great escape* had occurred. Although his involvement in the historic escape was a prided family story, the role that gardening had played in his time in the camp was new information.

Throughout his internment, tending gardens was a cherished source of activity in the face of the endemic boredom in the POW camp. In short, the act of planting, growing, and harvesting vegetables was, much like storytelling by prisoners in the novel and film *Kiss of the spider woman*, a way to distract from the harsh realities of the circumstances and to fill the void of purposeless time. In so doing, gardening became a means for preserving wellbeing. Seeds supplied by the Red Cross were also a tangible link to life back at home. The knowledge that the 'mother' plants which had produced the seeds had grown them in the soil of 'Old Blighty,' was a source of solace. During the final months of the war when the Reich was collapsing and reliable food delivery became an issue, produce from the gardens prevented starvation. The gardens also played an integral role in the escape itself. Because the camp was built on sandy sediment, the presence of different-coloured soil deposited in a pile would have been a tell-tale sign of the tunnelling activity. And so, the continual cultivation of the gardens provided camouflage in which to discretely scatter and mix in the excavated soil amongst the tomatoes, peas, and lettuce plants. Finally, after learning of the executions of the recaptured escapes – an infamous war crime – all of whom were of course known to my grandfather, he assuaged his anger and grief by narrowing his world-view to nothing more than focusing on tending the garden as a reprieve from the pain of loss. Never has the concept of horticultural therapy been so sorely needed.

Book

Restorative commons: creating health and well-being through urban landscapes, edited by Lindsay Campbell and Anne Weisen, arose from a 2007 conference run by the New York Academy of Medicine. The Foreword by noted physician Oliver Sacks lays out the theme of the tome through emphasising the importance of the restorative and healing power of gardens and nature. Included here is his coining of the term 'hortiphilia' to mean the special desire deeply instilled within us to interact with, manage, and tend nature. The book is richly illustrated with almost a hundred colour photographs, some spread across two pages. Written by a multidisciplinary group of experts, the book is focused on therapeutic design and is composed of sixteen articles, more than half of which explicitly dealing with urban agriculture. 'Commons' are defined as places that are publicly accessible, non-excludable, and managed through shared governance. They are considered to be 'restorative' if they contribute to the wellbeing of individuals, communities, and the landscape by providing a sense of salubrity. As explained in one chapter, commons, which were traditionally a central field shared by a community for grazing or growing, become transformative when they apply the restorative qualities of nature to urban landscapes to enhance human and environmental wellbeing.

The history of creating city parks as a catalyst for inspiring human health and societal wellbeing is reviewed, with the point being made that *restorative* landscapes need more than simply the provision of open space. Active stewardship is required to enhance the restorative elements that open spaces can offer. Moreover, stewardship implies responsibility. This is an important aspect of the experience of place and provides a way in which individuals can contribute to larger system of civic meaning. Gardens, in this light, become important expressions of both self and community.

Other urban agriculture-themed chapters (a) highlight gardens as dynamic and productive ecosystems; (b) deal with the challenge of designing gardens for those with special needs, balancing peaceful spaces for contemplation and didactic spaces for education, exploration, and horticultural therapy; and (c) introduce the concept of gardens as living memorials or 'memoryscapes.' One case study concerns a jail-to-street rehabilitation programme to train inmates for horticulture careers once they are released. Another chapter focuses on youth empowerment through urban agriculture, the director stating that he doesn't care if the troubled teams become farmers, it being simply enough that they grow up with a sense of caring for something. An important feature of community gardens is that they encourage imaginative play and risk-taking in an unstructured, natural environment. This is something that is deemed critical for creative development, and which is absent from traditional playgrounds and fields designed for organised sports. Community gardens therefore foster neighbourhood pride at the same time as providing one of the essential ingredients of a healthy childhood. As one contributor states: 'I believe that this 'transforming the dangerous abandoned space into a flourishing garden' story is archetypal; it is a metaphor for the personal transformation many gardeners felt while engaged in the creative process of building and maintaining community gardens' (p. 136).

Sites

7.2.1 Natural Growth Project (London, UK)

The organisation 'Freedom from Torture' has been in operation since 1992 and provides survivors of torture and their families with medicinal and social care and practical assistance through psychological and physical therapy (www.freedomfromtorture.org). From the start, therapists incorporated different approaches to make rehabilitation more holistic, including therapies involving art, music, and especially horticulture (Linden and Grut 2002). The latter was deemed especially important because for some of the damaged clients the simple act of being inside an enclosed room with a therapist asking them questions could be enough to trigger flashbacks to prison interrogations. However, 'when one works outdoors in nature, immediately, a different dynamic is established. In nature, we are connected to a wider world; the torture experience becomes a part of life, rather than the only thing a client is thinking about. The healing process can begin' (www.resettlement.edu). In addition to its local rehabilitation work, the organisation investigates and raises awareness about human rights abuse around the world.

Freedom from Torture's volunteers and paid staff who provide medical consultation are based in a building referred to as the 'London Centre.' Given that their clients suffer from the extreme limits of inhumanity, careful attention was paid to the building's design. Brutal and angular architecture is replaced by gently curving corridors. Consultation rooms are physically much larger than needed in order to be less threatening to those who may have been confined in tiny cells. In 2007, a fundraising campaign allowed for a small therapy garden to be created behind the building, which was opened to clients three years later as a healing sanctuary. 'Natural Growth Project,' the horticultural therapy wing of the organisation, uses this onsite space as well as allotment gardens for their rehabilitation work. Extremely fragile clients receive psychotherapy as individuals or in small groups in the secure space at the London Centre. Other clients meet weekly with psychotherapists and project workers at prearranged times to work on communal plots in the allotments. Those who are considered to be most robust have their own set of keys and can visit the allotments on their own. In all cases, the wellbeing of those in recovery is monitored, including reductions in nightmares, flashbacks, and other PTSD symptoms. Participants in the programme develop a sense of ownership and belonging to the garden, in that 'a small kind of homeland is created, from which clients can venture forth emboldened in their lives.'

Nothing of particular architectural note is observed on the walk from the Underground station along Isledon Road until the Freedom from Torture's signature building, the London Centre, is reached. There, the modern limestone building follows the gentle curve of the road, progressing to a backyard which contains attractive wood-stained sheds that peek out over top of the arching wooden fence [e7.2.1.1]. From the building, a small courtyard is accessed which functions as a secluded therapeutic space. The water feature consists of a holding tank and a small babbling channel crossed by a tiny footbridge in a patio with plantings of conifers and flowers and seating benches (Figure 7.37) [e7.2.1.2]. A sign directs one to the garden [e7.2.1.3], which is entered through a walkway in a lovely wooden building whose shape and the presence of potted bushes in the portico is reminiscent of pavilions in Japanese gardens [e7.2.1.4]. A small glasshouse is the first structure encountered [e7.2.1.5], peering into whose windows reveals plastic trays of seedlings and potted plants (Figure 7.38) [e7.2.1.6]. One area of the garden is focused on aesthetics, containing a stone-laid path, seating benches, and plantings of ornamental trees and flowers (Figure 7.39) [e7.2.1.7]. The largest area is devoted to edible plots in raised wooden boxes and elevated troughs, behind which are several storage sheds (Figure 7.40) [e7.2.1.8]. The storage sheds are themselves beautiful structures [e7.2.1.9] and feature rain barrels to collect rooftop runoff [e7.2.1.10]. A secluded area with a bench and sundial is provided for quiet contemplation [e7.2.1.11].

Located four kilometres away from the London Centre building are the Collindale Allotments [e7.2.1.12], with plots tended by the Natural Growth Project's clients (had I had access to the gated site to be able to see the gardens, my positioning of this project in the sequence of those reviewed in this chapter would have, I'm sure, been higher). In addition to growing vegetables, the trees, shrubs, and flowers planted by survivors are acts laden with meaning (Mistiaen 2002). Some individuals grow flowers that their mothers used to love or plants that remind them of their home countries. Others create remembrance gardens as living memorials to loved ones whose remains were never recovered. The idea is to use nature as a medium to allow those subjected

Figure 7.37.

Figure 7.38.

Figure 7.39.

Figure 7.40.

to unspeakable acts to achieve a sense of healing through regeneration. Digging in the soil and removing weeds become symbolic acts of burying or expunging trauma. The processes of growth and renewal in the garden provide hope for a promising future. As Linden and Grut (2002: 11) state, 'the art of making things grow is in itself an important symbol of healing.' Surveys of participants show that they believe the mix of therapy and doing something purposeful in the garden helps them heal by giving them peace and a reprieve from the burden of painful memories. As one client phrased it, the allotments are 'the hospital with a blue sky' (Linden and Grut 2002: 9). Five additional benefits were identified by participants in the Natural Growth Project (Anonymous 2015):

> Connect with others (feel less isolated and able to make and meet friends; meet others in the community especially at the allotment; re-learn and re-experience trusting others).
>
> Be more active (enjoy doing physical work; strength and stamina improved; take fewer painkillers; feel relaxed and less tense).

Take notice of the world (gain a sense of achievement – see the results of their work; are able to take pleasure in the world around them; re-kindles a sense of hope).

Keep learning (learn new skills and re-visit existing skills and knowledge; gain confidence about learning in a new country, circumstances and language).

Give to others (some are volunteering in green community projects; support each other).

One torture survivor taking part in the Natural Growth Project remarked that 'there is no evil in the garden'. The reality, however, is different, as shown in a sad postscript that underscores the deep and lasting damage that can mark the lives of victims of violence and those with whom they interact. Ten months after my visit, the 80-year-old allotment secretary and treasurer was found strangled with an extended cord from a lawn mower, her body dumped in a tool shed. The murderer, who was sentenced to a minimum of 19 years, turned out to be a man who had been involved in the allotment therapy program, first as a victim of torture himself, and then later as a volunteer councillor who helped others (Kleeman 2019).

Upon arrival I follow the attractive wooden fence that curves around the property. Partway along I spy a sign announcing the application of anti-climbing paint as a deterrent to those contemplating scaling the fence. To a long-time and passionate rock climber like myself, however, such expressions always invoke feelings of aspiration rather than dissuasion. I furtively glance about to ensure that I am alone, drop my daypack, and make a series of upward moves. I soon reach a point of where I am confident that, paint notwithstanding, I could, even with the absence of my old EB climbing shoes, make it to the top. I quickly drop back down to the sidewalk with a loud ka-thunk just as a pedestrian rounds the corner. I pick up my pack and nod to him, hoping my friendliness will remove the puzzled look across his face. A short while later I enter the site, doing so legally via the front door.

I linger in the patio beside the small water channel, listening to the babble of the water that echoes off the walls surrounding the atrium. I think of Henry David Thoreau, one who certainly recognised the therapeutic benefit of water, and who wrote that 'he who hears the rippling of the rivers in these degenerate days will not utterly despair' (France 2001). I try to fathom the unimaginable misery suffered by those who have been tortured, and sincerely hope that the sound of the water as well as the act of participating in tending the adjacent gardens may have brought them a reprieve from their pain, even if it be so but only temporarily.

7.2.2 Forty Hall Vineyard (London, UK)

'Forty Hall Vineyard' is a non-profit, multiple award-winning social enterprise project located in the northeast London Borough of Enfield. It is also the only commercial vineyard in London, and the first there since the Middle Ages (www.fortyhallvineyard.com). The vineyard is located on the Forty Hall Farm, a 69-hectare organic farm and local food hub that is managed by Chapel Manor College, London's only specialised environmental college. The area is an important one in terms of rural heritage (Laviscio 2020) in that it was once part of a medieval hunting forest which was later converted into Tudor, Jacobean, and then Georgian estates, with many of the remaining manor houses being Grade 1 to 3 Listed Buildings (Carter 1995). The Forty Hall Estate contains a seventeenth-century manor house with farm buildings, remnants of a sixteenth-century deer park, a lime grove, water gardens, eighteenth-century pleasure grounds, and nineteenth-century picturesque tree plantings, all subject to protection by the national organisations Heritage England and Natural England. As described by Laviscio (2020), 'the complete complex was constructed, over the centuries, to compactly reproduce all the features of the country greater residences, wherein each part is related to the other by functional and symbolic relationships to guarantee recreation, food consumption, and landscape control'.

The modern Forty Hall Farm runs a market garden that produces food for the onsite shop and café and local farmers' market, as well as providing horticultural training and employment. A CSA programme exists with members receiving weekly 'veg bags' at various pick-up locations around Enfield. The financially self-sustaining enterprise also raises rare breeds of sheep, pigs, and cattle, and has an orchard of over one hundred fruit trees and three hundred fruiting hedge rows, including several local heritage varieties. The site is a good demonstration of the heritage potential of peri-urban agrarian landscapes. The fact that 8 acres are devoted to vegetable production used in the café goes 'in some way [to] restoring the ancient relationship between the house, its farm building, and the surrounding agricultural land' (Loviscio 2020). The shop sells fresh vegetables, eggs, free-range meat, and flowers, in addition to products from local producers such as ales and ciders, pickles and chutney, and of course, the wine produced from the grapes grown in the onsite vineyards. It is the latter which are the reason why the project is included herein.

The vineyard occupies four hectares of the Forty Hall Farm and is dedicated to producing the best quality, certified organic English wine. The vineyard was started in 2009 with one hundred volunteers planting the first hectare. By 2018, the picked harvest of grapes totalled more than twenty tonnes. Along the way, challenges included crop losses due to bad weather, disease, and bird damage. Pests are controlled biologically with no synthetic fungicides, herbicides, or fertilisers being used. Instead, grapes are grown in chemical-free soil which makes for a stronger natural resistance to disease and also allows for the wine to reflect the *terroir* of the unique characteristics of place, soil, and microclimate. Horse-driven vineyard equipment was imported from Italy to help reduce use of the diesel-powered tractor, something that contributed to the vineyard receiving a local green business enterprise award. Grapes are harvested and driven to a regional winemaker for vinification. Chardonnay, Pinot Noir, and Ortega are produced. Their Bacchus vine was a gold medal winner in 2019 and is described with the sort of purple prose customary for the profession: 'An easy-drinking, elegant Bacchus with notes of nettles, pea-shoots and elderflower. Dry and pithy;' and 'The wine is fresh, light, bright and clean with flavours of crisp apples, and aromas of gooseberry and blackcurrant. A great aperitif and food-match for shellfish, seafood and oriental dishes'. Profits from sales of wine are put back into the project to deliver health and wellbeing benefits to the local community. And it is this element of the project that makes Forty Hall Vineyard an important example of regenerative agrourbanism, one worthy of its awards for both urban food production and for social enterprise creation.

The borough of Enfield has a high rate of depression, with more than thirty thousand of its residents living with anxiety and mental health disorders. As such, there is a high demand for community-based services to address the problem. By offering volunteering opportunities to work in a safe and secluded area 'with stunning views over London and abundant wildlife,' the vineyard venture is 'based on the theory, now backed by research, that being involved in purposeful work in the outdoors with others promotes physical and psychological healing and resilience'. The eco-therapy program is targeted to individuals with mental health support needs, adults with learning disabilities, young people on the autistic spectrum, those experiencing social isolation and stigma, ethnic minority women suffering mental distress, those suffering from bereavement, unemployment, anxiety, or low self-esteem, as well as refugees and asylum seekers, families experiencing child or parent abuse, and other marginalised groups.

Online videos show dozens of people socialising while they tend the vineyard. Tasks include grape picking, sorting and cleaning, and winter pruning. In addition to vineyard management, other volunteer projects have included beekeeping, clay oven building, nature walks, willow weaving workshops for seniors, and wildlife surveys. Volunteers report the following benefits: 68% felt more positive, 84% felt more connected to nature, 90% felt more socially responsible, 58% felt physically fitter, 62% had improved mental health and wellbeing, 65% felt more confident, 55% had greater self-esteem, and 88% felt that their range of friendships had increased. The vineyard's website contains testimonials culled from the organisation's social impact reports, such as: 'Living with schizophrenia, I have found working at the vineyard has been an important part of my journey back to health. Green therapy at the vineyard worked for me and I have seen improvements in others. Coming to the vineyard gives structure to my week and gets me out of the house and into the fresh air. I enjoy working and socialising with the other volunteers – good for body and soul!' Employment and training programmes allowed participants to regain their health to allow them to expand their social networks and enable them to take the first steps to re-enter the workforce with increased self-confidence, as for example: 'At that point, I was struggling with anorexia, which had taken control of my life. Each day had become incredibly painful mentally and physically. Those few hours working outside in the vineyard with the other volunteers [were] like a saviour to me.' At Forty Hall Vineyard, she discovered her 'passion for viticulture and horticulture', gaining hands-on vineyard experience and knowledge, which was furthered by completing a horticulture course at neighbouring Capel Manor College. 'Both my health and my career have made huge leaps forward.'

Benefits accruing to the Borough of Enfield from the Forty Hall Vineyard include:

- increased local opportunities for volunteering and skills development for vulnerable adults and people from deprived communities;
- greater community involvement in civic and environmental action;
- healthier, stronger, more resilient and more connected communities;
- enhanced local civic pride and sense of place (being part of a unique project in London and the UK);
- reduced stigma of residents with mental health support needs;
- greater understanding of mental health issues and how they affect us all;
- increased social cohesion as volunteers from a wide range of backgrounds work together.

Laviscio (2020), expressing an appreciation of the experiential qualities of edible landscapes that is rare in the professional literature on agrourbanism, describes Forty Hall Farm as 'a very attractive, rolling agricultural perspective, with open fields drained by a cross-cross of watercourses. The impression is of a very rural area even though we are on the edge of the town. Also, the smells and sounds reflect those of an agricultural and natural landscape.'

The walk to the estate from the public transit station in Enfield allows glimpses of the grandeur of the manor houses through the wrought iron gateways designed to keep out the lower classes. In contrast, sidewalk signs announcing that both shop and café are open [e7.2.2.1] signify public accessibility to the Forty Hall Estate. A map provides details of the site features [e7.2.2.2]. A short stroll through the landscape of stately trees and past the ornamental pond [e7.2.2.3] and through the impressive entrance gate [e7.2.2.4] leads to the manor built in the seventeenth-century for the Lord Mayor of London (Figure 7.41) [e7.2.2.5]. Inside, rooms showcase ornate furniture, marble fireplaces, decorated plaster ceilings, and white wainscoting [e7.2.2.6]. The farm shop is located in an adjacent building [e7.2.2.7], where one can purchase melons, pumpkins, other vegetables, and meat from the farm [e7.2.2.8], as well as jams and preserves, ciders and beers, and books and gift cards [e7.2.2.9]. Also for sale are various types of wine produced from the Forty Hall Vineyard located on the estate grounds (Figure 7.42).

Figure 7.41.

Figure 7.42.

If interested in the agricultural workings of site, it is necessary to leave behind the *flâneurs* enjoying the pleasure grounds about the manor and to move into the seldom visited farm proper, with its numerous outbuildings and machinery (Figure 7.43). Here, heritage breeds of animals, including goats peaking their heads out from a barn [e7.2.2.10], are accompanied with information panels about other types of livestock [e7.2.2.11]. Signs direct the way to the orchard and market garden [e7.2.2.12], the latter containing several hoop-houses (Figure 7.44).

Walking through the farm leads to the vineyard, where a sign describes its development, operating principles, and cultivation techniques, in addition to mentioning that it is hoped the site will produce fifteen thousand bottles per year [e7.2.2.13]. Rows of young vines stretch out across a large grass field, merging into woods in the middle distance, above which rise apartment towers on the horizon (Figure 7.45) [e7.2.2.14]. The vines are anchored by metal posts and guide wire, with their bases protected by plastic tubes [e7.2.2.15]. Amenities for the volunteers include a sheltered workspace, toilets, and what looks like might be an oven [e7.2.2.16].

Figure 7.43.

Figure 7.44.

Figure 7.45.

It has been four years since my last visit to Enfield and my tour of Valley Grown Salads (see Chapter 3.2). At the time of this, my present visit to the Forty Hall Farm and Vineyard, it is a bleak day in November 2016, one made bleaker upon learning later that evening of the results of certain national election 'across the Pond.' As I walk around the vineyard thinking of geopolitics and what changes the election might bring, I think upon those who have come here seeking therapy in the vineyard. I reflect too upon the wisdom of Cicero, who, when confronted by the political turbulence of his own day, remarked that as long as one has a garden and a library then that is enough. Society, I feared, might very well be in collective need of green therapy in order to help weather the potential stormy years ahead. Or, as I look out over the vines, if not that, at least the solace and enjoyment that can be found in partaking of the passion of oenophiles.

7.2.3 ReVision Urban Farm (Boston, MA, USA)

Located in the Dorchester district of greater Boston, the 'ReVision Urban Farm' is run by a community-based, non-profit organisation whose mandate is to 'provide fresh, affordable, locally grown produce to families in the city with limited incomes and transportation access, as well as to neighbours in the area' (www.vpi.org/revison). The unique connection between Victory Program's Revision Urban Farm and the ReVision Family Home has created a dynamic and innovative approach to ending homelessness. Begun in 1990 as a small garden beside its shelter for homeless mothers, the project has expanded to becoming a farm through purchase of three abandoned lots across the street. Gardens were created with soil imported from a local cemetery in addition to composted manure from a local zoo and the police department's horse stables (Rich 2012). Working with a motto that 'all things are possible when lives are given the proper nurturing, sunlight and time to grow,' the specific goals which underlie the farm and programmes of the ReVision project are: small-scale, green, economic development; community food security; and job training and education. Over its lifespan, the project has supplied food and shelter to hundreds of homeless families and has created healthy food options for thousands of people in the community.

It is the close interworking of the urban farm and the shelter for homeless women that makes this project inspirational. The shelter houses 22 women and their children and provides services for life-stabilisation, permanent housing, and vocation selection. Part of the goal of increasing self-sufficiency is facilitated by increasing nutritional well-being. This comes about through providing information about healthy eating and sustainable urban farming in addition to acquiring direct, hands-on experience by growing, harvesting, and consuming food from the gardens. By integrating agriculture and enterprise, the ReVision Farm teaches homeless mothers and youth practical job skills related to food production, marketing, and community outreach. This is accomplished by employing some of the shelter's residents in addition to local high school students as summer agriculture interns. These interns, in turn, run a programme, 'SPROUTS,' based on developing an 'age-appropriate gardening curriculum for pre-schoolers, teaching young children about fruits and vegetables, while reinforcing basic cognitive skills.' The webpage is filled with moving testimonials from summer interns.

The homeless shelter (Figure 7.46) is located in a densely inhabited neighbourhood of detached homes and abandoned lots [e7.2.3.1]. The urban farm occurs on three vacant lots sandwiched between houses across the street from the shelter [e7.2.3.2]. Rows of vegetables are planted directly in the soil due to an absence of industrial contamination on the site [e7.2.3.3]. A hoop-house is also used to sprout plants in raised beds and containers (Figure 7.47) [e7.2.3.4]. Site amenities include a lovely wooden pavilion (Figure 7.48) [e7.2.3.5] and a newly constructed stone amphitheatre for lectures and social events (Figure 7.49) [e7.2.3.6]. The original garden is situated opposite the pavilion and alongside the family home shelter (Figure 7.50). Attached to the rear of the home shelter is a multi-storey glass enclosure designed to house tilapia (Figure 7.51) [e7.2.3.7]. At the time of my visit this was no longer operational. The backyard contains a playground and small gardening plots for young children [e7.2.3.8] that are arranged around a ground mosaic spelling out the words: 'PLAY, LAUGH, LEARN, & GROW' (Figure 7.52).

Figure 7.46.

Figure 7.47.

Figure 7.48.

Figure 7.49.

Figure 7.50.

Figure 7.51.

A significant portion of ReVision's operating expenses are obtained from selling produce to the neighbourhood at their nearby roadside food stand (Figure 7.53) [e7.2.3.9] and from a stall at the local farmers' market. Additional revenue is generated through establishing a CSA to families residing in the city's food deserts. The latter is a means for a collective made up of such urban farms to ensure their viability by facilitating access to the larger market with little cost. Finally, ReVision Farm runs a seedling co-op of herbs, flowers, and vegetables for urban gardeners. These are grown using sustainable products and methods and marketed with the tagline of 'buy[ing] your transplants from growers you know and trust'.

Figure 7.52.

Figure 7.53.

I have situated myself on the top tier of the amphitheatre, the best location in which to observe the entire site. My most memorable experience of visiting this urban farm is defined by sensorial absence. It is early in the morning and despite the density of homes, there is a pervasive sense of peacefulness and silence that envelopes the landscape. I can hear neither cars, nor people, nor songbirds. Even the wind seems to be absent. It is as if all the hubbub of life except for that of the plants growing quietly about me has suddenly packed up and departed for a holiday. I lift my head to gaze across the street at the shelter where I imagine the lives of the residents therein to be anything but pacific. How remarkable it must be for them to find some reprieve in the serenity of this garden should they too be fortunate to be present in similar circumstances to what I am presently experiencing. I close my eyes and reminisce to circumstances of tribulation in my own life, and remember how focusing on the minutiae of work had provided solace. Again, I stare out past the growing plants to the shelter. If one is made homeless through the act of fleeing an abusive domestic situation, it is fair to assume that such individuals do not have the luxury of being able to turn off the unpleasant outside world by throwing themselves into purposeful activity of their jobs. In consequence, I imagine that participating in the physical and focused act of gardening this space must be a cherished refuge from life's misfortunes.

7.2.4 Earthworks Urban Farm (Detroit, MI, USA)

Although now situated within the urban framework of Detroit, at the time when the Franciscans originally established their monastery [e7.2.4.1] the site was located beyond the city limits. This allowed the Capuchins to practice their long history of raising livestock and growing their own food. In 1928, they opened the 'Capuchin Soup Kitchen' to feed the hungry and care for the poor, which ninety years on, remains a fixture in the neighbourhood (Figure 7.54) [e7.2.4.2]. Beginning in 1997, operations have physically expanded with the development of a cluster of gardens on vacant lots surrounding the seminary, and with an expanded Catholic mission 'to address the systematic causes of poverty, broken relationships and a wounded Earth' (www.cskdetroit.org/earthworks). Inspired by the tradition of St. Francis [e7.2.4.3], the 'Earthworks Urban Farm' sets out to 'restore our connection to the environment and community.' The name 'Earthworks' was chosen for its symbolism as a bulwark against antagonistic forces and for reflecting the goods and services work provided by the Earth. The underlying belief is that a working farm is an ideal strategy for building a socially-just response to hunger. Furthermore, as the director states, 'The gardens are very much metaphors for enlivening our region. If we can envision what a little piece of property could be, then we should be able to envision what an old factory could be – perhaps a community-run packing house or lumber salvage operation from decommissioned buildings' (Rich 2012:104).

Figure 7.54.

It is a given that poor families often do not have access to easy transportation. In consequence, Earthworks decided to switch things around and bring fresh food from the farm to people by operating weekly markets at local health clinics. They then shifted focus from providing their produce to markets to providing meals in their onsite facilities with that same grown produce. Their aim is to 'serve the poor who suffer from the lack of basic human needs: material needs, such as food, clothing, and household necessities; psychological needs, such as motivation, self-esteem, and rehabilitation; and social needs, such as support and meaningful relationships.' A chalkboard in the kitchen dining room posts the health benefits of their garden-grown food. Other ventures include a mobile market that distributes fresh food to the elderly in the neighbourhood, and a programme in which Earthworks staff and volunteers help with the building of gardens at senior care facilities. A monthly programme for discussing topics related to food literacy, policy, justice, security, and access to nutritionally adequate diets meets at community potluck evenings. Information is shared with the community through open-house events and classes are taught on urban gardening and beekeeping skills. The aim of these is to 'introduce people to the pleasures and benefits of gardening in Detroit' in order to build a 'just, beautiful food system for all.' Farm operations are helped by the mobilisation of enthusiastic volunteers and a relationship with students at a Detroit university (Pothukuchi 2011).

Earthworks is the oldest certified organic farm in Detroit and uses regenerative agricultural practices designed to give more life and health to the soil than what is taken from it. Pesticides are not used and compost and green manure replace the need for fertilisers. Water is conserved and fields are left fallow to allow for the recovery of the soil. And in the spirit of Saint Francis (France 2020), the gardens are blessed twice a year.

A youth programme was established in partnership with a local church. 'Growing Healthy Kids' focuses on 'nutrition and wholesome activities, including growing, cooking and eating homegrown food,' in addition to environmental stewardship and cultural awareness for 5- to 11-year-olds living within a three-kilometre radius of the site. After receiving a healthy snack, children spend time planning the flower and vegetable gardens, painting garden signs, cooking pancakes, making flavoured popcorn, extracting honey, and starting seed transplants in the greenhouse. An annual International Food Night has children sampling fare from around the world and getting their 'passports' stamped. Teenagers become involved in the 'Youth Farm Stand' where they participate in the planning, growing, harvesting, preparing, packaging, and marketing of produce from the Earthworks Farm. To increase revenue and spread the news about the programme, Earthworks has started to produce value-added products such as canned tomatoes, jams, picked beets, honey, and beeswax hand balm.

Earthworks is located in a section of Detroit characterised by derelict houses, vacant lots, and isolated, still-occupied homes [e7.2.4.4]. Abandoned properties were reclaimed by Earthworks through a year-long process of removing debris and importing loads of compost to rebuild the soil. Seven gardens spread over twenty city lots within a two-block radius of the Soup Kitchen are now tended, each identified by a prominent wooden sign (Figure 7.55) [e7.2.4.5]. One of the gardens poignantly occurs on a lot that had once had a small grocery store; a strong lesson in food deserts. Several sites, such as the Garden of Unity and the Youth Farm Stand Garden, also have more fulsome signage headed with slogans such as 'Growing community and self-reliance' in the case of the former (Figure 7.56), and 'Young leaders, growing together' for the latter [e7.2.4.6]. Most of the vegetables that are regularly found in Michigan are grown in the gardens [e7.2.4.7], in addition to some specialised culinary and medicinal herbs. There is also a community orchard of fruit trees fronted by a lovely sidewalk flower garden [37.2.4.8]. At the largest site, in a field alongside a brick building, rows of crops are irrigated with plastic tubing, with some covered by gauze to impede egg-laying insects (Figure 7.57) [e7.2.4.9]. Here too, nestled against the building, are found vast heaps of decomposing organic matter that are guarded by the art installation of a compost monster [e7.2.4.10]. During the last decade, Earthworks has also operated several hoop-houses [e7.2.4.11] which produce over a three hundred thousand vegetable seedlings for their own gardens as well as for hundreds of families, community, and school gardens in the area (Figure 7.58) [e7.2.4.12]. At several locations along the street in which many of the food gardens occur, the efforts of the growers have spilled out in the form of curbside plantings of flowers in efforts toward neighbourhood beautification and biodiversity fostering (Figure 7.59). These benefits are described in a two-page, self-guided tour brochure that can be picked up in the soup kitchen.

Figure 7.55.

Figure 7.56.

Figure 7.57.

Figure 7.58.

Figure 7.59.

Having once walked from La Verna to Assisi, I find while visiting the scattered gardens and hoop-houses of the Capuchins in Detroit, that my thoughts wander to their founder, Saint Francis, and to the stories told of him in relation to gardens (a subject covered in my book derived from that pilgrimage – France 2020). I pause to listen to the chatter of several birds dashing in and out of nearby bushes and think that Francis might disapprove. For example, the Rose Garden at his much-loved Porziuncola was the scene of a miracle that demonstrated, yet again, the unseemly side of the patron saint of ecology through his domination of animals; in this case chastising and silencing a crying lamb lest it interfere with his prayers. Then, while spying a cabbage growing in another garden, I remember the bizarre story of Francis at the hermitage of Montecasale, where he tested the obedience of two boys who wished to become members of his Order. The Saint tells them to plant cabbage in the garden with the roots pointing upward. One boy states the obvious, that vegetables cannot grow upside down; the other boy does as directed. The result is that the latter individual is admitted into the brotherhood whereas the former is sent back to the fields with Francis' blessing to become a good farmer. A quick check reveals that the vegetables in front of me are planted the right way around. Later, I spot a hoe leaning against the side of a hoop-house and remember that there is another garden story from the hagiography that is worthy of rumination. Once when he was asked what he would do upon learning that he would expire before sunset, the Little Poor Man of Assisi responded, returning to his task of tending a row of vegetables, that he would simply finish hoeing his garden.

7.2.5 Detroit Dreams (Detroit, MI, USA)

This final case study overview is different from others described in this book as it describes two promising, but as yet unrealised, projects. These occur in Detroit, that city of broken dreams and incredible potential, where many observers consider that agrourbanism's lauded benefits are both most sorely needed and most thoroughly tested.

The decline and blight of Detroit results from a perfect storm of outside forces and internal problems which has led to physical abandonment and degradation on a scale that is unsurpassed (Vogel 2008). The net result of this has been threefold. Since 1950, the city has lost half of its population. As a result, there are between forty to sixty-five thousand vacant or abandoned lots, which together comprise an incredible one-third of the 360-square-kilometre surface area of the city. And in 2013, Detroit became the largest city in United States history to formally declare bankruptcy.

Many have been the voices calling for urban agriculture to step up to the plate to turn the tide (Ciadella 2008; Goodspeed 2009; Lachance 2004) and change the dialogue from Detroit being a city of despair to it becoming a city of hope (Bogg 2011; Gallagher 2010). The Detroit Agricultural Network (DAN) – in operation since the mid-1990s – were one of the first groups to realise that the one asset that Detroit has over all other cities is vast amounts of space in which to grow food. Their motto, 'growing people and communities,' reflects the belief that a network of community gardens could serve to transform neighbourhoods. The city developed a framework in 2012 to guide future land-use planning, in which urban agriculture was to play a major role (Viljoen *et al.* 2015). For some, community gardens are seen as an important component toward rebuilding and revitalising Detroit (Gopakumar 2005). According to urban planners (Northup 2008), however, it is doubtful that given Detroit's impoverished population, that it can recover through grassroots initiatives alone. The alternative is to look toward large-scale, top-down projects that might offer the creative solutions that the city so desperately needs.

One initiative that has raised considerable attention – as witness to links to dozens of newspaper articles on the webpage, including Mantyk 2010 and Kavanaugh 2010 – at the same time as generating considerable notoriety, is 'Hantz Farms,' which boldly champions itself as being 'Detroit's saving grace'. The company's dream is to rejuvenate Detroit by returning the city to its agrarian roots through creating the world's largest urban farm on the underutilised land. This is very much an agri-business model in which, in addition to growing edible crops, the farm would also grow Christmas trees and other mercantile products. The proposal is to have large blocks of the city ceded to the corporation to run as an economically viable urban farm which would create many green economy jobs while providing services to the neighbourhood. Research and educational initiatives would be created to address economic development, environmental quality, neighbourhood stabilisation, and consumer understanding of agriculture and natural resource management. Beneficial by-products would include land consolidation, blight removal, conservation of city services, and neighbourhood beautification. Ultimately, Detroit would become a major centre for exporting technology on how agrourbanism can contribute to feeding the world. Critics, however, schooled in the bottom-up idealism of the community garden movement (see Chapter 4.1), are adamantly opposed to what they regard as little more than a corporate land-grab. Hantz insists that it is of course a land-grab, for without land, agriculture cannot take place (Cockrall-King 2012). Critics claim that rather than fitting into neighbourhoods, the large commercial operation of Hantz Farms will be a force of domination and destruction. Should urban agriculture, they ask, be the end result of community redevelopment or the catalyst for it? And certainly there are many benefits accruing from urban agriculture that have nothing to do with revenue generation. Such opposition from grassroots community garden organisations, is not new to Detroit, as for example the umbrage directed to the corporately sponsored Lafayette Greens, one of the most visually stunning agrourbanism projects anywhere in the world (see Chapter 5.2). Hantz Farms has a legion of supporters just as it has a host of detractors (Hodgson *et al.* 2011), with its director being regarded as either a visionary or a profiteer (Cockrall-King 2012). Ladner (2011) is of the latter mind, referring to the Hantz's 'Ayn-Rand-infused approach to urban agriculture' (p. 96) that 'completely misses the vital self-empowerment payoffs of combing community farming with a higher social purpose' (p. 97). Others from academe, politicians, the business community, farmers' markets, and practitioners of agrourbanism in Detroit, believe that Hantz Farms has the potential to become an important contributor to the city's urban agriculture regime. As one prominent and level-headed worker in the field opined to me, there is simply so much vacant land in the city that there are numerous opportunities to experiment with dozens of different models in order to see which ones work and which do not.

Cognisant of the criticism levelled against the previous large-scale enterprise, two other landscape proposals have been advanced that pay particular attention to the social concerns of Detroit's underprivileged citizenry. Both proposals focus on the lower east side, an area of the city where, with the exception of a few blocks of nicely tended properties [e7.2.5.1], the urban devastation is starkly apparent. Here, properties go through the transition from abandonment [e7.2.5.2], through progressive dilapidation (Figure 7.60) [e7.2.5.3], to final removal by the city, leaving behind vast open spaces (Figure 7.61) [e7.2.5.4]. These spaces are empty of new construction with the exception of the absurd legislated requirement to have sidewalk interchanges to accommodate wheelchairs for the blind (Figure 7.62) [7.2.5.5]. This despite the fact that few people, sighted or otherwise, actually live there. In this area, too, are the derelict ruins of former markets, schools, and apartment blocks (Figure 7.63) [e7.2.5.6], some of whose structural integrity is compromised by the selective and illegal removal or 'mining' of metal posts and beams for resale to scrap yards [e7.2.5.7]. Some people still do live in the area, in an almost medieval feudal society (Vogel 2008). Attempts have been made over the years to begin community gardens (Figure 7.64) [e7.2.5.8]. An interesting public art project is based on collecting debris from the ruined homes in the neighbourhood and displaying it as a giant scar across the vacant property lots (Figure 7.65).

For more than two decades, the Detroit Collaborative Design Center at the University of Detroit Mercy has explored conceptual options for regenerating a portion of the lower east side of the city (Detroit Collaborative Design Center (DCDC) 2011; Lystra *et al.* 2010; Vogel 2008). The central spine of the area is Bloody Run Creek, named for the bodies of British soldiers that had coloured the water during a 1763 battle at the location. Like many cities at the end of the nineteenth-century, Detroit attempted to combat epidemic diseases by building a piped system to carry combined sewer overflow and stormwater runoff underground, in this case into the

Figure 7.60.

Figure 7.61.

Figure 7.62.

Figure 7.63.

Figure 7.64.

Figure 7.65.

Figure 7.66.

conveniently buried Bloody Run Creek. Today, ironically, the only visible section of the creek is an above-ground stretch through a cemetery (Figure 7.66). As well as hundreds of vacant lots, the area is characterised by the skeletal remains from the automotive industry and rail corridors (see Chapter 3.2). Significantly, urban farming and guerilla gardening are widely practiced in the Bloody Run area (Vogel 2008: 197):

> Vacant single-family lots are used for high-density food producing gardens, whereas the consolidation of vacant land is used for farming on a large scale [including pasturage for horses]. Alfalfa is a popular crop because there is a market for hay and because alfalfa removes toxins from contaminated soil and purifies the land in a relatively short period of time. Much of the farming and gardening efforts illegally utilise city-owned land, although the city has generally ignored this violation.
>
> Evidence of the growth and importance of farming is the increasing use of the Michigan State University Extension Service and importance of the 4H Club. When quizzed about the popularity of farming, a resident of the Bloody Run Creek area stated, 'Our parents were share croppers in the South. Farming is in our blood.' The ability to be self-sufficient and grow your own food has attracted a hardy class of individuals, both black and white, to the inner city.

The redevelopment concept, called 'Adamah' in its first few iterations, is based on integrating cultural and environmental factors into a community that encompasses the watershed of Bloody Run Creek, whose proposed 'daylighting' and restoration would become the *genius loci* of the area. In addition to a suite of green planning and design ideas, all of which have a ring of real world authenticity about them in contrast to some of the far-fetched concepts of several pie-in-the-sky projects conceived of elsewhere (see France 2016), a major focus of the Bloody Run proposal is on urban agriculture (Vogel 2008). These include:

- Being rooted in the original French farming tradition of the area with long, narrow farms perpendicular to the Detroit River and wind turbines replacing the original windmills.
- Some of the 'walk to work' created jobs would be market gardening, crop farming, and fish and mollusc aquaculture.
- Reused buildings for single-family dwellings and local farmers' markets and repurposed abandoned lots for cultivation would form the organisational core of the new community.
- The daylighted creek would be used for irrigation.

These ideas are based on recognising that, in the absence of an otherwise healthy economy (Lystra *et al.* 2010), urban agriculture is playing an increasingly important supportive role in the lives of Detroit's residents. Given all this, it is disappointing that the original Adamah vision was deemed to be unrealistic by the city's planning department.

The most recent version of the project, developed by the DCDC and supported by, among others, the Kresge Foundation, is the Bloody Run Creek Greenway Redevelopment Project (DCDC 2011). The master plan covers an area of more than a thousand hectares, touching upon the Eastern Market (see Chapter 3.2) and the envisioned RecoveryPark site (below). The new master plan continues to champion agrourbanism in the form of farming, research, food processing, and markets, and expands upon the continuous productive urban landscape (CPUL) ideas of Viljoen (2008). The plan restores the creek and situates the food system activities on the 40% of the vacant land that is publicly owned. The plan proposes to address the following potentials:

- What if innovative ecological technologies and practices formed new landscapes of production?
- What if the neighbourhood became a public heritage parkway and welcoming entry into the City?
- What if native species filtered water to the Detroit River?
- What if phytoremediation gardens regenerated the site?
- What if abandoned buildings became ecological destinations?
- What if bike trails highlighted the City's historic landmarks?
- What if wetlands and urban life were combined?
- What if a neighbourhood took part in producing its own energy, food, and culture?

The project's vision is to improve the quality of life for citizens by transforming a marginalised landscape through private and public funding into a healthy landscape that is equally productive of food products and services, ideas, sustainable energy, social exchange, and especially jobs, while at the same time as providing opportunities for recreation, culture, education, and ecology. It is estimated that the projected economic impacts of developing the project would be hundreds of permanent jobs and millions of dollars in subsequent tax revenue. The master plan includes a timeline of future steps in order to implement a demonstration project.

A conceptually and physically sympatric project to the Bloody Run Creek Greenway Redevelopment Project is RecoveryPark. A major difference between the two proposals is that RecoveryPark is primarily job creation-based whereas the former is largely housing creation-based (G. Wozniak, pers. com. 2012). The underlying idea is that it is impossible to consider recovering the landscape of Detroit without first recovering the socio-economic wellbeing of its citizens. And therein lies the challenge. For it is always much easier to restore a physically broken landscape than it is to restore the hopes, dreams, and aspirations of the broken lives of its residents (France 2007, 2008). Creating jobs can be the key, for putting dollars directly into the pockets of the disfranchised is the best strategy toward creating a self-sustaining community (Wozniak, pers. com.). And that is the major hurdle that RecoveryPark addresses. With an unemployment rate of over thirty percent and with more than twenty percent of those being released from state prisons moving to Detroit each year – many of whom have limited education and diminished employment skills – there is a desperate need for jobs in the city. Indeed, Detroit has the highest unemployment rate in the United States, close to three times the national average (Wicke 2012).

RecoveryPark is envisioned as a decade-long, multi-million dollar community redevelopment project on the east side of Detroit, an area characterised by occupied properties that stand stalwart cheek-and-jowl with derelict building ruins and vacant lots (Figure 7.67). Its vision is to become a model that integrates the strengths of the natural environment and the existing neighbourhoods to demonstrate the true potential of urban agriculture as an economic engine. The project is intended to help revitalise the city through innovative job creation that will benefit the community of individuals who are in recovery from addiction, prison, job loss, and hopelessness. The aim is therefore to recover people by enabling them to assume control of their own lives in a non-judgmental environment that is supportive over the long-term in 'transform[ing] addicted

Figure 7.67.

and other impacted individuals into happier, healthier and more self-sustaining citizens. As an extension, changed lives translate into seeds being planted for transforming Detroit into more diverse and vibrant communities whose brightest days lie in the future. What we practice and believe today will become tomorrow's reality' (www.recoverypark.org; accessed in 2016). In short, RecoveryPark is based on the concept of recovering landscape and economy as a means for recovering spirit of the dispossessed. This will be brought about by job creation, small business incubation, productive land use transformation, building rehabilitation, increased tax revenue, and lower social service costs. With a slogan of 'Renewal – Growth – Recovery', the modus operandi to bring this about is through the planned repurposing of land and lives with a focus on urban farming and food production. In this respect, RecoveryPark serves as an important reminder that in its original definition the word 'agri-culture' is about much more than just food.

Rather than attempting to transform a single large square of vacant land, RecoveryPark plans to use an amoeba approach to mould itself around existing structures, taking advantage of the surrounding context. Hoop-house orchards will be built in alleys rather than main roads, and productive hedge barriers of fruit trees will be used to shield residents from the farms. Jobs will be created in farming and aquaculture that do not require professional degrees. A big emphasis will be placed on indoor farming which can generate far more jobs per hectare than field crops. And once this is up and working, expansion of indoor cultivation will enable packaging and distribution jobs to follow. The plan is to get food products into every store in the city, not just at the nearby Eastern Market. Because the purpose is to produce food not as an end itself but rather as a means to produce employment, high-tech/high-efficiency growing options will be avoided in place of more traditional, labour-intensive methods. It is important to remember that the history of employment in Detroit is one that is sensitive to issues of technological progress given that it was the auto industry's shift to robotics in the 1980s that killed thousands of jobs, not the arrival of Japanese imports to America.

More than ninety organisations are partnered with RecoveryPark. When I visited the intended site in 2012, several projects were at the planning and development stage, including one to renovate and repurpose an abandoned factory into an organic (hormone and chemical free) tilapia farm. When it reaches peak production, it is thought that this facility, 'AquaNourish Farms,' might be able to provide two million kilograms of fresh fish to the southeast Michigan region while creating jobs for fifty Detroit residents. Future plans for this facility include a bio-digester to produce energy from waste and an algae farm to clean the water. Another plan that was in the works at the time of my visit, was for an abandoned and dilapidated former market (Figure 7.63) to be transformed into a horse stable and large-animal veterinary practice.

The strategy of RecoveryPark is to first bring in a concentration of jobs and then to focus on establishing new housing. After this, it is hoped that private development will follow on its own accord. In this respect, philanthropic contributions will only be needed as seed money. So in addition to urban agriculture and jobs, RecoveryPark's long-term goal is to position itself as a magnet to attract other commercial and housing development. Viljoen and Bohn (2015: 136) concluded that RecoveryPark 'does offer a new model, for large-scale urban agriculture opportunities with potential efficiencies of scale, while retaining a focus on neighbourhood and individual farm workers. The challenge of setting up such a large project, and learning about its operation, can only be judged in the future.'

While attending the 'Urban Agriculture Summit' in Toronto a few months after my guided tours of Detroit's lower east side in 2012, I became disturbed by the hagiography heaped upon one prominent urban agriculture individual. The sycophancy in the lecture theatre filled with upwards of three hundred 'worshippers' was palpable. This was exacerbated by a film shown by this individual of his exploits wherein frequent reference was made to his winning a Macarthur Genius award. That evening I had a conversation with one of this individual's senior collaborators who admitted concern that this was developing into a very troubling situation of a personality cult wherein the messenger had become inflated with his own self-importance such that his critical message was now at risk of being forgotten under an avalanche of arrogance that was starting to alienate many of the coworkers as well as outside observers like myself. The next day, during a session in which the RecoveryPark project was presented to a much more modest gathering of several dozen attendees, I was struck by the injustice of the previous rock-star hype and the understated but remarkable promise offered by RecoveryPark spokespeople. During the question and discussion session that followed, I hopped on a soapbox and stated that one of the luxuries of being a university professor with a travel grant meant that I had the opportunity to visit and critically compare and contrast dozens of agrourbanism projects around the world. And it is through this process, that it was my opinion that if only half of the proposals advanced by the RecoveryPark project meet with fruition, then it will still be the most interesting, most integrative, most ambitious, and most promising of all the case studies I had visited.

Afterword: Fast forward to a decade later and both these 'Detroit Dreams' projects are still waiting to be actualised to their planned potential. Online searches reveal a dearth of information about the Bloody Run Creek Greenway Redevelopment Project following a flurry of news reportage from 2012 coincident to the release of its conceptual plan. RecoveryPark on the other hand became a media darling in the interim, garnering coverage in such august outlets as *The Wall Street Journal* and *The New York Times*. That said, its current website is a scaled down version in terms of information from what had once been posted there. All that is now shown is a timeline heralding accomplishments that have occurred since the time of my visit to Detroit up to early 2021. These include: negotiations with the city for urban agriculture permissions and signing leases for vacant land, acquisition of warehouse facilities, a pilot farm and hydroponics test facility, acquisition of office space, establishment of a clean-up and anti-dumping programme, a handful of hoop-houses, environmental and water quality surveys, expansion to additional farm sites, obtaining funding, creation of a master plan, hiring of employees, construction of an indoor hydroponics system, sales to local restaurants, launching of a housing support platform, and securing commitment for future funding for the construction of a large glass greenhouse along with a promise of debt financing from the United States Department of Agriculture. The bad news is that various growing facilities were vandalised several times. But that is minor compared to the controversy within which RecoveryPark now finds itself embroiled (DeVito 2021a,b,c): a messy soap-opera with defamation accusations and lawsuits involving charges of dishonesty and professional mismanagement, leading to the temporary or possibly permanent collapse of what one disgruntled employee referred to as 'a brilliant idea'. In conclusion, this is a demonstration that sometimes the travails of agrourbanism practitioners, just like those of environmentalists, can be frustratingly akin to the labours of Sisyphus.

7.2.6 Case study lessons

- Whether expressed and enacted explicitly (Natural Growth Project, Forty Hall Vineyard) or implicitly (ReVision Urban Farm), horticultural therapy can be a powerful tool in the healing of broken lives.
- Horticultural healing can occur in a variety of different physical settings, including sequestered gardens (Natural Growth Project), adjacent or proximal plots (ReVision Urban Farm, Earthworks Urban Farm), onsite fields (Forty Hall Vineyard), and offsite community gardens (Natural Growth Project).
- Societal healing through regenerative agrourbanism occurs by closely tying the growing of food to other facilities such as counseling offices (Natural Growth Project), homeless shelters (ReVision Urban Farm), soup kitchens (Earthworks Urban Farm), and onsite shops or food stands (Forty Hall Vineyard, Revision Urban Farm).
- Agrourbanism contributes to social enterprise development through entrepreneurial training and educational programmes (Forty Hall Vineyard, Revision Urban Farm, Earthworks Urban Farm).
- Healing provided by regenerative agrourbanism projects benefit from inclusion of beauty in the form of heritage landscapes (Forty Hall Vineyard), decorative plantings (Earthworks Urban Farm), and amenities (Natural Growth Project, ReVision Urban Farm).
- Participants seeking solace through urban agriculture can achieve this through private time spent in a quiet garden (Natural Growth Project) or through solidarity as part of a group tending a working farm (Forty Hall Vineyard).
- Regenerative agrourbanism calls for planners and practitioners to dream large and to reach far (Bloody Run Creek Greenway Redevelopment Project, RecoveryPark), always keeping in mind the inspiring questions posited by William Morris – 'Dreamer of dreams, born out of my due time, Why should I strive to set the crooked straight?' and of Robert Browning – 'Ah, but a man's reach should exceed his grasp. Or what's a heaven for?'

References

Anonymous. 2007. Brownfields success in New England. Land & Community Revitalization fact sheet.
Anonymous. 2011. Green Barn program report 2010-2011. The Stop Community Food Centre.

Anonymous. 2015. Case study: green therapy for survivors of torture. Using ecotherapy to work with people who have experienced trauma. Ecominds.

Anthropoulou, T., S. Nikolaidou, M. Partalidou and M. Petrou. 2018. The emergence of municipal allotment gardens in Greece in times of crisis: governance challenges for new urban gardening practices. In: Soulara, C-T., C. Perrin and E. Valette (eds.). Toward sustainable relations between agriculture and the city. Springer, Berlin, Germany.

Apostolopoulou, E. and P. Kotsila. 2012. Community gardening in Hellinikon as a resistance struggle against neoliberal urbanism: spatial autogestion and the right to the city in post-crisis Athens, Greece. Urban Geography 43: 293-319. https://doi.org/10.1080/02723638.2020.1863621

Beaton, B. and S. Todd. 2015. Reclaiming the ruins: a case study of graffiti heritage interpretation at the Evergreen Brickworks in Toronto. In: Lovata, T.R. and E. Olton (eds.) Understanding graffiti. Routledge, London, UK.

Bogg, G.L. 2011. Detroit: place and space to begin anew. Available at: www.thenation.com.

Campbell, L. and A. Wiesen (eds.) 2009. Restorative commons: Creating health and well-being through urban landscapes. USDA Forest Service, Washington, DC, USA.

Campigotto, R.M. 2010. Farmers' markets and their practices concerning income, privilege, and race: a case study of Wychwood Artspace Barns in Toronto. Ph.D. Thesis, University of Toronto, Canada.

Carroll, A. 2016. Brownfields as sites for urban farms. In: Brown, S., K. McIvor and H. Snyder (eds.). Saving seeds in the city. Springer, Berlin, Germany.

Carter, V. 1995. Forty Hill and Bulls Cross. Heritage Walks No. 1. Enfield Preservation Society, Enfield, UK.

Ciaidella, J. 2008. Debating Detroit's redevelopment. Available at: www.planetizen.com.

Cobb, T.D. 2012. Reclaiming our food: how the grassroots food movement is changing the way we eat. Storey Publishing, North Adams, MA, USA.

Cockrall-King, J. 2012. Food and the city: urban agriculture and the new food revolution. Prometheus Books, Buffalo, NY, USA.

De la Salle, J. and M. Holland. 2010. Agricultural urbanism: handbook for building sustainable food & agriculture systems in 21st century cities. Green Frigate Books/Libri, Farringdon, UK.

Detroit Collaborative Design Center. 2011. The Bloody Run Creek greenway redevelopment project: a transformative destination for the 21st century. DCDC, University of Detroit Mercy, Detroit, MI, USA.

DeVito, L. 2021a. Can Detroit's RecoveryPark recover? Detroit Metro Times, 19 May.

DeVito, L. 2021b. RecoveryPark strikes back. Detroit Metro Times, 26 May.

DeVito, L. 2021c. Former RecoveryPark farm associate accuses owner of intending to 'flip' properties given by Detroit for profit. Detroit Metro Times, 23 July.

Digiacaomo, R. 2017. After two decades, a Philadelphia urban farm grapples with growing pains. Civil Eats, 22 August.

Dodds, R. and L. Jolliffe. 2016. Experiential tourism: creating and marketing tourism attraction experience. In: Sotiriadis, M. and D. Gursoy (eds.). The handbook of managing and marketing tourism experiences. Emerald Group Publications, Bingley, UK.

Flannery, J.A. and K.M. Smith. 2015. Evergreen Brickworks. Eco-landscape design. Springer, Berlin, Germany.

France, R. L. (ed.) 2001. Reflecting heaven: Thoreau on water. Houghton Mifflin, Boston, MA, USA.

France, R.L. (ed.) 2002. Handbook of water sensitive planning and design. Lewis/CRC Press, Boca Raton, FL, USA.

France, R.L. 2003. Wetland design: principles and practices for landscape architects and land-use planners. W.W. Norton, New York, NY, USA.

France, R.L. (ed.) 2007. Healing natures, repairing relationships: new perspectives on restoring ecological spaces and consciousness. Libri, Farringdon, UK.

France, R. (ed.) 2008. Handbook of regenerative landscape design. CRC Press, Boca Raton, FL, USA.

France, R.L. 2011. Veniceland Atlantis: the bleak future of the world's favorite city. Libri, Farringdon, UK.

France, R.L. 2016. Designing urban agricultural forms: history, education, proposals and projects. In: France, R.L. (ed.) Integrated urban agriculture: precedents, practices, prospects. Libri, Farringdon, UK.

France, R.L. 2020. Waymarking Italy's influence on American environmental imagination while on pilgrimage to Assisi. Cambridge Scholars Publishing, Newcastle upon Tyne, UK.

Gallagher, J. 2010. Reimagining Detroit: opportunities for redefining an American city. Painted Turtle, Detroit, MI, USA.

Gavrilidou, E., H. Kleinmann, E. Ourelidou and S.G.Z. Zafeiropoulos. 2015. Urban agriculture in Thessaloniki: an academic project meets reality. Ri-Vista 02:60-70.

Goodspeed, R. 2009. Detroit and the limits to urban decline. Available at: www.goodspeedupdate.com.

Gopakumar, G. 2005. Case study of community gardens in Detroit, Michigan. Detroit Agriculture Network.

Gorgolewski, M., J. Komisar and J. Nasr. 2011. Carrot city: Creating places for urban agriculture. Monacelli Press, New York, NY, USA.

Hanson, D. and E. Marty. 2012. Breaking through concrete: building an urban farm revival. University of California Press, Berkeley, CA, USA.

Herrington, S. and D.M. Lopes. Neo-picturesque. In: Bicknell, J., J. Judkins and C. Korsmeyer (eds.) Philosophical perspectives on ruins, monuments, and memorial. Routledge, London, UK.

Hodgson, K., M.C. Campbell and M. Bailkey. 2011. Urban agriculture: growing healthy, sustainable places. American Planning Association, Chicago, IL, USA.

Hurdle, J. 2008. Where industry once hummed, urban garden finds success. The New York Times, 20 May.

Irvine, S. 2012. Evergreen Brick Works: an innovative and sustainability case study. Technology Innovation and Management Review, July: 21-25.

Johnson, L. 2010. City farmer: Adventures in urban food growing. Greystone Books, Vancouver, BC, Canada.

Kamvasinou, K. 2017. Temporary intervention and long-term legacy: lessons from London case studies. Journal of Urban Design 22: 187-207.

Karagiorgas, A. 2016. From zero to 'hero'! In: Keshavarz, N. and R. Fox-Kämper (eds.) Urban allotment gardens in European cities: future, challenges and lessons learned. European Science Foundation, Strasbourg, France.

Kavanaugh, K. 2010. John Hantz: the man has a plan, but does Detroit have a farming future? Available at: www.modelmedia.com.

Kleeman, J. 2019. Murder on the allotment: 'she was everything that he wanted to be.' The Guardian, 13 April.

Lachance, J.D. 2004. Supporting urban agriculture: a proposed supplement to the City of Detroit master plan of policies. City of Detroit Dept. Planning and Development, Detroit, MI, USA.

Ladner, P. 2011. The urban food revolution: changing the way we feed cities. New Society Publishers, Gabriola Island, Canada.

Laviscio, R. 2020. Is urban agriculture an opportunity to preserve landscape systems? In: Scazzosi, L. and P. Brandoini (eds.). AgriCultura: urban agriculture and the heritage potential of agrarian landscapes. Springer, Berlin, Germany.

Linden, S. and J. Grut. 2002. The healing fields: working with psychotherapy and nature to rebuild shattered lives. Freedom from torture. Frances Lincoln Publishers Ltd., London, UK.

Lystra, M. S. Vogel and D. Pitera. 2010. Adamah redux: Urban agi[culture] in Detroit's lower east side. School of Architecture, University of Detroit Mercy, Detroit, MI, USA.

Mantyk, E. 2010. World's largest urban farm slated for Detroit. Available at: www.theepochtimes.com.

Mistiaen, V. 2002. Thyme is the healer. The Guardian, 2 October.

Nairn, M. and D. Vitiello. 2010. Lush lots: everyday urban agriculture. Harvard Design Magazine 31: 1-8.

Nasr, J., J. Komisar and M. Gorgolowski. 2015. Urban agriculture as ordinary urban practice. In: Viljoen, A. and K. Bohn (eds.) Second nature urban agriculture: designing productive cities. Routledge, London, UK.

Northrup, S. 2008. Detroit should give the status-qo a go. Available at: www.planetizen.com.

Novakovics, S. 2019. 10 projects that defined a decade of Canadian architecture. The Azure Magazine, 18 December.

Pothukuchi, K. 2011. Building sustainable, just food systems in Detroit: reflections from SEED Wayne, a campus-community collaborative. Sustainability 4: 193-198.

Rich, S.C. 2012. Urban farms. Abrams Books, New York, NY, USA.

Sepgupta, U., G. Cape, S. Irvine and P. Bertrand. 2008. Evergreen Brickworks: Planning for success in a triple, bottom-line enterprise. In: Stoner, A.F. and C. Wankel (eds.) Global sustainability initiatives: new models and new approaches. Information Age Publishing, Charlotte, NC, USA.

Spirn, A.W. 1998. The language of landscape. Yale University Press, London, UK.

Spirn, A.W. 2005. Restoring Mill Creek: landscape literacy, environmental justice and city planning and design. Landscape Research 30: 395-413.

Thwaites, K., A. Mather and I. Simkins. 2013. Socially restorative urbanism: the theory, process and practice of experiemics. Routledge, London, UK.

Travalinie, K. and C. Hunold. 2010. Urban agriculture and ecological citizenship in Philadelphia. Local Environments 15: 581-590.

Viljoen, A. (ed.) 2008. Continuous productive urban landscapes: designing urban agriculture for sustainable cities. Elsevier, Amsterdam, the Netherlands.

Viljoen, A. and K. Bohn. 2015. Second nature urban agriculture: designing productive cities. Routledge, London, UK.

Viljoen, A., J. Schlesinger, K. Bohn and A. Drescher. 2015. Agriculture in urban design and spatial planning. In: de Zeeuw, H. and P. Drechsel (eds.) Cities and agriculture: developing resilient urban food systems. Earthscan, London, UK.

Vogel, S. 2008. Detroit (re)turns to nature. In: France, R.L. (ed.) Handbook of regenerative landscape design. CRC Press, Boca Raton, FL, USA.

Walker, J. 2012. In defense of the land: The Mill Creek Farm and Brown Street Community Garden. Available at: www.millcreekurbanfarm.org.

Walker, J. and J. Rosen. 2006/08. Mill Creek Farm annual reports.

Wicke, R. 2012. Detroit: one chance for re-birth? www.greenhouseonline.org.

Wissinger, M. 2016. Agriculture along the transect: case studies. University of Miami School of Architecture and DPZ Partners, Miami, FL, USA.

Yannacopoulou, B. 2012. Greeks reclaim the land to ease the pain of austerity. The Ecologist, 10 January.

www.ingramcontent.com/pod-product-compliance
Lightning Source LLC
Chambersburg PA
CBHW061200240926
51846CB00006B/1067